T0238961

Terrestrial Fluids, Earthquakes and Volcanoes: The Hiroshi Wakita Volume I

Edited by
Nemesio M. Pérez
Sergio Gurrieri
Chi-Yu King
Kenneth A. McGee

2006

Birkhäuser Verlag
Basel · Boston · Berlin

Reprint from Pure and Applied Geophysics
(PAGEOPH), Volume 163 (2006) No. 4

Editor's

Nemesio M. Pérez
Environmental Research Division
Instituto Tecnológico y de Energias
Renovables
Polígono Industrial de Granadilla s/n
38611 Granadilla, Tenerife
Canary Islands
Spain
e-mail: nperez@iter.es

Chi-Yu King
Earthquake Prediction Research, Inc
381 Hawthorne Ave.
Los Altos, CA 94022
USA
e-mail: Chiyuking@aol.com

Sergio Gurrieri
Istituto Nazionale di Geofisica e
Vulcanologia
Sezione di Palermo
V. Ugo La Malfa, 153
90146 Palermo
Italy
e-mail: peppo@pa.ingv.it

Kenneth A. McGee
Project Chief, Volcano Emissions Project
U.S. Geological Survey
Cascades Volcano Observatory
1300 SE Cardinal Court, # 100
Vancouver, Washington 98683
USA
e-mail: kenmcgee@usgv.gov

A CIP catalogue record for this book is available from the Library of Congress, Washington
D.C., USA

Bibliographic information published by Die Deutsche Bibliothek:
Die Deutsche Bibliothek lists this publication in the Deutsche Nationalbibliographie; detailed
bibliographic data is available in the internet at <http://dnb.ddb.de>

This work is subject to copyright. All rights are reserved, whether the whole or part of the material is concerned, specifically the rights of translation, reprinting, re-use of illustrations, recitation,
broadcasting, reproduction on microfilms or in other ways, and storage in data banks. For any
kind of use, permission of the copyright owner must be obtained.

ISBN 3-7643-7580-9 Birkhäuser Verlag, Basel – Boston – Berlin

© 2006 Birkhäuser Verlag, P.O.Box 133, CH-4010 Basel, Switzerland
Part of Springer Science+Business Media
Printed on acid-free paper produced from chlorine-free pulp TCF ∞

ISBN 10: 3-7643-7580-9
ISBN 13: 978-3-7643-7580-5

9 8 7 6 5 4 3 2 1

PURE AND APPLIED GEOPHYSICS
Vol. 163, No. 4, 2006

Contents

Pure appl. geophys. 163 (2006) 629–631
0033–4553/06/040629–3
DOI 10.1007/s00024-006-0052-z

© Birkhäuser Verlag, Basel, 2006

Introduction

Terrestrial Fluids, Earthquakes and Volcanoes: The Hiroshi Wakita Volume I is a special publication to honor Professor Hiroshi Wakita for his scientific contributions. This volume consists of 17 original papers dealing with various aspects of the role of terrestrial fluids in earthquake and volcanic processes, which reflect Prof. Wakita's wide scope of research interests. Volume II is planned for publication later this year. Both volumes should be useful for active researchers in the subject field, and graduate students who wish to familiarize them with it.

Professor Wakita co-founded the Laboratory for Earthquake Chemistry in 1978 and served as its director from 1988 until his retirement from the university in 1997. He has made the laboratory a leading world center for studying earthquakes and volcanic activities by means of geochemical and hydrological methods. Together with his research team and a number of foreign guest researchers that he attracted, he has made many significant contributions in the above-mentioned scientific fields of interest. This achievement is a testimony to not only his scientific talent, but also his enthusiasm, his open mindedness, and his drive in obtaining both human and financial support.

The seventeen papers in this volume are arranged into two groups. The first group of seven papers deal with the movement of terrestrial fluids related to earthquakes. The paper by King *et al.* gives an updated general review of this field with special emphasis on implications for earthquake prediction. The paper by Kitagawa *et al.* uses data from a borehold strainmeter and other crustal-deformation sensors to infer the occurrence of a 1-mm creep event on the Yasutomi fault in Japan. The significance of the work is that this would be the first evidence of an aseismic-slip event on this inland fault. The paper by Ohno *et al.* describes anomalous water level changes at two wells associated with seismic swarm activity off Izu Peninsula in March, 1997. Coseismic water-level drops followed by gradual postseismic water-level rises were observed and the postseismic signatures were explained by the authors in terms of the horizontal pressure diffusion due to the pressure gradient in the aquifer induced by coseismic static strain. The paper by Song *et al.* describes anomalous chloride and sulphate in the water at the springs of Chiayi area in west-central Taiwan as likely precursors to earthquakes. The observed anomalies are explained as a reflection of stress/strain-induced pressure changes in subsurface water systems. Another paper from Taiwan is presented by Yang *et al.*, who describe observed CO_2/CH_4 variations in bubbling gases from a mud pool at an active fault zone of southwestern Taiwan. The observed variations were detected a few days to a few weeks before the earthquakes and correlated well with those of local magnitude > 4.0 and local intensities > 2. The paper by WAILA *et al.* describes the detection of radon precursory signals for some earthquakes of magnitude > 5 that occurred in N-W Himalaya. The authors describe also the precursory nature of observed helium anomalies related to earthquake activity in the region, suggesting that these observations are a strong indicator in favor of geochemical precursors for earthquake prediction studies. The paper by TOUTAIN *et al.* is the last of the first group of contributions; it demonstrates that coseismic transient anomalies in groundwater are evidenced by modeling the mixing fuction *F* characteristic of the groundwater dynamics in the Ogeu (western French Pyrénées) seismic context, and demonstrates that groundwater systems in such environments are unstable systems that are highly sensitive to both rainfall inputs and microseismic activity.

The earthquake-related papers are followed by a paper by SANO *et al.* on helium isotope systematics, describing and discussing the geographical distribution of helium isotope ratios observed in the Chugoko district of southwestern Japan. This paper is then followed by nine papers dealing with observations related to volcanic processes. The paper by Chiodini *et al.* describes an anomalous and vigorous submarine gas emanation at Panarea (Aeolian Islands, Italy) and discusses its origin and implications for regional volcanic surveillance. The paper by GRASSA *et al.* details the chemical and isotopic composition of dissolved gases in several warm and thermal springs of Sicily, suggesting the action of several processes to explain the high CO_2 contents dissolved in the thermal water springs. The paper by SATO *et al.* reports on the observation of some geochemical changes in spring waters associated with the 2000 eruption of the Miyake-jima volcano in Japan. Sulphate content and isotopic values measured in the water samples and leachate of the ejecta allowed the authors to conclude that the observed increases on the sulphate concentration in the spring waters were mainly derived from mixing with leachate of the ejecta.

The remaining six papers deal with diffuse degassing of carbon dioxide in active volcanic systems from Hawaii, Japan, Italy and Central America. This geochemical approach has recently become a useful tool for volcano monitoring research. New data on diffuse carbon dioxide emission from Mt. Fuji are presented by NOTSU *et al.*, who also describe a conceptual model for the temporal evolution of non-visible vs. visible degassing during inter-eruptive, pre-eruptive, eruptive, and post-eruptive phases of active volcanoes. The paper by MCGEE *et al.* reports an interesting study carried out at Puhimau thermal area, Hawaii, and discusses the role of CO_2 degassing reaching the East Rift Zone (ERZ) and its contribution to Kilauea's CO_2 output. They conclude that Puhimau Thermal Area is not currently a significant contributor to the total CO_2 budget of Kilauea. An additional paper on diffuse CO_2 degassing at Kilauea is reported by GIAMMANCO *et al.*, who studied the relationship between soil CO_2 emissions and both tectonic and volcano-tectonic structures on the summit and the lower East Rift Zone of Kilauea volcano. The paper by HERNÁNDEZ *et al.* reports the first study of diffuse CO_2 degasssing at Showa-Shinzan, a parasitic volcanic dome of Usu volcano (Japan) where the observed low CO_2 efflux values and light carbon isotopic signatures of the soil CO_2 gas suggested a poor contribution of deep-seated degassing near or within the volcanic hydrothermal system of Showa-Shinzan. The paper by PÉREZ *et al.* shows the results of continuous monitoring of soil CO_2 efflux for the purpose of detecting precursory signatures of short-term unrest at San Miguel volcano, El Salvador, Central America. This work demonstrates the sensitivity of this geochemical approach in detecting early signs of volcanic unrest in areas of low degassing rate. Finally, the paper by CAMARDA, *et al.* describes a new method to measure *in situ* soil gas permeability; tests on Vulcano showed very poor correlation between CO_2 flux and shallow soil permeability.

The guest editorial team would like to thank all the contributors and all reviewers involved, who are listed below: A. Aiuppa, G. Chiodini, B. Christenson, W. D'Alessandro, A. Eff-Darwich, W. Evans, C. D. Farrar, C. Federico, F. Goff, F. Grassa, S. Gurrieri, P. A. Hernández, G. Igarashi, C.-Y. King, L. Mastin, K. Notsu, N. Osskarson, N. M. Pérez, Y. Taran, J.-P. Toutain, Y. Sano, N. Sturchio, J. Varekamp, and N. Varely. Special thanks are due to Pedro A. Hernández for his considerable assistance to the team, and to Renata Dmowska, without whose marvellous support this special volume would not have been possible.

Nemesio M. Pérez
Environmental Research Division
Instituto Tecnológico y de Energías
Renovables (ITER)
Tenerife, Canary Islands
Spain

Sergio Gurrieri
Istituto Nazionale di Geofisica e
Vulcanologia
V. Ugo La Malfa, 153
90146 Palermo
Italy

Chi-Yu King
Earthquake Prediction Research, Inc.
381 Hawthorne Ave.
Los Altos, CA 94022
USA

Kenneth McGee
Cascades Volcano Observatory
U.S. Geological Survey
1300 SE Cardinal Court, #100
Vancouver, WA 98683
USA

Pure appl. geophys. 163 (2006) 633–645
0033–4553/06/040633–13
DOI 10.1007/s00024-006-0049-7

© Birkhäuser Verlag, Basel, 2006

❘Pure and Applied Geophysics

Earthquake-induced Groundwater and Gas Changes

Chi-Yu King,[1] Wei Zhang,[2] and Zhaocheng Zhang[2]

Abstract—Active faults are commonly associated with spatially anomalously high concentrations of soil gases such as carbon dioxide and Rn, suggesting that they are crustal discontinuities with a relatively high vertical permeability through which crustal and subcrustal gases may preferably escape towards the earth's surface. Many earthquake-related hydrologic and geochemical temporal changes have been recorded, mostly along active faults especially at fault intersections, since the 1960s. The reality of such changes is gradually ascertained and their features well delineated and fairly understood. Some coseismic changes recorded in "near field" are rather consistent with poroelastic dislocation models of earthquake sources, whereas others are attributable to near-surface permeability enhancement. In addition, coseismic (and postseismic) changes were recorded for many moderate to large earthquakes at certain relatively few "sensitive sites" at epicentral distances too large (larger for larger earthquakes, up to 1000 km or more for magnitude 8) to be explained by the poroelastic models. They are probably triggered by seismic shaking. The sensitivity of different sites can be greatly different, even when separated only by meters. The sensitive sites are usually located on or near active faults, especially their intersections and bends, and characterized by some near-critical hydrologic or geochemical condition (e.g., permeability that can be greatly increased by a relatively small seismic shaking or stress increase). Coseismic changes recorded for different earthquakes at a sensitive site are usually similar, regardless of the earthquakes' location and focal mechanism. The sensitivity of a sensitive site may change with time. Also pre-earthquake changes were observed hours to years before some destructive earthquakes at certain sensitive sites, some at large epicentral distances, although these changes are relatively few and less certain. Both long-distance coseismic and preseismic changes call for more realistic models than simple elastic dislocation for explanation. Such models should take into consideration the heterogeneity of the crust where stress is concentrated at certain weak points (sensitive sites) along active faults such that the stress condition is near a critical level prior to the occurrence of the corresponding earthquakes. To explain the preseismic changes, the models should also assume a broad-scaled episodically increasing strain field.

1. Introduction

Groundwater and gas changes (other than seismic oscillation), such as in water level, color, and smell in wells, have been observed since ancient times in many seismic regions. A systematic search for such changes to uncover earthquake-related, and especially pre-earthquake, changes with scientific instruments did not begin, however, until the 1960s. Earlier results, obtained before 1980 mainly in China,

[1]Earthquake Prediction Research, Inc., 381 Hawthorne Ave., Los Altos, CA, 94022, U.S.A
[2]Center for Analysis and Prediction, China Seismological Bureau, Beijing, 100036, China

Japan, the former Soviet Union, and the United States, were reviewed by KING (1986), THOMAS (1988), MA, *et al.* (1990), IGARASHI and WAKITA (1995), and ROLOEFFS (1988, 1996), among others. This paper gives a brief review of more recent results.

2. Earlier Results

Hydrologic and geochemical parameters studied include water level/pressure, temperature, electric conductivity at wells, and flow rate at springs, concentration of various ions and dissolved gases, and components of soil gas in shallow holes. Earthquake-related changes were reportedly observed before and after many destructive earthquakes, especially at certain relatively few "sensitive" sites. The sensitive sites were usually found to be situated along active faults, especially at intersections of faults, and may be at unexpectedly large epicentral distances (many source dimensions, or hundreds of km, away for large events). This pattern was difficult to understand by many seismologists, who were accustomed to using simple dislocation models of earthquake sources that assume a homogeneous and linearly elastic medium. Because of this difficulty as well as the poor quality of some studies which sometimes failed to take proper account of background noises caused by such environmental variables as rainfall, barometric pressure, temperature, and earth tide, many geophysicists dismissed the earthquake relatedness of such changes (e.g., GELLER, 1997). Nevertheless, as pointed out by KING (1986) and MA *et al.* (1990), the difficulty may be partly due to unrealistic modeling of earthquakes occurring in an inhomogeneous and non-elastic crust by such simple dislocations. They suggested that if the crustal heterogeneity and a large-scale, episodically increasing tectonic stress field are taken into consideration, the observed features can be reasonably understood. This suggestion is confirmed by recent observations, as reviewed below.

In a related group of studies, many active faults, including some buried ones, were found to show spatial anomalies, such as higher concentration of various terrestrial gases (radon, helium, hydrogen, mercury vapor, carbon dioxide, isotope ratios, etc.) in groundwater and soil air (e.g., IRWIN and BARNES, 1980; KING, 1986; MA *et al.*, 1990).

3. Recent Observations

Observations of earthquake-related groundwater and gas changes were made more carefully since 1980, many by using continuously-recording instruments of better sensitivity and reliability and by giving proper consideration to background noise, such as changes caused by rainfall, barometric-pressure variation, and solid-earth tide. The results, however, still show largely the same characteristic features as observed earlier: (1) Sensitivity of monitoring sites, even only meters apart, can be greatly different; (2) at near-field sites, some recorded earthquake-related changes are approximately consistent with poroelastic dislocation models of earthquakes; (3) however, "far-field" co- and postseismic changes recorded for many moderate and large earthquakes at some "sensitive sites" are too large (up to about 1000 km or more for magnitude 8) to be explained by the poroelastic models. Being recorded at larger distances for larger earthquakes, they are probably triggered by seismic shaking; (4) the recorded premonitory changes are relatively few and uncertain; (5) sensitive sites are usually located on or near active faults and characterized by some near-critical hydrologic or geochemical condition (e.g., permeability that can be greatly changed by a slight shaking or stress change); (6) the earthquake-related changes recorded for different earthquakes at a sensitive site are usually similar, regardless of the earthquake's location and focal mechanism; and (7) the sensitivity may change with time as the crustal stress changes (e.g., SILVER and VALLETTE-SILVER, 1992; ZHANG, 1994; ROELOFFS, 1988, 1996; WAKITA, 1996; TOUTAIN and BAUBRON, 1999; KING et al., 1999, 2000; KING and IGARASHI, 2002; MONTGOMERY and MANGA, 2003).

A notable example of the long-distance seismic-wave triggering effect of large earthquakes is the observation of numerous earthquakes triggered by the 7.3 earthquake at Landers, California, in 1992. Many earthquakes were triggered minutes to hours later in geothermal and volcanic areas across the western United States at distances of up to 1250 km, about 10 fault lengths, from the Landers epicenter (HILL et al., 1993). No significant earthquakes were triggered, however, in the nearby San Andreas Fault segments outside the aftershock area. Such a long-distance triggering phenomena cannot be explained by the small static stress changes of the earthquake expected from elastic-dislocation models, but can be attributed to seismic-wave-triggered fluid movement in and near some nearly critically loaded faults in the triggered area. Indeed, various hydrologic phenomena, including gas bubbling, increased spring discharge, and groundwater-level changes, were observed after the earthquake in many of such areas (ROELOFFS, 1996). Since Landers earthquake, numerous additional moderate-to-large earthquakes have been found to have dynamically triggered distant earthquakes and volcanic eruptions (HILL et al., 2002); one such example is the Hector Mine earthquake, which was followed by an abrupt increase in seismicity at distances of 110–270 km at sites within the Salton Trough to the south (GOMBERG, et al., 2001). It is the observation of such long-distance triggering of events that has helped to remove some geophysicists' skepticism about the possibility of far-field hydrologic and geochemical (as well as some other kinds of) anomalies (including precursors), and to demonstrate the inadequacy of the simple elastic dislocation models in explaining anomaly occurrences.

Another example of earthquake-related changes in crustal-fluid movement is provided by the study of eruption-interval changes at a nearly periodic geyser located in proximity to the San Andreas Fault north of San Francisco, California (SILVER and VALLETTE-SILVER, 1992). Co- and preseismic changes were detected for three earthquakes (including the magnitude 6.9 Loma Prieta event in 1989) at epicentral distances of 100–200 km. Such changes may possibly be due to permeability changes caused by seismic shaking and an episodically increasing stress field, respectively.

In Parkfield, where an earthquake-prediction experiment that includes a water-level study at a dozen wells has been in progress for about two decades, steplike coseimic water-level drops were recorded at four wells at the time of the magnitude 5.8 Kettleman Hills earthquake in 1985 approximately 35 km away (ROELOFFS and QUILTY, 1997). Such coseismic changes are of the correct sign and roughly the expected size from the poroelastic dislocation model, based on the well's strain sensitivity as inferred from earth tidal response (QUILTY and ROELOFFS, 1997; see also WAKITA, 1975). This is in contrast to coseismic changes at long distances near active faults recorded elsewhere (larger changes than expected and may be of a different sign, see IGARASHI and WAKITA, 1991 and KING *et al.*, 1999). A detailed study of the Kettleman Hills data also shows small water-level changes beginning three days before the earthquake in two of the wells. In addition, there is a shallow well at which coseismic water-level rises were observed at the time of three local and five distant earthquakes with different azimuth and focal mechanisms (as distant as 730 km for a magnitude 7 event) (ROELOFFS, 1998). These rises, like those observed by the above-mentioned other authors, cannot be explained by poroelastic response to coseismic static strain changes. No pre-earthquake changes were observed by Roeloffs at this well.

In Japan, coseismic water-level drops were recorded at two wells monitored by the Geological Survey of Japan in Shizuoka prefecture in response to earthquakes up to 740 km away; some of the larger changes were also preceded by smaller preearthquake changes (e.g., MATSUMOTO, 1992). Similar coseismic and sometimes pre-earthquake changes have been observed in groundwater temperature and discharge also (e.g., MOGI, *et al.*, 1989; FUJIMORI *et al.*, 1995). The University of Tokyo has maintained a network of about a dozen continuously monitored hydrologic and geochemical stations along the nearby Pacific Coast since 1978 (WAKITA, 1996). One of the stations in Izu Peninsula recorded a clear radon-concentration change before the nearby Izu-Oshima-Kinkai earthquake in 1978. Similar preseismic changes in groundwater level, temperature and strain were observed at three other sites, all at epicentral distances of 25–50 km beginning a month before the event. Also, over the years, only two of the stations (one for radon and one for water level) have been found to be sensitive to distant earthquakes (such as a magnitude 8.1 event 1200 km away). The sensitivity was found to vary with time (WAKITA, 1996), probably due to local stress or permeability variation.

Many co- and preseismic changes in groundwater level/discharge, geochemistry, radon concentration in groundwater and atmosphere, and strain were observed at locations up to 220 km away from the magnitude 7.2 Kobe earthquake in 1995, in spite of the fact that the earthquake occurred beyond the Japanese intensive-study area of Tokai (e.g., SATO, et al., 2000). Most of the preseismic changes were recorded within the ultimate aftershock zone beginning three months before the event (IGARASHI et al., 1995; KING et al., 1995; TSUNOGAI and WAKITA, 1995; YASUOKA and SHINOGI, 1995; SUGISAKI et al., 1996). In a questionnaire survey conducted shortly after the earthquake, KOIZUMI et al., (1996) found postearthquake water level/discharge changes at many sites (among many more sites without such changes) at epicentral distances up to about 300 km. The distribution pattern of these changes cannot be explained satisfactorily by volume strain changes derived from dislocation models of the earthquake but may be caused by permeability enhancement (SATO et al., 2000). Also FUJIMORI et al. (1995) observed coseismic seep discharge increases at the Rokko-Takao site near Kobe in response to earthquakes up to several hundred km away; they also observed a gradual discharge increase over a two-month period preceding the Kobe earthquake, which ruptured to within 20 km of the discharge point and produced a coseismic discharge increase of more than one order of magnitude.

KING et al. (1999) studied water-level data continuously recorded over a period of ten years at a closely clustered set of 16 wells within 400 m of a fault in Tono, central Japan. They observed co- and postseismic changes for many moderate local and distant earthquakes (longer distance for larger earthquakes up to more than 1000 km, at locations where the calculated coseismic strain is as low as a few 10^{-8}). They found that the changes recorded at wells on different sides of the fault show very different features. The difference is attributable to a seismically induced permeability increase of the normally impermeable fault-gouge zone, which had been sustaining a considerable hydraulic gradient across the fault. One of the wells showed coseismic changes for more than 20 local and distant earthquakes, including a magnitude 8 event about 1200 km away. All the coseismic changes (water-level drops) and subsequent recoveries have similar shapes, irrespective of the location and focal mechanism of the earthquakes. This well is located on the higher-groundwater-pressure side of the fault. The high sensitivity is attributable to this well's tapping a high-permeability aquifer connected to one of the high-permeability fracture zones that bracket the impermeable fault-gouge zone. Such a structure is observable in an underground tunnel through the fault zone, and is commonly observed elsewhere (see EVANS et al., 1997). Because of the large hydraulic gradient across the gouge zone, a relatively small seismic shaking (or tectonic stress increase) may have introduced some quickly healable fissures in the gouge zone and resulted in a significant but temporary permeability increase which allowed water seepage across the fault and thus caused the observed changes. Water-level at this well was also found to have shown pre-earthquake drops for several earthquakes. Some of the pre-earthquake drops were later found to be due to human activities (e.g., drilling of holes by other people across the fault, resulting in a cross-fault water flow), however one of them, beginning about six months before a magnitude 5.8 local earthquake (the largest near the well since 1983) 50 km away in 1997 may be truly premonitory in nature (KING *et al.*, 2000). The sensitivity of the well showed a temporary decrease during a one-year period after the earthquake, similar to what was observed by WAKITA (1996), presumably due to a temporary relaxation of local stress to a subcritical level, as a result of the magnitude 5.8 earthquake.

Earthquake-related permeability increase is also invoked to explain post-earthquake (including the magnitude 7.1 Loma Prieta event in 1989) flow rate increases and water-chemistry changes observed at some springs and streams in California (KING *et al.*, 1994; ROJSTACZER and WOLF, 1992; ROJSTACZER *et al.*, 1995) as well as at many springs near the epicenter of the1995 Kobe earthquake in Japan (SATO *et al.*, 2000). Some changes that began before certain local earthquakes may be premonitory in nature (KING *et al.*, 1994). Coseismic excess stream flows were observed also for several other earthquakes, including the magnitude 7.5 Hebgen Lake event in 1959 and the magnitude 7.3 Borah Peak event in 1983 (MUIR-WOOD and KING, 1993). The changes in the latter case, however, were attributed to the expulsion of water from the depth due to elastic compression.

In Taiwan, coseismic water-level changes of up to 11.1 m were recorded at 157 of 179 monitored wells located in an alluvial fan 2 to 50 km from the seismogenic fault of the magnitude 7.3 Chi-Chi earthquake (the largest in about 100 years) in 1999 (CHIA *et al.*, 2001). LEE *et al.* (2002) found that the polarities of most of the observed (near field) water-level and some river-discharge changes are in good agreement with those of the static volumetric strain calculated from a well-constrained dislocation model of the seismogenic fault.

Since the mid-1960s, China has established an extensive network of earthquake-related hydrologic and geochemical monitoring stations in areas of earthquake risk, especially along major active faults. The network covers large areas of the China mainland with 330 hydrogeochemical and 250 groundwater-level monitoring stations (ZHANG and LI, 1994). Various effective observation measures have been adopted to obtain reliable continuous data. Studies of possible earthquake precursors and their mechanisms towards earthquake prediction have been carried out from various approaches, both in the field and in the laboratory. The resultant data showed the usefulness of monitoring such parameters as ^{222}Rn, H_2, He, CO_2, H_2S, total gas, Hg^0, F^-, Cl^-, water level, conductivity, temperature and flow (ZHANG, 1994). Since 1966, numerous strong earthquakes of magnitude 7 or above have occurred in the monitoring areas of China: Bohai, 7.4, 1969; Haicheng, 7.3, 1975; Longlin, 7.4, 1976; Tangshan, 7.8, 1976; Songpan, 7.2, 1976; Langcan, 7.6, 1988; Gonghe, 7.0, 1990; and Mongliang, 7.3, 1995. Valuable raw data and practical experience have been accumulated. Many earthquake-related anomalies were recorded; those observed in epicentral areas (near field) were found to have different characteristics from those recorded at large distances (far field). The anomalies in near field are closely related to the process of earthquake preparation, whereas those in the far field are due to stress concentration along active faults, especially at their intersections and bends, under the action of regional stress field. Many spike-like anomalies were observed, and are thought to be useful for predicting earthquake time, whereas the spatial and temporal extent of anomalies are thought to be useful for earthquake magnitude. Some recent studies also found that radon-concentration changes in escaped gases from groundwater showed better correlation with solid-earth tides than in dissolved gases and were more sensitive to earthquake occurrences (SHI and ZHANG, 1995; ZHANG and ZHANG, 1996). These findings are similar to observations by WOLLEN-BERG (1985) in California that groundwater radon at certain sites may respond to minute crustal stress changes, with amplitudes comparable to tidal amplitude (about 0.001 MPa).

A series of publications (8 volumes so far), entitled Earthquake Cases in China, have been published in China, summarizing more than 2000 earthquake-related changes recorded prior to about 200 earthquakes of magnitude 5 and above during 1966–99 (ZHANG et al., 1988–2000; CHEN et al., 2002–03). About 30% of these changes are hydrologic and geochemical. These data are the result of some systematic and standardized studies and are of particular importance to understanding the seismogenic processes in future studies. However, they are currently reported only in Chinese, as are many co- and post-earthquake changes.

In the Koyna-Warna region in western India where reservoir-triggered earthquakes have been studied for four decades, CHADHA et al. (2003) recorded four cases of coseismic and some preseismic water-level changes of several centimeters at a network of wells, for earthquakes of magnitude of 4.3–5.2 at epicentral distances up to 24 km during 1997–2000.

Recent geochemical studies at certain volcanic-hydrothermal systems in the Central American graben detected some earthquake-related CO_2-efflux changes (SALAZAR et al., 2002). Continuous monitoring of CO_2 flux in volcanic areas, where deeply penetrating faults abound, can be easily done and could be useful because CO_2, being a major volcanic gas that has low solubility in silicate melts, can readily migrate to the surface and be influenced by earthquake-related stress changes (SILVER and WAKITA, 1996; PEREZ and HERNANDEZ, 2005).

Many gas-geochemical surveys were conducted across active faults worldwide. Most of the results show fault-related spatial anomalies of such soil air components as CO_2, He, H_2, Hg^0, and ^{222}Rn, suggesting that active faults are crustal discontinuities with a higher vertical permeability than surrounding rocks, thus providing preferential pathways for crustal and subcrustal gases to escape towards the surface (e.g., IRWIN and BARNES, 1980; SUGISAKI et al., 1983; ZHANG et al., 1988; KLUSMAN, 1993; KING et al., 1996). Based on measurements of ^{222}Rn, CO_2, and O_2 in soil air, KING et al. (1996) attributed the ^{222}Rn anomalies observed along several fault segments in California to an upward flow of soil air in the high vertical-permeability fracture zones of the faults. LEWICKI and BRANTLEY (2000) found similar fault-related CO_2 flux anomalies along 12 of 16 San Andreas fault-crossing transects at five sites in Parkfield. Based on carbon-isotope data, they concluded the measured CO_2 to be of biogenic origin, and not deeply derived. Geochemical studies of deeper fluids in wells and springs along the San Andreas in central and south-central California by KENNEDY *et al.* (1997) exhibited high $^3He/^4He$ ratios, suggesting mantle origin of the fluids, which may contribute to the fault-weakening high-fluid pressures at seismogenic depths; they estimated an upward flow rate of about 1 to 10 mm/year, which may be too small to significantly affect the soil-air movement.

4. Mechanisms for Earthquake-related Hydrologic and Geochemical Changes

The earth's crust contains numerous pores and fractures filled with water, gas and other fluids that have different chemical compositions at different places. When the crust is deformed in the tectonic process of earthquake generation, certain transient movements (fissure opening, fault creep, etc.) may be induced in the crust (BERNARD, 2001), and fluids in the fault and other weak zones may be forced to migrate to different locations and thus cause the observed hydrologic and geochemical changes at these locations.

Several inter-related mechanisms were proposed to explain the hydrologic and geochemical changes in the epicenter area and at the distant sensitive sites: Dilatancy due to the increasing number and size of microcracks; increased upward flow of deep-seated fluids to the monitored aquifers or shallow holes; squeezing of gas-rich pore fluids out of the rock matrix into the aquifers; mixing of water from other aquifers through tectonically created fissures in the intervening barriers (permeability enhancement); increased rock/water interactions; and increased gas emanation from newly created crack surfaces in the rock to the pore fluids (e.g., NUR, 1974; KING, 1986; THOMAS, 1988; MUIR-WOOD and KING, 1993; KING and MINISSALE, 1994; ROJSTACZER *et al.*, 1995). For ^{222}Rn, which has a short half-life of 3.8 days, the anomalous changes are likely dependent on corresponding changes of some carrier gases, such as CO_2. KING (1986) demonstrated how an upward flow of soil air could perturb the subsurface ^{222}Rn profile to cause the anomalies that he had observed. The ^{222}Rn pulses observed in China and elsewhere may be caused by the release of pockets of relatively high or low soil air. For a brief review of the origin and migration of fluids in the crust, see KING and IGARASHI (2002).

Some of the explanations were tested by laboratory studies of audio and ultrasonic vibrations and pressure solution of crushed rock samples , and by field studies of underground explosions, large-scale hydraulic fracturing, and ground-water pumping (e.g., BRACE and ORANGE, 1968; GIARDINI *et al.*, 1976; HONDA *et al.*, 1982; KITA *et al.*, 1982; KING, 1986; KING and LUO, 1990; ZHANG, 1994). Most of the laboratory studies involve the pre-failure phenomena of the volume increase of the specimen and stable sliding along a plane of weakness. When a rock specimen is subjected to increasing uniaxial or triaxial compression, it commonly begins to display nonelastic volume increase (dilatancy) due to the occurrence of micro-cracks at a stress level somewhat below the failure level. The same phenomena may presumably occur in the crust, but to date has not been observed directly.

However, a 'self-organized critical state' (BAK and TANG, 1989), including dilatancy and other related phenomena (permeability increase, acoustic emission, etc.), may possibly have occurred only at the hypocenter and to lesser degrees at some relatively few widely scattered sensitive sites, where the local stress had reached certain subcritical level before the corresponding earthquake; and thus it is hard to detect. To account for such a possibility, one needs to invoke imhomogeneity of the crust, and in the case of pre-earthquake changes a broad-scaled and episodically increasing stress field (KING, 1986; MA et al., 1990). Such a field also may be evidenced by common observation of multiple earthquake occurrences within a short-time interval over a broad region. Under such a condition, strain may concentrate along some faults and weak zones, especially at their intersections and bends, where voids and fluids abound and the stress is near a critical level, thus causing local dilatancy and related hydrologic and geochemical (as well as electric, magnetic, and animal behavior) anomalies at/near the time of the earthquakes. This scenario may explain the observation that the sensitivity of a sensitive site may change with time, because the sensitivity depends on how close the pre-earthquake local stress is to the critical level. The same scenario may also explain volcanic unrest and eruptions triggered by earthquakes over surprisingly long distances (HILL et al., 2002).

5. Discussion and Conclusions

Hydrologic and geochemical parameters, like other geophysical parameters, have been studied extensively during the past several decades, mainly in search of possible premonitory changes useful for earthquake prediction. Nevertheless, relatively few significant precursors have been recorded and the results are still inconclusive. The enormous complexity of the challenge has not been met by a matching amount of funding and manpower. Not knowing where the next destructive earthquakes are going to strike, it has been extremely difficult to deploy a sufficient number of reliable instruments at appropriate locations long enough to record sufficient background data and to catch a significant number of possibly true precursors. To date few significant earthquakes have occurred in several areas of intensive global studies at anticipated times. Maintaining monitoring efforts over long periods of time is difficult because of some inevitable problems, such as funding uncertainty, personnel change, and instrument failure. Other difficulties arise from the inaccessibility of earthquake sources, inhomogeneity of the crust (and the resultant problem of sensitive vs. insensitive sites), and the lack of realistic models that can allow for the imhomogeneity.

In spite of the above-mentioned difficulties, recent studies have shown credible premonitory changes as well as many co- and postseismic changes, some at previously unexpectedly extended epicentral distances. Although the long-distance changes were recorded only at relatively few sensitive sites, their reality can no longer be denied. This phenomenon calls for more realistic models for explanation. The sensitive sites are usually located at structurally weak zones, characterized by certain near-critical stress conditions where some local property, such as permeability, can be greatly changed by seismic shaking and possibly by a small stress increment. Permeability in a fault zone may be enhanced by such processes as faulting, fracturing, microcracking, and brecciation, and reduced by such processes as gouge formation, microcrack healing, hydrothermal cementation of fractures, and solution precipitation (HICKMAN *et al.*, 1995; PARRY, 1998; MORROW *et al*, 2001). The sensitivity of a sensitive site may change with time, depending on how close the pre-earthquake stress is to the critical level for new fissure production.

Because of the current lack of sufficient data to meet the enormous challenge, it seems too early to conclude at this time whether earthquakes are predictable or not in general terms, as GELLER (1997) did. To overcome the above-mentioned difficulties, a prediction effort should: adopt a multistage, multisite, multidisciplinary approach; use sensitive telemetered instruments of good long-term stability and time resolution; deploy them at a sufficient number of sensitive locations long enough to recognize normal background noise and to catch enough target earthquakes; develop sufficiently realistic geophysical and geochemical models for the heterogeneous crust; and use appropriate statistical methods and relevant environmental monitoring for objective data analysis.

Acknowledgments

We would like to thank N. M. Perez and J.-P. Toutain for their benificial reviews.

REFERENCES

BAK, P., and TANG, C. (1989), *Earthquakes as self-organized critical phenomena*, J. Geophys. Res. *94*, 15635–15637.

BERNARD, P. (2001), *From the search of 'precursors' to the research on 'crustal transients'*, Tectonophysics *338*, 225–232.

BRACE, W. F. and ORANGE, A.S. (1968), *Electrical resistivity changes in saturated rocks during fracture and frictional sliding*, J. Geophys. Res. *73*, 1433–1445.

CHADHA, R. K., PANDEY, A. P., and KUEMPEL, H. J. (2003), *Search for earthquake precursors in well water levels in a localized seismically active area of reservoir triggered earthquakes in India*, Geophys. Res. Lett. *30*, 10.1029/2002GLO16694.

CHEN, Q.-F., ed. (2002, 2002, 2003), *Earthquake Cases in China 1992–94, 1995–96, 1997–99*, Seismological Press, Beijing (in Chinese).

CHIA, Y., WANG, Y.-S., CHIU, J.J., and LIU, C. W. (2001), *Changes of groundwater level due to the 1999 Chi-Chi earthquake in the Choshui river alluvial fan in Taiwan*, Bull. Seismol. Soc. Am. *91*, 1062–1068.

EVANS, J.P., FORSTER, C. B., and GODDARD, J. V. (1997), *Permeability of fault-related rocks, and implications for hydraulic structure of fault zones*, J. Struct. Geol. *19*, 1393–1404.

GELLER R.J. (1997), *Earthquake prediction: A critical review*, Geophys. J. Int. *131*, 425–450.

GIARDINI, A. A., SUBBARAYUDU, G. V., and MELTON, C. E. (1976), *The emission of occluded gas from rocks as a function of stress: its possible use as a tool for predicting earthquakes*, Geophys. Res. Lett. *3*, 355–358.

GOMBERG, J., REASONBERG, P. A., BODIN, P., and HARRIS, R. A. (2001), *Earthquake triggering by seismic waves following the Landers and Hector Mine earthquakes*, Nature *411*, 462.

HICKMAN, S., SIBSON, R., and BRUHN, R. (1995), *Introduction to special section: Mechanical involvement of Fluids in Faulting*, J. Geophys. Res. *100*, 12831–12840.

HILL, D.P. *et al.* (1993), *Seismicity remotely triggered by the magnitude 7.3 Landers, California, earthquake*, Science *260*, 1617–1623.

HILL, D.P., POLLITZ, F., and NEWHALL, C. (2002), *Earthquake-volcano interactions*, Physics Today, November, 41–47.

HONDA, M., KURITA, K., HAMANO, Y., and OZIMA, M. (1982), *Experimental studies of H_2 and Ar degassing during rock fracturing*, Earth Planet. Sci. Lett. *59*. 429–436.

IGARASHI, G. and WAKITA, H. (1991), *Tidal responses and earthquake-related changes in water level of deep wells*, J. Geophys. Res. *96*, 4269–4278.

IGARASHI, G., SAEKI, T., TAKAHATA, N., SUMIKAWA, K., TASAKA, S., SASAKI, Y., and SANO, Y. (1995), *Groundwater radon anomaly before the Kobe earthquake in Japan*, Science *269*, 60–61.

IGARASHI, G. and WAKITA, H. (1995), *Geochemical and hydrological observations for earthquake prediction in Japan*, J. Phys. Earth *43*, 585–598.

IRWIN, W.P. and BARNES, I. (1980), *Tectonic relations of carbon dioxide discharge and earthquakes*, J. Geophys. Rev. *85*, 3115–3121.

KENNEDY, B.M., KHARAKA, Y.K., EVANS, W.C., ELLWOOD, A., DePAOLO, D.J., THORDSEN, J., AMBATS, G., and MARINER, R.H. (1997), *Mantle fluids in the San Andreas fault system*, California, Science *278*, 1278–1281.

KING, C.-Y. (1986), *Gas geochemistry applied to earthquake prediction: An overview*, J. Geophys. Res. *91*, 12269–12281.

KING, C.-Y. and IGARASHI, G. (2002), *Earthquake-related hydrologic and geochemical changes*, Int. Handbook of Earthq. Engg. Seismol. *81A*, 637–645.

KING, C.-Y. and LUO, G. (1990), *Variations of electrical resistance and H_2 and Rn emission of concrete blocks under increasing uniaxial compression*, Pure Appl. Geophys. *134*, 45–56.

KING, C.-Y. and MINISSALE, A. (1994), *In search of earthquake-related hydrologic and chemical changes along Hayward fault*, Appl. Geochem. *9*, 83–91.

KING, C.-Y., BASLER, D., PRESSER, T. S., EVANS, W. C., WHITE, L. D., and MINISSALE, A. (1994), *In search of earthquake-related hydrologic and chemical changes along Hayward fault*, App. Geochem. *9*, 83–91.

KING, C.-Y., KOIZUMI, N., and KITAGAWA, Y. (1995), *Hydro-geochemical anomalies and the 1995 Kobe earthquake*, Science *269*, 38–39.

KING, C.-Y., KING, B.-S., EVANS, W. C., and ZHANG, W. (1996), *Spatial radon anomalies on active faults in California*, Appl. Geochem. *11*, 497–510.

KING, C.-Y., AZUMA, S., IGARASHI, G., OHNO, M., SAITO, H., and WAKITA, H. (1999), *Earthquake-related water-level changes at 16 closely clustered wells in Tono, central Japan*, J. Geophys. Res. *104*, 13073–13082.

KING, C.-Y., AZUMA, S., OHNO, M., ASAI, Y., HE, P., KITAGAWA, Y., IGARASHI, G., and WAKITA, H. (2000), *In search of earthquake precursors in the water-level data of 16 closely clustered wells at Tono, Japan*, Geophys. J. Int. *143*, 469–477.

KITA, I., MATSUO, S., and WAKITA, H. (1982), *H_2 generation by reaction between H_2O and crushed rocks: An experimental study in H_2 degassing from the active fault zone*, J. Geophy. Res. *87*, 10789–10795.

KLUSMAN, R. W., (1993), *Soil Gas and Related Methods for Natural Resource Exploration*, John Wiley and Son, New York .

KOIZUMI, N., KANO, Y., KITAGAWA, Y, SATO, T., TAKAHASHI, M., NISHIMURA, S., and NISHIDA, R. (1996), *Groundwater anomalies associated with the 1995 Hyogo-ken Nanbu earthquake*, J. Phys. Earth *44*, 373–380.

LEE, M., LIU, T.-K., MA, K.-F., and CHANG, Y.-M., (2002), *Coseismic hydrological changes associated with dislocation of the September 21, 1999 Chichi earthquake, Taiwan*, Geophys. Res. Lett. *29*, 17, 1824.

LEWICKI, J.L. and BRANTLEY, S.L. (2000), *CO_2 degassing along the San Andreas fault, Parkfield, California*, Geopyhs. Res. Lett. *27*, 5–8.

MA. Z., FU, Z., ZHANG, Y., WANG, C., ZHANG, G., and LIU, D. , *Earthquake Prediction: Nine Major Earthquakes in China* (1966–76) (Seismological Press, Beijing 1990).

MATSUMOTO, N. (1992), *Regression analysis for anomalous changes of ground water level due to earthquakes*, Geophys. Res. Lett. *19*, 1193–1196.

MONTGOMERY, D. R. and MANGA, M. (2003), *Stream flow and water well responses to earthquakes*, Science *300*, 2047–2049.

MORROW, C.A., MOORE, D.E., and LOCKNER, D.A. (2001), *Permeability reduction in granite under hydrothermal conditions*, J. Geophys. Res. *106*, 30551–30560.

MUIR-WOOD, R. and KING, G. C. P. (1993), *Hydrologic signatures of earthquake strain*, J. Geophys. Res. *98*, 22035–22068.

NUR, A. (1974), *Matsushiro, Japan, earthquake swarm: Confirmation of the dilatancy-fluid diffusion model*, Geology *2*, 217–222.

PARRY, W.T. (1998), *Fault-fluid compositions from fluid inclusion observations and solubilities of fracture-sealing minerals*, Tectonophysics *290*, 1–26.

PEREZ, N.M. and HERNANDEZ, P. A. (2005), *Earthquake monitoring and prediction research in active volcanic areas by means of diffuse CO_2 emission studies*, Geophys. Res. Abs. *7*, 10152.

QUILTY, E. and ROELOFFS, E. (1997), *Water level changes in response to the December 20, 1994 M4.7 earthquake near Parkfield, California*, Bull. Seismol. Soc. Am. *87*, 310–317.

ROELOFFS, E. A. (1988), *Hydrologic precursors to earthquakes: a review*, Pure Appl. Geophys. *126*, 177–209.

ROELOFFS, E. A. (1996), *Poroelastic techniques in the study of earthquake-related hydrologic phenomena*, Adv. Geophys. *37*, 135–195.

ROELOFFS, E. A. (1998), *Persistent water-level changes in a well near Parkfield, California due to local and distant earthquakes*, J. Geophys. Res. *103*, 869–889.

ROELOFFS, E. A. and QUILTY, E. (1997), *Water level and strain changes preceding and following the August 4, 1985 Kettlemean Hills, California, earthquake*, Pure Appl.Geophys. *149*, 21–60.

ROJSTACZER, S. and WOLF, S. (1992), *Permeability changes associated with large earthquakes: An example from Loma Prieta, California*, Geology *20*, 211–214.

ROJSTACZER, S., WOLF, S., and MICHEL, R. (1995), *Permeability enhancement in the shallow crust as a cause of earthquake-induced hydrological changes*, Nature *373*, 237–239.

SALAZAR, J.M.L., PEREZ, N.M., HERNANDEZ, P.A., SORIANO, T., BARAHONA, F., OLMOS, R., CARTAGENA, R., LOPEZ, D. L., LIMA, R. N., MELIAN, G., GALINDO, I., PADRON, E., SUMINO, H., and NOTSU, K. (2002), *Precursoy diffuse carbon dioxide degassing signature related to a 5.1 magnitude earthquake in El Salvado, Central America*, Earth Planet. Sci. Lett. *205*, 81–89.

SATO, T., SAKAI, R., FURUYA, K., and KODAMA, T. (2000), *Coseismic spring flow changes associated with the 1995 Kobe earthquake*, Geophys. Res. Lett. *27*, 1219–1222.

SHI, R. and ZHANG, W. (1995), *The correlation between radon variation and solid-earth tide change in rock-groundwater system – the mechanical foundation for using radon change to predict earthquake*, J. Earthq. Prediction Res. *4*, 423–430.

SILVER, P. G. and VALLETTE-SILVER, N. J. (1992), *Detection of hydrothermal precursors to large northern California earthquakes*, Science *257*, 1363–1368.

SILVER, P. G. and WAKITA, H. (1996), *A search for earthquake precursors*, Science *273*, 77–78.

SUGISAKI, R., IDO, M., TAKEDA, H., ISOBE, Y., HAYASHI, Y., NAKAMURA, N., SATAKE, H., and MIZUTANI, Y. (1983), *Origin of hydrogen and carbon dioxide in fault gases and its relation to fault activity*, J. Geology *91*, 239–258.

TOUTAIN, J.-P. and BAUBRON, J.-C. (1999), *Gas geochemistry and seismotectonis: A review*, Tectonophys. *304*, 1–27.

WAKITA, H. (1975), *Water wells as possible indicators of tectonic strain*, Science *189*, 553–555.

WAKITA, H. (1996), *Geochemical challenge to earthquake prediction*, Proc. Natl. Acad. Sci. USA *93*, 3781–3786.

WOLLENBERG, H. A. (1985), *Radon-222 in groundwater of the Long Valley caldera*, Pure Appl. Geophys. *122*, 34–36.

ZHANG, W. (1994), *Research on hydrogeochemical precursors of earthquakes*, J. Earthq. Prediction Res. *3*, 170–182.

ZHANG, W. and LI, X. (1994), *A survey of the hydrogeochemical observation network for earthquake prediction in China*, J. Earthq. Res. in China *8*, 377–386.

ZHANG, W., LUO, G., XIN, Y., and WEI, J. (1988), *A gas geochemical method in detecting active faults*, J. Earthq. Res. in China *2*, 537–541.

ZHANG, W. and ZHANG, X. (1996), *A contrast study on escaping and dissolved radon's micro-behavior characteristics in groundwater*, J. Earthq. Prediction Res. *5*, 148–153.

ZHANG, Z., ed. (1988, 1990, 1990, 1999, 2000), *Earthquake cases in China 1966–75, 1976–80, 1981–85, 1986–88, 1989–91*, Seismological Press, Beijing (in Chinese).

(Received: May 15, 2004; revised: December 1, 2005; accepted: December 4, 2005)

To access this journal online:
http://www.birkhauser.ch

Pure appl. geophys. 163 (2006) 647–655
0033–4553/06/040647–9
DOI 10.1007/s00024-006-0041-2

© Birkhäuser Verlag, Basel, 2006

⌐ Pure and Applied Geophysics

Groundwater-level Changes Due to Pressure Gradient Induced by Nearby Earthquakes off Izu Peninsula, 1997

MASAO OHNO,[1,4] TSUTOMU SATO,[2] KENJI NOTSU,[1] and HIROSHI WAKITA,[†,1]
and KUNIO OZAWA[3]

Abstract—Anomalous water level changes were observed at two wells associated with seismic swarm activity off Izu Peninsula on March, 1997. These are coseismic water level drops followed by gradual post-seismic water level rise at the time of large earthquakes during the swarm activity. The post-seismic water level rises, which can be fitted by an exponential function with a time constant of about six hours, are explained in terms of the horizontal pressure diffusion due to the pressure gradient in the aquifer induced by the coseismic static strain.

Key words: Groundwater, water level, earthquake, strain.

Introduction

Changes in water levels and discharge rates associated with earthquakes have frequently been reported (see e.g., ROELOFFS, 1996). These changes are often discussed in relation with static strain changes due to earthquakes. For example, WAKITA (1975) found that the distribution of groundwater-level rises and drops associated with an earthquake was consistent with the area of contraction and dilatation expected by the faulting. There are also changes that cannot be explained in terms of coseismic static strain change. ROJSTACZER and WOLF (1992) investigated increased discharge in stream after the 1989 Loma Prieta earthquake, and concluded that the earthquake increased rock permeability and temporarily enhanced ground-water flow rates.

[†]Now at Gakushuin Women's College.

[1]Laboratory for Earthquake Chemistry, Graduate School of Science, The University of Tokyo, Tokyo, 113-0033, Japan.
[2]Geological Survey of Japan, AIST, Tsukuba, 305-8567, Japan.
[3]Shizuoka Prefecture Office, Shizuoka, 420-8601, Japan.
[4]Graduate School of Social and Cultural Studies, Kyushu University, Ropponmatsu, Fukuoka, 810-8560, Japan. E-mail: mohno@scs.kyushu-u.ac.jp

ROELOFFS (1998) showed that the coseismic water level rises in a well in California represent diffusion of abrupt coseismic pressure increases in an area in the aquifer.

In this paper we show the anomalous water-level changes of two wells which are located in Ito City on the east coast of Izu Peninsula, Japan (Fig. 1) during the swarm activity in March, 1997. Coseismic drops of water levels are interpreted to be due to static strain changes that are significant as the epicenters are close. Postseismic changes are attributed to the pressure diffusion caused by a pressure gradient in the aquifer.

Observation Wells

Seismic activities in the area off the east coast of Izu Peninsula became high after the Izu-Oshima-Kinkai Earthquake (M7.0) occurred on January 14, 1978. The

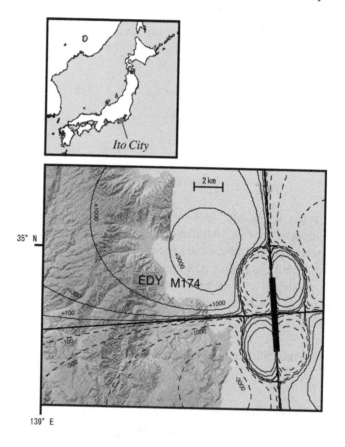

Figure 1
Location of the observation wells (red crosses). Solid and dotted lines show coseismic static strain, dilatation and contraction, respectively, calculated for the fault (thick line) of E2 earthquake using a code by OKADA (1992): The attached figures indicate values in nano-strain.

largest activity occurred in July 1989, accompanied by a submarine eruption off Ito City on July 13 (see IDA and MIZOUE, J. Phys. Earth, Special Issue, 1991). Three days prior to the submarine eruption, four old hot spring wells were found to start to spout suddenly (NOTSU et al., 1991), two of which we will investigate in this paper: EDY well (Matsubara No.136 well) and M174 well (Matsubara No. 174 well). Both wells, separated by 500 m from each other, are about 85 m deep, and are considered to be tapped into an identical aquifer that extends around this area. No data are available on the structure of the wells. The observation systems are described by OHNO et al. (1999). After the time interval of recording the water levels was changed to every second, a variety of water level changes was observed in relation to earthquakes. OHNO et al. (1997) compared water-level fluctuations and strain fluctuations caused by large remote earthquakes, and estimated the hydraulic parameters of the aquifer. Changes in water levels that were synchronized to the burst of seismicity were observed, accompanied by 1996 swarm activity (OHNO et al., 1999). KUNUGI et al. (2000) focused on small (less than one centimeter) changes related to nearby earthquakes and analyzed the underdamped response of the well.

After the swarm activity in 1989, major activities arose in May-June in 1993, September-October in 1995, and October in 1996. On March 3 in 1997, seismic swarm activity resumed. 9334 earthquakes were observed in total at Kamata by JMA (Japan Meteorological Agency), which was similar in number to the activities in 1993 and 1995. In Figure 2, we show the water level changes of the two wells. Associated with this activity, anomalous water temperature change was observed in another well in Ito City (UNIVERSITY OF TOKYO, 1997). Radon changes (UNIVERSITY OF TOKYO, 1997) and water level changes (KOIZUMI et al., 1999) were also observed around Ito City.

Coseismic Change

In the changes of water level in Figure 2, the most striking are the two drops associated with the two large earthquakes: M 5.0 earthquake at 23:09 on March 3 (hereafter referred to as E1) and M 5.7 earthquake (E2) at 12:51 on March 4. The water level of EDY well dropped 150 cm in response to the occurrence of E1; it dropped more than 147 cm after E2, although we could not measure the exact value because the water level dropped below the sensor. At M174 well the level dropped by 138 cm after E1 and by 161 cm after E2.

Here, we will calculate the volumetric strain changes due to the earthquakes, and compare the water level changes to them. The fault parameters during the swarm activity are estimated from gravity change (YOSHIDA et al., 1999) and geodetic data (CERVELLI et al., 2001), respectively. They analyzed the accumulated change through the period of swarm activity, but did not give the parameters for E1 or E2. Here, we used the parameters of left lateral fault that AOKI et al. (1998) estimated from geodetic data for E2, and calculated the volumetric strain change (Fig. 1). The calculated strain

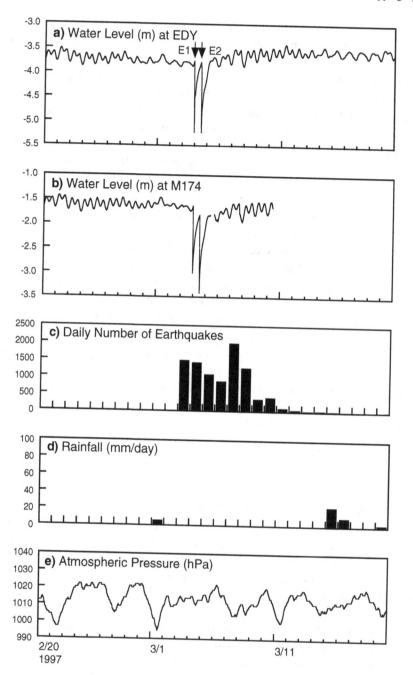

Figure 2
Water level at a) EDY and b) M174, together with the daily number of earthquakes (JMA), rainfall and atmospheric pressure in Ito City.

steps are 1.2×10^{-6} strain for both EDY and M174. The strain sensitivity of the wells was estimated to be 5 mm/nano-strain for EDY (see Appendix in KUNUGI et al., 2000), and that of M174 well was estimated to be 60% of that of EDY (see OHNO et al., 1999). The water level drops calculated from the strain step are 6 m (EDY) and 3.6 m (M174), respectively. These values are larger than the observed changes, but are consistent in order. A possible cause for this discrepancy is that we applied a uniform half-space model in calculating strain caused by such a local earthquake. It is also notable that the calculated strain is sensitive to slight change in the fault parameters as seen in the dense contours around the well in Figure 1.

Post-seismic Change

Another remarkable feature is the changes after these drops; the water levels rose about 1.5 m in c.a. 12 hours, which we refer to as post-seismic changes. To analyze these post-seismic changes, we removed the tidal component and the response of barometric pressure with a computer program BAYTAP-G (TAMURA et al., 1991). The results are shown in Figure 3, in which each post-seismic change is fitted by an exponential function: the fitting function $W(t)$ at time t is denoted as,

$$W(t) = A(1 - e^{t/\tau}), \tag{1}$$

where τ is the time constant and A is the amplitude at the initial time ($t = 0$). It is remarkable that the time constants are nearly the same for all cases; they are as short as about six hours.

To test if the water level rises are due to aquifer unconfinement, we inferred the degree of aquifer confinement from the response of water wells to atmospheric loading by following ROJSTACZER (1988). The calculated barometric efficiency and its phase are almost constant for as long a period as 10 days and do not indicate any diffusion due to unconfinement which might induce the observed quick level change.

In these wells water level rises of $20 \sim 40$ cm were observed associated with the activity in 1993 and 1995 (UNIVERSITY OF TOKYO, 1994, 1996; SATO et al., 1997). Based on a questionnaire to owners of hot springs in Ito City, SATO et al. (1997) proposed that the water level rise was induced by the enhancement of permeability due to shaking, which made hot springs in this area turbid. However a rise of three meters, as the total of the change for E1 and E2, is too large to be attributed to the enhancement of permeability compared to the rises observed in the previous activities.

How can such a quick and large water level change occur? KÜMPEL (1992) and ROELOFFS (1998) discussed water level changes due to the diffusion of coseismic pressure increases in an area apart from the well. ROELOFFS (1998) demonstrated that coseismic water level changes in a well in California could represent diffusion of abrupt coseismic pore pressure increase within several meters of the well, produced by a mechanism similar to that of liquefaction. KÜMPEL (1992) discussed stress diffusion due to the invasion of pressurized pore fluid, in which the shape of the source region was approximated by a sphere. The influence of local inhomogeneities was also discussed. In the present study, it is characteristic that the static strain change caused by E2 has significant gradient around the wells as seen in Figure 1, because the earthquake occurred quite close to the wells. In response to this strain gradient, horizontal pressure gradient should be induced in the aquifer around the wells. In the present study, we will examine water level change caused by horizontal diffusion due to the pressure gradient induced in the aquifer. We will calculate the time constant of water level change due to horizontal diffusion at the present aquifer, and compare it to the observation.

Here we consider the case of one dimension because the strain distribution around the wells is almost one-dimensional in the direction of south to north as seen in Figure 1. The diffusion of pressure u is governed by the equation,

$$\frac{\partial u}{\partial t} = c_h \frac{\partial^2 u}{\partial x^2}, \tag{2}$$

where c_h is the horizontal hydraulic diffusivity. The induced pressure diffuses following a decaying exponential function, whose time constant τ is represented as,

$$\tau = \frac{L^2}{4\pi^2 c_h} \tag{3}$$

in which L denotes the wavelength of sinusoidal pressure distribution (see e.g., RILEY *et al.*, 1997).

OHNO *et al.* (1997) estimated the parameters of aquifer: K (hydraulic conductivity) $=0.01$ m/s, S (storativity) $=0.0001$, and b (thickness of aquifer) $=5$ m. Then, the hydraulic diffusivity c_h

$$c_h = \frac{K}{S \cdot b} \tag{4}$$

is calculated to be 20 m^2/s. Although information regarding the extent of the aquifer is poor, the scale of aquifer (La) is presumed to be $1 \sim 3$ km from the distribution of hot springs in this area in SATO *et al.* (1992). Hot springs distribute in the alluvial plain (the green colored area around the wells in Fig. 1). The length L in equation (3) is the spatial scale of the pressure distribution induced in the aquifer by strain. In the present case, the pressure in the aquifer decreases monotonously from south to north as the strain increase from south to north around the wells (Fig. 1). If we approximate this pressure distribution by a part of a sinusoidal curve as

$$u(x) = a \cos\frac{\pi x}{La} \quad (0 < x < La), \tag{5}$$

where a is a constant, then $L=2La$. Therefore, the time constant τ is calculated to be $1.4 \sim 10$ hours, which is consistent with the observed time constant of about six hours.

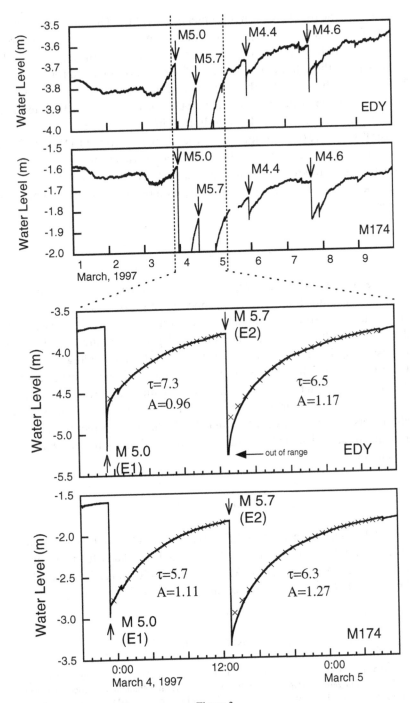

Figure 3
Water levels at every minute after removing the components of barometric response and tidal components.
Part of the upper diagrams is enlarged to lower diagrams. Crosses in lower diagrams show the fitted
functions. Each time constant (τ) is given in hours, and amplitude (A) in meters.

E1 occurred at nearly the same hypocenter as that of E2. The focal mechanism of E2 is similar to E1, but it has a component of vertical movement (Japan Meteorological Agency). It is probable that pressure gradient was induced in the aquifer by E1 in the same way as E2, although the parameters of fault are not obtained for E1.

The differences between the observations and the fitted functions which are found immediately after the occurrence of earthquakes in Figure 3 can be fitted again by exponential functions with considerably shorter time constant. For example, after we subtract the fitted curve ($\tau = 6.5$ hour) from the water-level rise of EDY for E2, then the residual can be fitted again by another exponential function ($\tau = 0.7$ hour). These changes with shorter time constants are considered to be related to smaller spatial scale, which may be local pressure gradients due to inhomogeneity of physical properties of the aquifer.

Conclusion

We have investigated the response of the water level of wells to local earthquakes. The water levels, which dropped in response to coseismic static strain of local earthquakes, rose exponentially with a time constant of ca. six hours. If we consider pressuré diffusion due to the horizontal gradient induced in the aquifer by local coseismic static strain, the time constant of water level change is calculated to be $1.4 \sim 10$ hours, which is consistent with the observed ones. This phenomenon is considered to be characteristic in cases that a significant earthquake occurs quite close to a well whose aquifer has high diffusivity. In such cases as the present study, we will understand the observed water level changes correctly by considering pressure diffusion, which will aid further investigation of the mechanism of the seismic activities.

Acknowledgments

We thank F. Tanabe, the owner of EDY well, and the office of Ito City for allowing us to use the well for observation, and Y. Tamura and Y. Okada for allowing us to use computer programs. We are grateful to M. Inoue and T. Mori for assistance in observation. We acknowledge S. Gurrieri, C. Federico, and J.-P. Toutain for critical comments.

REFERENCES

AOKI, Y., KATO, T., SEGALL, P., and CERVELLI, P. F. (1998), *Focal process of March, 1997, off-Ito swarm activity derived from geodetic data*, Abstract of Japan Earth and Planetary Science Joint Meeting, 290.
CERVELLI, P., MURRAY, M. H., SEGALL, P., AOKI, Y., and KATO, T. (2001), *Estimating source parameters from deformation data, with an application to the March 1997 earthquake swarm off the Izu Peninsula, Japan*, J. Geophys. Res. *106*, 11217–11237.

IDA, Y., and MIZOUE M. (editors) (1991), *Special issue: Seismic and volcanic activity in and around the Izu Peninsula and its tectonic implications*, J. Phys. Earth *39*, 1–460.

KOIZUMI, N., TSUKUDA, E., KAMIGAICHI, O., MATSUMOTO, N., TAKAHASHI, M., and SATO, T., (1999), *Preseismic changes in groundwater level and volumetric strain associated with earthquake swarms off the east coast of the Izu Peninsula, Japan*, Geophys. Res. Lett. *26*, 3509–3512.

KÜMPEL, H.-J. (1992), *About the potential of wells to reflect stress variations within inhomogeneous crust*, Tectonophysics *211*, 317–336.

KUNUGI, T., FUKAO, Y., and OHNO, M. (2000), *Underdamped responses of a well to nearby swarm earthquakes off the coast of Ito City, central Japan, 1995*, J. Geophys. Res. *105*, 7805–7818.

NOTSU, K., WAKITA, H., IGARASHI, G., and SATO, T. (1991), *Hydrological and geochemical changes related to the 1989 seismic and volcanic activities off the Izu Peninsula*, J. Phys. Earth *39*, 245–254.

OHNO, M., WAKITA, H., and KANJO, K. (1997), *A water well sensitive to seismic waves*, Geophy. Res. Lett. *24*, 691–694.

OHNO, M., SATO, T., NOTSU, K., WAKITA, H., and OZAWA, K. (1999), *Groundwater-level changes in response to bursts of seismic activity off the Izu Peninsula, Japan*, Geophys. Res. Lett. *26*, 2501–2504.

OKADA, Y. (1992), Internal deformation due to shear and tensile faults in a half-space, Bull. Seismol. Soc. Am. *82*, 1018–1040.

RILEY, K. F., HOBSON, M. P., and BENCE, S. J. (eds.), *Mathematical Methods for Physics and Engineering*, (Cambridge University Press 1997).

ROELOFFS, E. A. (1996), *Poroelastic techniques in the study of earthquake-related hydrologic phenomena*, Adv. Geophys. *37*, 135–195.

ROELOFFS, E. A. (1998), *Persistent water level changes in a well near Parkfield, California, due to local and distant earthquakes*, J. Geophys. Res. *103*, 869–889.

ROJSTACZER, S. and WOLF, S. (1992), *Permeability changes associated with large earthquakes: An example from Loma Prieta, California*, Geology *20*, 211–214.

SATO, T., TAKAHASHI, M., and OZAWA, K. (1997), *Hydrologic changes associated with the swarm activities off the east coast of Izu Peninsula in October, 1996*, Rep. Cood. Comm. Earthq. Pred. *57*, 367–372.

TAMURA, Y., SATO, T., OOE, M., and ISHIGURO, M. (1991), *A procedure for tidal analysis with a Bayesian information criterion*, Geophys. J. Int. *104*, 507–516.

UNIVERSITY OF TOKYO (1994), *Groundwater changes associated with the seismic swarms off the east coast of Izu Peninsula (May 26 – June 3, 1993)*, Rep. Cood. Comm. Earthq. Pred. *51*, 430–432.

UNIVERSITY OF TOKYO (1996), *Groundwater changes associated with the seismic swarms off the east coast of Izu Peninsula*, Rep. Cood. Comm. Earthq. Pred. *55*, 354–356.

UNIVERSITY OF TOKYO (1997), *Groundwater changes associated with the seismic swarms off the east coast of Izu Peninsula (March, 1997)*, Rep. Cood. Comm. Earthq. Pred. *58*, 318–319.

YOSHIDA, S., SETA, G., OKUBO, S., and KOBAYASHI, S. (1999), *Absolute gravity change associated with the March 1997 earthquake swarm in the Izu Peninsula, Japan*, Earth Planets Space *51*, 3–12.

WAKITA, H. (1975), *Water wells as possible indicators of tectonic strain*, Science *189*, 553–555.

(Received: June 8, 2003; revised: January 15, 2005; accepted: February 2, 2005)

To access this journal online:
http://www.birkhauser.ch

© Birkhäuser Verlag, Basel, 2006

Pure appl. geophys. 163 (2006) 657–673
0033–4553/06/040657–17
DOI 10.1007/s00024-006-0048-8

❘Pure and Applied Geophysics

Detection of Aseismic Slip on an Inland Fault by Crustal Movement and Groundwater Observations: A Case Study on the Yamasaki Fault, Japan

Yuichi Kitagawa,[1] Naoji Koizumi,[1] Ryu Ohtani,[1] Kunihiko Watanabe,[2] and Satoshi Itaba[2,3]

Abstract—To understand the detailed process of fault activity, aseismic slip may play a crucial role. Aseismic slip of inland faults in Japan is not well known, except for that related to the Atotsugawa fault. To know whether aseismic slip does not occur, or is merely not detected, is an important question. The National Institute of Advanced Industrial Science and Technology constructed an observation site near Yasutomi fault, a part of the Yamasaki fault system, and has collected data on the crustal strain field, groundwater pressures, and crustal movement using GPS. In a departure from the long-term trend, a transient change of the crustal strain field lasting a few months was recorded. It indicated the possibility of an aseismic slip event. Furthermore, analyses of data from the extensometers at Yasutomi and Osawa observation vaults of Kyoto University, as well as GPS data from the Geographical Survey Institute (GEONET), revealed unsteady crustal strain changes. All data could be explained by local, left-lateral, aseismic slip of the order of 1 mm in the shallow part of the Yasutomi fault.

Key words: Borehole strainmeter, contractional strain, Yamasaki fault system, aseismic slip, groundwater.

Introduction

Aseismic slip may play various roles according to the time and place where the slip occurs. An example is the case in which aseismic slip causes a stress concentration, accelerating the occurrence of a large earthquake. Another example is the case in which aseismic slip relieves enhanced stress due to large earthquake

---
[1]Tectono-Hydrology Research Group, Geological Survey of Japan, National Institute of Advanced Industrial Science and Technology, Site C7, 1-1-1, Higashi, Tsukuba, Ibaraki, 305-8567, Japan.
E-mail: y-kitagawa@aist.go.jp; koizumi-n@aist.go.jp; ohtani-ryu@aist.go.jp
[2]Research Center for Earthquake Prediction, Disaster Prevention Research Institute, Kyoto University, Gokasho Uji,, Kyoto, 611-0011, Japan. E-mail: watkun@rcep.dpri.kyoto-u.ac.jp; itaba@rcep.dpri.kyoto-u.ac.jp
[3]Tectono-Hydrology Research Group, Geological Survey of Japan, National Institute of Advanced Industrial Science and Technology, Site C7, 1-1-1, Higashi, Tsukuba, Ibaraki, 305-8567, Japan.
E-mail: itaba-s@aist.go.jp

occurrence. In order to understand the process of fault activity, it is necessary to know whether or not stable or unsteady aseismic slip occurs, and to know additional characteristics of the fault besides the seismic slip. For the San Andreas fault in the USA, unsteady aseismic slip (creep) is well known, and creep events have been studied by use of creep meters, groundwater levels, and dilatometers (e.g., ROELOFFS *et al.*, 1989, WESSON ,1988). Many creep events on the San Andreas fault involve total slip of a few mm or less and durations of a few days or less. In Japan, with the recent advance of continuous GPS monitoring, knowledge of aseismic slip on faults near the plate boundary has been accumulating (e.g., OZAWA *et al.*, 2002) and the detection of aseismic slip expected to occur in the lower crust has been attempted.

In the past, it was believed that no aseismic slip occurred on the inland faults in Japan (MATSUDA, 1969). At present, cases of the Atotsugawa fault are the only known examples as mentioned below. TADA (1998) found possible creep events of 1–1.5 mm/year occurring on the central part of the Atotsugawa fault by geodetic measurements. ITO and KUWAHARA (1999) detected a possible creep event of 0.5 mm using borehole strainmeters. HIRAHARA *et al.* (2003) also suggested possible creep events of 1.5 mm/year detected by the clustered GPS monitoring sites around the Atotsugawa fault.

The National Institute of Advanced Industrial Science and Technology (AIST) has been carrying out multi-parameter observations near active faults, in order to study interrelationships between the process of stress and strain accumulation at the active faults, seismicity, and groundwater level changes (TSUKUDA *et al.*, 2000). The Yasutomi station (YST) was sited near the Yasutomi fault, a part of the Yamasaki fault system, to monitor the crustal strain field with an Ishii-type three-component

(a)   **(b)**

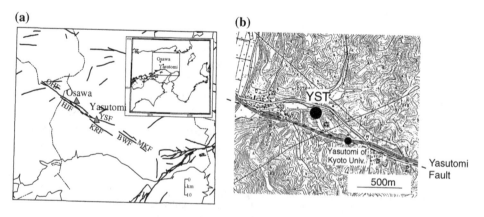

Figure 1
Locations of the observation stations. (a) The Yamasaki fault system consists of several faults. Solid triangles show the Yasutomi and the Osawa observation stations. OHF: Ohara Fault, HJF: Hijima Fault, YSF: Yasutomi Fault, KRF: Kuresakatoge Fault, BWF: Biwakou Fault, MKF: Miki Fault. (b) The map shows the Yasutomi area. Large solid circle is the Yasutomi station (YST) of AIST. Small one is the Yasutomi station of Kyoto University.

borehole strainmeters, groundwater pressure, and crustal movement with GPS monitoring, respectively, since 1998 (KOIZUMI *et al.*, 2000). The trend of the crustal strain field started to show a slight change in April 2002, and an anomalous rapid contraction of the crustal strain field was observed in June 2002. Thereafter all three components of the borehole strainmeter maintained a contractional trend until

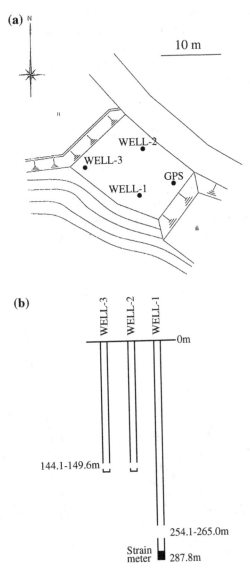

Figure 2
Layout of the observation wells at YST. (a) Plan view of three wells located within 10 m of each other. (b) Elevation view of three wells of which perforated intervals are shown as open parts.

mid-August, 2002. Altogether the contractional trend endured for a few months. This anomalous phenomenon had not been recorded since the observation of the crustal strain field at YST began in 1998, and is interpreted as a result of an aseismic slip (creep). Previously, little creep was known on the Yamasaki fault system (KISHIMOTO and OIKE, 1985).

It is possible that the phenomenon is a form of activity on the Yamasaki fault. To understand inland faults in Japan, the study of the characteristic appearance of this phenomenon is thought to be important. In this paper, we analyzed the phenomenon utilizing data from the borehole strainmeters, groundwater pressures, and GPS monitoring at YST, and the extensometers at the Yasutomi and Osawa observation vaults of Kyoto University.

Observations

The Yamasaki fault system is a left-lateral strike-slip fault. The total length of the Yamasaki fault system is about 80 km and its displacement rate is about 0.3 meters per 1000 years (THE RESEARCH GROUP FOR ACTIVE FAULTS OF JAPAN, 1991). The AIST observation site YST is located within 200 meters of the well-defined displacement zone of the Yasutomi fault (Fig. 1). At YST, three boreholes were

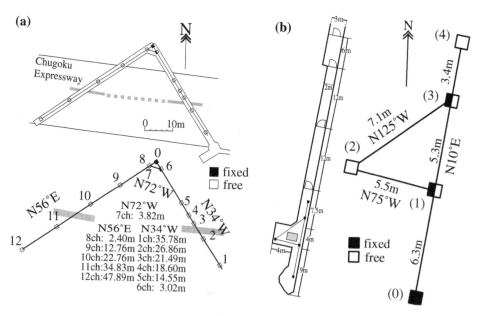

Figure 3
Layout of the extensometers at (a) the Yasutomi and (b) the Osawa observation vaults of Kyoto University.

drilled within 10 meters of each other (Fig. 2). Groundwater pressures were measured in three boreholes, and in the deepest borehole the crustal strain field was measured by use of an Ishii-type multi-component borehole strainmeter that includes three-component strainmeter. At WELL-1 of YST, an Ishii-type multi-component borehole strainmeters was set at 287.8 m depth, and the groundwater pressure was measured through a perforated interval between depths of 254.1–265.0 meters. At WELL-2 and WELL-3, the groundwater pressures were measured through perforated intervals between depths of 144.1–149.6 meters. The

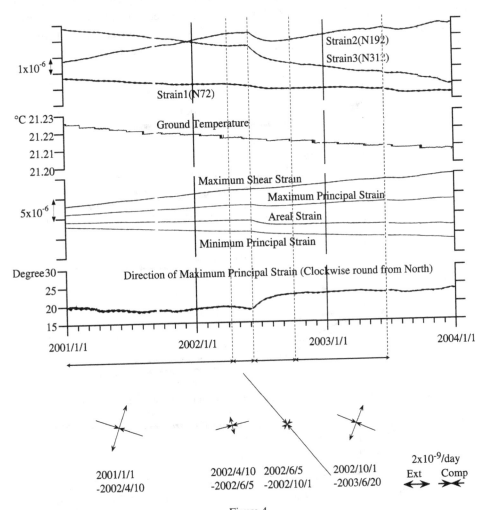

Figure 4

Crustal strain at YST. The uppermost graph shows raw data from the borehole strainmeters. The upper-middle graph is the temperature of the borehole strainmeters. The lower-middle and lowermost graphs show the results of strain analyses of raw data. The strain field during each term is shown pictorially at the bottom.

continuous GPS monitoring utilizes a 5-meter high antenna. GPS observation conditions at YST are not very favorable because a small mountain on the south side of YST obstructs the view of satellites (OHTANI *et al.*, 2003).

The Yamasaki Fault Observation Station of the Research Center for Earthquake Prediction, Disaster Prevention Research Institute, Kyoto University, was located about 500 meters southeast of YST (Fig. 1b). At the L-shaped Yasutomi observation vault, excavated under the Chugoku Expressway (Fig. 3a), the crustal extension has been observed by twelve sensors. In the vault, a clear fracture zone passes between sensors Nos.2 and 3 and between Nos.10 and 11. WATANABE (1991a) noted that from 1976 to 1987, no left-lateral strike-slip had been detected and the long-term strain variations with time intervals of a few years were recognized. Therefore, it is probable that the enitre L-shaped vault exists within the fracture zone of the

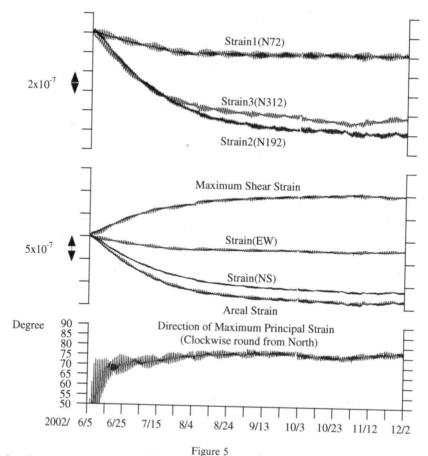

Figure 5

Crustal strain at YST after removal of the long-term trend. The uppermost graph shows the results after the trend values are subtracted from those for January 2002 to March 2002. The middle and lowermost graphs show the results of the strain analysis of the data in the uppermost graph.

Yasutomi fault (WATANABE, 1991a). Geodetic measurements by FUJIMORI et al. (1996) indicated that the Yasutomi fault had displacements not only on the main fracture zone, but also within a wide shear zone.

The Osawa observation vault is located near the Hijima fault. The crustal extension is observed along five lines (Fig. 3b). There is a clear fracture zone in the 0–1 comp (ITABA et al., 2001).

Results

Crustal Strain Field Observed with Borehole Strainmeters

Figure 4 shows the crustal strain data recorded with the Ishii-type three-component borehole strainmeters at YST. Until April 2002, the crustal strain components in N72E-N252E and N312E-N132E directions showed stable and contractional long-term trends, and the crustal strain component in N192E-N12E showed a stable and extensional long-term trend, except for the beginning of the observation and coseismic strain steps. This trend agrees well with the crustal strain field for most of western Japan (GEOGRAPHICAL SURVEY INSTITUTE, 1987). On April 10, 2002, the trend showed a slight change. The strain rate decreased and the crustal strain field changed to NNW-SSE extension and ENE-WSW contraction. This trend is also within the crustal strain field for most of western Japan (GEOGRAPHICAL SURVEY INSTITUTE, 1987). Since June 5, 2002, the crustal strain field has shown a large contraction. This anomalous contraction had the maximum contraction in the NW-SE direction and was very different from the previous trend. The change during the first half of June 2002 was the largest, and the amount of change gradually became small and had almost ceased in September 2002.

In order to accurately assess the change in crustal strain since June 5, 2002, the anomalous change needs to be separated from the raw data by estimating the long-term trend and removing it from the raw data. Each trend is estimated by a linear fit to each of the three components of strain data collected from January 2002 to March 2002. The anomalous change of each component after June 5, 2002 is calculated from each raw data value by subtracting each trend value (Fig. 5). The resulting anomalous changes are similar to exponential functions. The total quantity of the anomalous change was derived by the result of strain analysis (Table 1). The accumulated crustal strain anomaly was an overall contraction with the maximum contraction in the NNW-SSE (N10–15°E) direction and the minimum contraction in the ENE-WSW (N75–80°E) direction.

Groundwater Pressures

Figure 6 shows the groundwater pressures at WELL-2 and WELL-3. Because at three boreholes the water head potentials are higher than the ground level, the

Table 1

Comparison of observed crustal strain change at YST with that calculated from the slip model shown in Figure 11

	Observation	Model
EW component	-0.24×10^{-6}	-0.23×10^{-6}
NS component	-1.15×10^{-6}	-1.17×10^{-6}
Areal strain	-1.39×10^{-6}	-1.40×10^{-6}
Maxmum shear	1.02×10^{-6}	0.97×10^{-6}
Direction	N77°E	N83°E

Direction means the direction of the maximum principal strain.

groundwater pressures as pore water pressures were measured in sealed boreholes. At WELL-1, the groundwater pressure was not accurately measured due to imperfect sealing, consequently the data from WELL-1 have not been used in this report. At WELL-2 and WELL-3, the pressures increased after 1999 as the boreholes were sealed. When an inspection took place in 2000, the boreholes were left open until they were sealed again in 2001. After the sealing they showed recovery of pressures. The

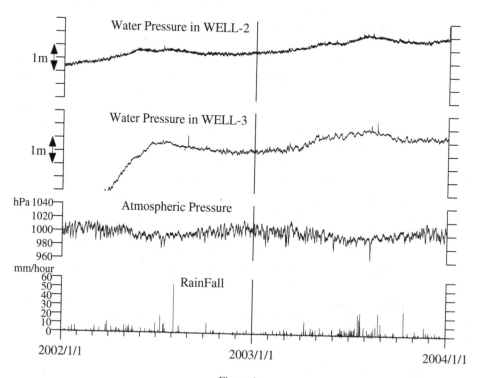

Figure 6
Groundwater pressures of WELL-2 and WELL-3, atmospheric pressure, and rainfall at YST. The groundwater pressure is converted into the height of the water, that is, groundwater level.

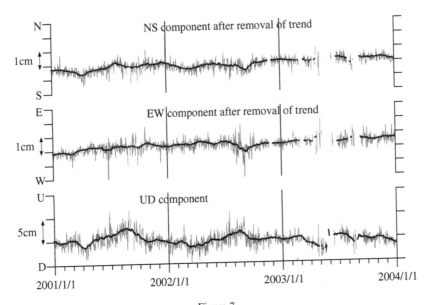

Figure 7

Relative baseline length change of YST by GPS (to Tsukuba as a reference point). The gray lines are those for daily data after the removal of trends. NS component has a southward trend of 0.45 cm per year. EW component has an eastward trend of 2 cm per year. Solid lines show 31-day moving averages.

differences in the timing of increases corresponded to the times when each borehole was sealed.

The depths of the perforated intervals of WELL-2 and WELL-3 are the same and it is thought that the same stratum has been monitored judging from the core data. However, the groundwater pressures at WELL-2 and WELL-3 differ considerably in the following manner, and it seems that the pressures are not well linked. From the tidal responses, the strain sensitivity of WELL-2 is 1.3–1.5 cm per 10^{-8} strain, while that of WELL-3 is 0.4–0.6 cm per 10^{-8} strain. The coefficient of the barometric response of WELL-2 is 0.3–0.5 cm per hPa in a range of 0.1–0.6 cycles per day and slightly smaller in low frequency. The coefficient of barometric response of WELL-3 is 0.5 cm per hPa at 0.1 cycles per day and 0.8 cm per hPa at 0.6 cycles per day, generally decreasing as frequency decreases. These results suggest that the aquifers are not well confined, particularly at WELL-3.

The groundwater pressure at WELL-2, after the resealing in 2001, rapidly increased until the middle of 2001, and thereafter slowly increased until the latter half of May 2002 when it started to decrease slowly. The groundwater pressure at WELL-3 rapidly increased just after the resealing in November 2001, and thereafter greatly increased until the middle of June 2002 followed by a slow decrease. The pressures thus increased after resealing, however because the observation was not for a long term, their annual fluctuations due to rainfall were not clearly understood. Therefore,

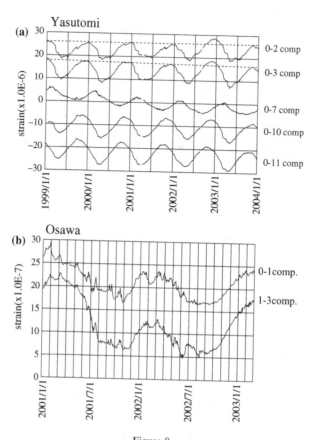

Figure 8
Extensometer results at (a) Yasutomi and (b) Osawa observation vaults of Kyoto University.

in the present circumstance it is difficult to distinguish changes due to crustal strain from the increases after resealing and the annual fluctuations due to rainfall. Considering strain sensitivities from tidal responses, a pressure increase of several tens of cm of water would have been expected to accompany the contractional strain of 1.4×10^{-6}, but in fact no clearly related increases were detected. In June 2002, the pressures rather slowly decreased, and therefore it may be possible that at the depths of the perforated intervals of WELL-2 and WELL-3, there were changes related to extensional strain. However, due to poorly confined aquifers, a transient pressure change may be unmaintainable for a long term.

GPS Monitoring

Figure 7 shows the relative baseline length change of YST to Tsukuba. For details of the results by GPS monitoring, refer to OHTANI *et al.* (2003). Since the

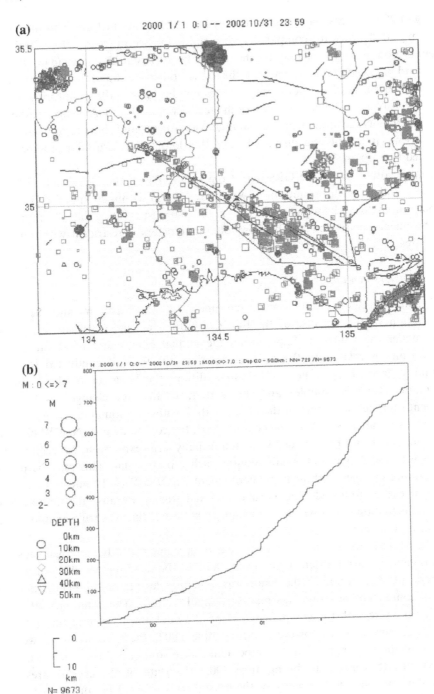

Figure 9
(a) Epicenters of earthquakes around Yamasaki fault from January 2000 to October 2002 and (b) accumulated number of earthquakes around the southeastern part of the Yamasaki fault system shown as an enclosed area in Figure 9 (a).

spring of 2002, an eastward migration may have slowed about 1 cm. This means that YST migrated westward compared with a stable long-term trend before 2002. This phenomenon may be related to the anomalous crustal strain change. Relative westward migration is consistent with the left-lateral strike slip of the Yasutomi fault. However, because of the poor geographical conditions of GPS monitoring at YST and the problem in the performance of the receiver made by AOA company, the GPS data clearly include annual and semiannual components and short-term fluctuations. At present, it is not conclusive whether this anomalous crustal phenomenon is detected by GPS at YST.

On the other hand, the GPS data from GEONET of the Geographical Survey Institute, the observation stations of which are located within 20 km of YST, do not display anomalous phenomena (OHTANI *et al.*, 2003), suggesting that the anomalous crustal phenomenon detected by borehole strainmeters is restricted within a small local area.

Extensometers in Vaults

Figure 8 shows the data from extensometers at the Yasutomi and the Osawa observation vaults. Because of the thin overburden of the Yasutomi observation vault under the Chugoku Expressway, the annual component of strain is large ($\sim 10^{-5}$) and is related to thermal deformation due to heat conducted from the ground surface (WATANABE, 1991a). In addition, the behavior of the Yasutomi observation vault is complex and shows frequent step-like changes ($\sim 10^{-7}$) and frequent short-term changes of the order of 10^{-7}, without anomalous changes of the borehole strainmeters at YST. Since May 2002, the components in N34°W (0–2 and 0–3 components in Fig. 8) manifested an unusually large expansion, the beginning of which was about one and a half months earlier than usual. This may be due to temperature changes, although the components in N56°E (0–10 and 0–11 components in Fig. 8) followed their normal seasonal trends. Because the amplitude of annual fluctuation varies widely, it remains to be seen if this was truly an anomalous crustal movement.

The Osawa observation vault is located in comparatively homogeneous tuff breccia, and its overburden in the deep vault is about 45 meters. The annual air temperature fluctuation in the vault was generally less than 0.1 degree and the annual component of strain was relatively small ($\sim 10^{-6}$). The component 0–1 was usually in phase with the component 1–3 on the tidal and long-term (more than a few days) movements. However, during June 2002, these components were in phase on tides, whereas the components were out of phase on long-term movements. It seems that during June 2002, the state of the crustal stress was anomalous or the inhomogeneity of the crustal stress existed because of a fracture zone. The cause and the relevance of these results to the crustal strain anomaly at YST remain to be solved.

Seismicity

We investigated whether or not earthquakes were related to the crustal strain anomaly at YST. Figure 9 shows the distribution of epicenters around the Yamasaki fault system since 2000. The preliminary reports of hypocenter data from Kyoto University and the Japan Meteorological Agency were used. A period of seismic activity began in the mid- 2001. The crustal strain field changes observed by the borehole strainmeters were not related to the active seismicity. In addition, on April 10, 2002, the seismicity became temporarily active. Hypocenters of this seismicity were located about 10 km southeast of YST and near Kuresakatoge fault rather than the Yasutomi fault (Fig. 1a). This seismicity occurred at the time when the trend of crustal strain observed by the borehole strainmeters began to change. Figure 10 shows the seismicity near YST. There was little seismicity from January 2002 to May 2002. On June 2, 2002, a small earthquake (M1.1) occurred along the Yasutomi fault ESE of YST, followed by a small earthquake (M1.3) on July 1, 2002, along the Yasutomi fault WNW of YST. Considering the hypocentral distances to YST, it was concluded that seismic slips by these two earthquakes did not affect the crustal strain at YST. It is possible, however, that source changes in crustal stress affected simultaneously both the crustal strain and seismicity.

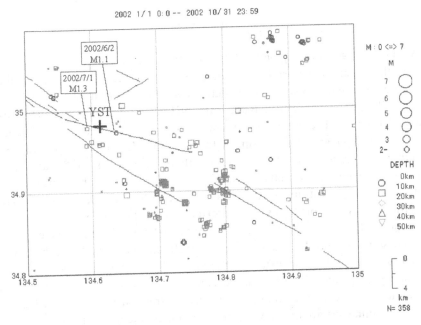

Figure 10

Epicenters of earthquakes near YST (marked in +) from January 2002 to October 2002.

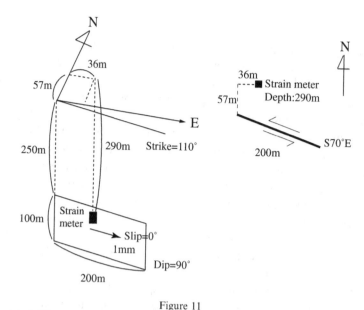

Figure 11
Slip model. (a) Solid rectangle corresponds to the position of the borehole strainmeters. Open parallelogram shows the slip area. (b) Horizontal view. Solid square designates the borehole strainmeters. Thick line shows the projected slip area.

Discussion

An unsteady aseismic slip on the Yasutomi fault is considered as a strong possible cause of the remarkable crustal strain change since June, 2002. Our model of the slip is proposed on the following three restricting conditions. (1) The slip is a perfect left-lateral strike slip on a vertical plane with N70°W strike, the same strike as the Yasutomi fault. (2) The borehole strainmeters at YST are located within 200 meters of the vertical slip plane of the Yasutomi fault. (3) The slip-model-induced changes in groundwater pressures at YST (at the depth of 150 meters), GPS data (on the surface), and extensometer data in the Yasutomi and Osawa observation vaults of Kyoto University (just under the surface) are below the level of background fluctuations. It is possible that GPS and extensometer data show anomalous results. At the moment, however, it cannot be determined whether these anomalies were real, and their sizes have not been determined. Therefore, we assumed that there were no anomalies in the GPS and extensometer data. The slip model includes no time dependency. The model with a finite slip size is obtained by fitting the final anomalous strain changes. Because the crustal strain anomalies recorded by the borehole strainmeters are the only clear constraints on the modeling, the model is not unique.

Considering 1 mm of left-lateral slip on a rectangle with a vertical dimension of 100 meters and a horizontal dimension of 200 meters as an example (Fig. 11), the observed crustal strain changes are acceptable in size and sign (Table 1). From this model, 7×10^{-8} contractional crustal strains are expected at the perforated intervals of WELL-2 and WELL-3 (at the depths of 150 meters) of YST. Based on the strain sensitivity of WELL-2 and WELL-3, the expected changes of groundwater pressures are increases on the scales of 3–10 cm. Considering the aseismic slip lasted for three months, from June 2002 to September 2002, no wonders these long-term pressure changes were not detected, because the aquifers are poorly confined. In addition, the expanded areas were adjacent to the contracted areas according to the spatial distribution of the calculated strain change by the slip model, and consequently groundwater pressures increase and decrease in each contracted and expanded area. Consequently groundwater will flow from increased areas to decreased areas and the flow will gradually cancel the increases and decreases in groundwater pressure by the slip. Therefore, the flow is a possible cause of why long-term pressure changes were not detected. At ground surface, no more than 10^{-2} mm of displacement and no more than 10^{-8} strain changes are expected from the model. These expected changes cannot be detected by use of GPS at YST and extensometers at the Yasutomi observation vault, and it is consistent with the present interpretation. However, if further analysis of GPS data at YST indicates that eastward migration has actually slowed by 1 cm (OHTANI et al. 2003), then we may need to reconsider the slip model.

As stated above, models with a few mm slip can also explain the crustal strain change. A slip of a few mm corresponds to those of typical creep events on different faults. However, typical durations of creep events range from a few hours to a few days, while the duration of the phenomenon described here is considerably longer. Duration of up to a few months is similar to the postseismic deformation (e.g., NAKANO and HIRAHARA, 1997). A hypothesis that a short-term slip creates a long-term crustal strain because of inelastic characteristics of crustal rocks needs to be considered.

Conclusions

Rapidly contracting crustal strain since June 2002 detected by the borehole strainmeters at the YST site of AIST can be explained by the unsteady aseismic slip (creep). The slip model is interpreted as a small local event and is consistent with the absence of signals from the groundwater pressure sensors and GPS at YST, and extensometers in the Yasutomi and Osawa observation vaults. The amount of slip needed to explain the strain anomaly agrees well with that of typical creep events. The duration of the event, however, is unusually long and it may require other interpretations. Clear-cut data were obtained only from the borehole strainmeters, although the interpretation was constrained not only by the crustal strain changes but also by the fact that no changes were detected in other observations. In order to fully characterize fault activity, it is important to investigate various time and spatial scales of crustal movements by combining the measurements in the boreholes and vaults, and the geodetic and the GPS monitoring.

Acknowledgments

We recognize cooperation from the Yasutomi town office and related organizations regarding the establishment and management of the YST observation site. Makoto Takahashi, Norio Matsumoto, and Tomoko Ookawa of AIST, and Setsuro Nakao of Kyoto University, assisted us with the maintenance of the observation at YST. We used computer software, BAYTAP-G (ISHIGURO *et al.*, 1981; TAMURA *et al.*, 1991) and MICAP-Gv2.1 (NAITO and YOSHIKAWA, 1999; OKADA, 1992) for analyses of data. We are grateful to those who are mentioned above.

REFERENCES

FUJIMORI, K., YAMAMOTO, T., and OTSUKA, S. (1996), *Geodetic measurements at the Yasutomi-Usuzuku baseline network across the Yamasaki fault (1975–1995)*, Annuals, Disas. Prev. Res. Inst., Kyoto Univ., No.39, B-1, 303–309 (in Japanese).

GEOGRAPHICAL SURVEY INSTITUTE (1987), *Horizontal strain in Japan*, 75, Association for the Development of Earthquake Prediction (in Japanese).

HIRAHARA, K., OOI, Y., ANDO, M., HOSO, Y., WADA, Y., and OHKURA, T. (2003), *Crustal movements of the Atotsugawa fault -Do creeps occur?*, Earth Monthly 25, 59–64 (in Japanese).

ISHIGURO, M., AKAIKE, H., OOE, M., and NAKAI, S. (1981), *A Bayesian approach to the analysis of earth tides*, Proc. 9th Internat. Symp. on Earth Tides, pp. 283–292.

ITABA, S., MATSUO, S., ASADA, T., WADA, Y., WATANABE, K., and NEGISHI, H. (2001), *Extensometer observation at Osawa area on the Yamasaki fault*, Annuals, Disas. Prev. Res. Inst., Kyoto Univ., No.44, B-1, 185–190 (in Japanese).

ITO, H. and KUWAHARA, Y. (1999), *An episodic creep event observed by 3 component strainmeters along the Atotsugawa fault*, Abstracts of the 1999 Fall Meeting of the Seismological Society of Japan, P141.

KISHIMOTO, Y. and OIKE, K., (1985), *A summary of the Yamasaki fault*, Earth Monthly 67, 4–8 (in Japanese).

KOIZUMI, N., CHO, A., TAKAHASHI, M., and TSUKUDA, E. (2000), *Groundwater level observation of Geological Survey of Japan in and around the Kinki district, Japan for earthquake prediction research*, Proc. Hokudan Internat. Symp. and School on Active Faulting, pp. 183–186.

MATSUDA, T., (1969), *Active fault and large earthquake in order to explicate earthquake phenomena*, Kagaku 39, 398–407 (in Japanese).

NAITO, H. and YOSHIKAWA, S. (1999), *A program to assist crustal deformation analysis*, Zisin 2, 52, 101–103 (in Japanese).

NAKANO, T. and HIRAHARA, K. (1997), *GPS observations of postseismic deformation for the 1995 Hyogo-ken Nanbu earthquake*, Japan, Geophys. Res. Lett. 24, 503–506.

OHTANI, R., KITAGAWA, Y., KOIZUMI, N., and MATSUMOTO, N. (2003), *GPS-derived coordinate variation corresponding to a non-secular strain change at the Yasutomi station of the Geological Survey of Japan, AIST*, submitted to Bull. Geol. Surv. Japan (in Japanese).

OKADA, Y. (1992), *Internal deformation due to shear and tensile faults in a half-space*, Bull. Seismol. Soc. Am. *82*, 1018–1040.

OZAWA, S., MURAKAMI, M., KAIDZU, M., TADA, T., SAGIYA, T., HATANAKA, Y., YARAI, H. and NISHIMURA, T. (2002), *Detection and monitoring of ongoing Aseismic Slip in the Tokai Region, Central Japan*, Science *298*, 1009–1012.

ROELOFFS, E.A., BURFORD, S.S., RILEY, F.S., and RECORDS, A.W. (1989), *Hydrologic effects on water level changes associated with episodic fault creep near Parkfield, California*, J. Geophys. Res. *94*, 12,387–12,402.

TADA, T. (1998), *Crustal movements of the Atotsugawa fault*, Earth Monthly *20*, 142–148 (in Japanese).

TAMURA, Y., SATO, T., OOE, M., and ISHIGURO, M. (1991), *A procedure for tidal analysis with a Bayesian information criterion*, Geophys. J. Int. *104*, 507–516.

THE RESEARCH GROUP FOR ACTIVE FAULTS OF JAPAN, *Active Faults in Japan, Sheet Maps and Inventories*, rev. ed., (Univ. of Tokyo Press 1991) pp. 292–293 (in Japanese).

TSUKUDA, E., KOIZUMI, N., and KUWAHARA, Y. (2000), *Integrated groundwater monitoring studies in the intensified observation area and near major active faults*, Bull. Geol. Surv. Japan. *51*, 435–445 (in Japanese).

WATANABE, K. (1991a), *Strain variations of the Yamasaki Fault zone, Southwest Japan, derived from extensometer observations, Part 1. On the long-term strain variations*, Bull. Disas. Prev. Res. Inst., Kyoto Univ. *41*, 29–52.

WATANABE, K. (1991b), *Strain variations of the Yamasaki Fault zone, Southwest Japan, derived from extensometer observations, Part 2. On the short-term strain variations from strain steps*, Bull. Disas. Prev. Res. Inst., Kyoto Univ. *41*, 53–85.

WESSON, R. L. (1988), *Dynamics of fault creep*, J. Geophys. Res. *93*, 8929–8951.

(Received: July 29, 2003; revised: December 1, 2005; accepted: December 7, 2005)

To access this journal online:
http://www.birkhauser.ch

Pure appl. geophys. 163 (2006) 675–691
0033–4553/06/040675–17
DOI 10.1007/s00024-006-0046-x

© Birkhäuser Verlag, Basel, 2006

❘Pure and Applied Geophysics

Hydrogeochemical Anomalies in the Springs of the Chiayi Area in West-central Taiwan as Possible Precursors to Earthquakes

S. R. Song,[1] W. Y. Ku,[1] Y. L. Chen,[1,2] C. M. Liu,[1] H. F. Chen,[1] P. S. Chan,[1] Y. G. Chen,[1] T. F. Yang,[1] C. H. Chen,[1] T. K. Liu,[1] and M. Lee[3]

Abstract—Water samples from both hot and artesian springs in Kuantzeling in west-central Taiwan have been collected on a regular basis from July 15, 1999 to the end of August 2001 to measure cation and anion concentrations as a tool to detect major earthquake precursors. The data identify chloride and sulfate ion anomalies few days prior to major quakes and lasting a few days afterward. These anomalies are characterized by increases in Cl^- concentrations from 34.9% to 41.2% and 71.5% to 138.1% as well as increases in SO_4^{2-} concentrations from 232.7% to 276.8% and 100.0% to 155.1% above the means in both hot and artesian springs. The occurrence of these anomalies is probably explained first as stress/strain-induced pressure changes in the subsurface water systems which then generate precursory limited geochemical discharges at the levels of subsurface reservoirs. Therefore, finally leading to the mixing of previously separated subsurface water bodies occurs. This suggests that the hot and artesian springs in the Kuantzeling area are possible ideal sites for recording strain changes serving well as earthquake precursors.

Key words: Chloride ion, sulfate ion, hot and artesian springs, anomaly, earthquake precursor, Taiwan.

1. Introduction

On account of highly active seismicity and a major destructive earthquake of magnitude $M_L = 7.3$ which occurred in a densely-populated area in Taiwan on 21 September, 1999, a large-scale research program to monitor active faults and identify earthquake precursors was jointly initiated by the Central Geological Survey, MOEA-ROC and the Institute of Geosciences, National Taiwan University. In one subprogram, weekly measurements of cation and anion concentrations are made in both hot and artesian springs in Taiwan to establish background concentrations and to identify geochemical earthquake-related anomalies. The

---

[1]Institute of Geosciences, National Taiwan University, P.O. Box 13-318, Taipei 106, Taiwan. E-mail: srsong@ntu.edu.tw
[2]Institute of Applied Geosciences, National Taiwan Ocean University.
[3]Central Geological Survey, MOEA.

purpose of this subprogram is to evaluate potential sites at which regular monitoring systems should be set up in the future.

The destructive Chi-Chi earthquake with magnitude $M_L = 7.3$ occurred in west-central Taiwan, causing a total of about 80–90 km in length of surface ruptures along the Chelungpu fault, with the largest measured vertical offsets reaching as far as 5–8 m (CHEN *et al.*, 2001). The epicenter of the earthquake was located about 15 km east of the surface trace of the thrust fault at 120.82°E and 23.85°N and had a hypocenter depth of about 12 km (CHUNG and SHIN, 1999; MA *et al.*, 1999; KAO and CHEN, 2000), near the town of Chi-Chi in Nantou County in west-central Taiwan (Fig. 1). This earthquake became one of the largest inland events in the past century, causing the death of about 2,400, injuring another 10,000 and destroying more than 100,000 buildings. Numerous aftershocks, including one event of $M_L = 6.8$, were distributed around the main shock in a large area of central Taiwan (KAO and CHEN, 2000). Following the Chi-Chi earthquake, another large quake with magnitude $M_L = 6.4$ struck the Chiayi area on October 22, 1999 in west-central Taiwan. The epicenter, 2.5 km northwest of Chiayi City was located at 120.40°E and 23.51°N and had a hypocenter depth of about 12.1 km (Fig. 1) (CWB, 1999).

One important goal of geoscientists has long been the detection of valuable short-term precursors of earthquakes, and, indeed, many types of precursors, including chemical and hydrological changes in subsurface fluids prior to large earthquakes. Among these, gases involved in hydrothermal processes (Rn, He, CO_2, CH_4, H_2, Ar and N_2) and water chemistry (Cl^-, F^-, NO_3^- and SO_4^{2-}) are the most unambiguous precursors (HAUKSSON, 1981; KING, 1986; SUGISAKI *et al.*, 1996; TSUNOGAI and WAKITA, 1995, 1996; TOUTAIN *et al.*, 1997; SONG *et al.*, 2005). These geochemical and hydrologic anomalies are generally related to changes in groundwater circulating systems because of earthquake generation (THOMAS, 1988; SUGISAKI *et al.*, 1996; KING *et al.*, 1999). Thus, geochemical anomalies observed in groundwater have provided useful information for earthquake prediction in seismic countries (e.g., KOIZUMI *et al.*, 1985; BARSUKOV *et al.*, 1984/1985; GUIRU *et al.*, 1984/1985). Meanwhile, preceding the major 1995 Kobe earthquake ($M_L = 7.2$) and the 1996 Pyrenean earthquake ($M_L = 5.2$) (TSUNOGAI and WAKITA, 1995, 1996; TOUTAIN *et al.*, 1997), anomalies of ions in commercialized bottled groundwater and spring water, respectively, have been recently detected.

This paper contributes to this field by presenting the results of a two-year study investigating hydrochemical changes in hot springs by collecting water samples from both hot and artesian springs in response to earthquakes in the Chiayi area of west-central Taiwan. Furthermore, the possible mechanisms inducing the chemical changes in the respective subsurface water systems are discussed.

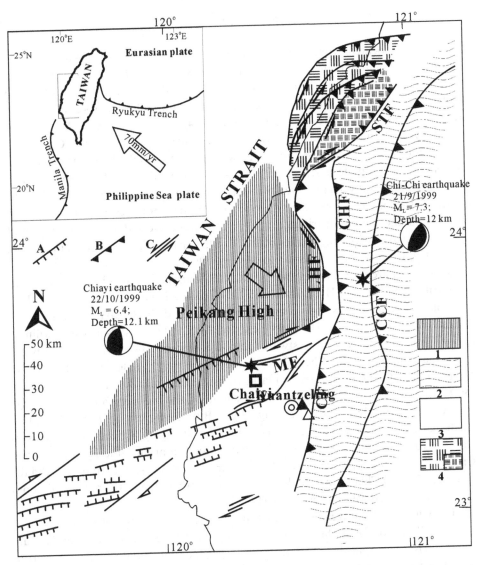

Figure 1

Regional structural sketch map of west-central Taiwan with the locations of the September 21 and October 22, 1999 earthquake epicenters and focal mechanisms. The legend in the lower right-hand corner indicates: 1: Pre-Tertiary basement; 2: Early Pleistocene tectonic belt; 3: Late Pleistocene tectonic belt; and 4: Escape blocks. The legend in the middle on the left-hand side indicates: A: Normal fault; B: Thrust fault; and C: Strike-slip fault. Shown on the map are: CCF: Chaochou-Chuchih fault; CF: Chukou fault; CHF: Chelungpu fault; LHF: Linnei-Hsinchu fault; MF: Meishan fault; and STF: Shihtan-Tuntzechiao fault (modified from BIQ, 1990; YANG *et al.*, 1994). Inset map shows the tectonics in the vicinity of Taiwan (modified from HO, 1986). Soild triangle: hot spring; open triangle: artesian spring.

2. Sites and Geological Background

Taiwan is located within the complexity of the oblique collision zone of the Eurasian continental plate and the Philippine Sea plate (Fig. 1). Presently, the Philippine Sea plate is moving WNW at about 70 mm per year (SENO and MARUYAMA, 1984), and it is believed the mountain-building process is still in progress (TSAI et al., 1981; YU and CHEN, 1994; YU et al., 1999). A dominant collision zone frequently inducing folding and fault thrusting, i.e., the Chelungpu thrust fault, may exist in west-central Taiwan. At the latitude of southern Taiwan, the Philippine Sea plate is riding up over the continental shelf of the South China Sea. Such active movements over the last 5 million years have been creating the island of Taiwan (Ho, 1986; TENG, 1987, 1990), and more recently, rapid crustal movements and widely distributed active structures have induced at least tens of large earthquakes with magnitudes over 7.0 in the last few hundred years (YU et al., 1997, 1999; CHANG et al., 1998; CHENG et al., 1999).

The Chiayi area is located in west-central Taiwan, and its pre-Tertiary basement high, called the Peikang High, is below. This Peikang High is an indentation block controlling the structures and seismic activities around the Chiayi area during the orogeny of Taiwan (Fig. 1) (LU, 1994; LU and MALAVIEILLE, 1994). The curvilinear active Chukou fault thrusts westward onto the Peikang High, with its northward extension connecting the Chelungpu thrust fault (Fig. 1) (BIQ, 1990; LU, 1994). South of the Peikang High there is a large transtension zone (BIQ, 1990; YANG et al., 1994), while in the middle, an active strike-slip fault, the Meishan fault, cuts through on the southern edge of the Peikang High (BIQ, 1990; LIN et al., 2000). Thus, earthquake epicenters in this area, one of the most highly active seismic areas of western Taiwan, distribute in a semicircular formation around the High (SHIN and CHANG, 1992). The epicenters of the September 21, 1999 Chi-Chi earthquake and the October 22, 1999 Chiayi earthquake are located about 35 km to the north and about 2.5 km to the northwest, respectively (Fig. 1).

The village of Kuantzeling is located in the southeastern part of the Chiayi area and is well-known for its hot spring spas. It is in the western foothills of Taiwan, where a passive margin shallow marine clastic sequence of the late Tertiary age crops out. Fossilferrous, fine-grained and little metamorphosed, the strata have been deformed by folding and faulting. The outpouring of hot spring waters is located on the axis of the Kuantzeling anticline, and a thrust fault, the Liuchungchi fault, cuts through it (HSU and WEY, 1983). The local geological structure, heat flow, silica geothermometry and Tritium data indicate that the hot springs may have come from a deep old water reservoir, over 2 km in depth rising along the fault fracture zone (CHEN et al., 2001). The temperature and pH value of the Kuantzeling hot springs are 79°C and 8.1, respectively. According to historical records, the perturbations of the hot spring system, i.e., the bursts of steam and increased outpouring, occurred a few days prior to, during and a few days after the 1964 Paiho earthquake, one of the most devastating earthquakes on Taiwan in the last hundred years, with magnitude 6.3 (CHENG et al., 1999) and its epicenter near the Kuantzeling area. Here, therefore, our attention was focused on the hot and artesian springs in this area, and the cations and anions in the water samples were regularly analyzed.

3. Sampling and Analysis

From July 15 to September 19, 1999 and for a period of about two months before the Chi-Chi earthquake, students from a local high school for a national science competition collected 9 samples from the Kuantzeling hot springs at different time intervals. Twenty days after the earthquake, the present researchers joined in and intensely collected one hot spring sample per day during two months and then decreased the sampling interval to one sample every three days and subsequently to one a week until the end of July 2001. Thus, a total of over 200 samples of hot spring water were collected for analyses. Meanwhile, one sample was also collected from the beginning of January 2000 to December 2000 every three days from an artesian spring located about 1 km southwest of the Kuantzeling hot springs. Later the sampling intervals decreased to one per week until the end of August 2001 for a total of about 170 samples of artesian spring water for analyses.

Dissolved anions and cations in both sets of samples were measured with an ion chromatographer (IC, Type Dionex DX-100) and an inductively coupled plasma-atomic emission spectrometer (ICP-AES, Type Jobin-Yvon JY-38plus), respectively. A sample from the same spring was measured after each sample analysis in order to enhance the precision of the measurements. Analytical uncertainties in the absolute concentrations were less than 3% for all of the anions and less than 5% for all of the cations. This study also analyzed the oxygen and hydrogen isotopes of the Kuantzeling hot springs from September 1999 to September 2000 using a Finnigan Delta Plus-Mass Spectrometer with precisions of about $0.1\%_0$ for the oxygen isotopes and $1\%_0$ for the hydrogen isotopes.

4. Results and Discussions

1. Temporal Variations in the Chemical Compositions

The temporal variations in the Cl^- and SO_4^{2-} concentrations of the water samples from the Kuantzeling hot springs from July 1999 to July 2001 are shown in Figs 2A and 2B, respectively. Chloride ion is the major anion in the hot springs and its average concentration reaches 2201 ppm, whereas that of the sulfate ion is about 33.6 ppm. Generally, the concentrations of chloride of in samples are almost constant, except on a few dates, but this is unlike those of sulfates, which are more

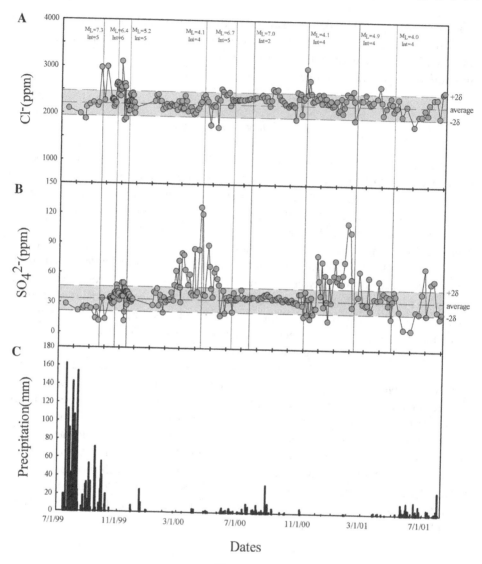

Figure 2
Temporal variations in (A) Cl^- and (B) SO_4^{2-} concentrations in the Kuantzeling hot springs. The average concentration (solid lines) and 2σ variation range (dashed lines) are also shown. The vertical lines represent the earthquakes with magnitudes and intensities greater than 4 that occurred in this area. (C) Daily amounts of precipitation obtained at the Kuantzeling area (Data from Central Weather Bureau of Taiwan).

fluctuant in two periods, i.e., from March 2000 to June 2000 and from December 2000 to February 2001. Two-sigma relative standard deviations (2σ) were calculated for those samples and are 12.0% (Cl^-) and 36.9% (SO_4^{2-}). We have considered the 2σ domains as representative of spring water background values, which may have resulted from water-rock interactions in the deep circulations of the hot spring reservoirs, sampling heterogeneity and analytical uncertainties. Except for the chloride and sulfate ions, all of the cations and anions vary within the 2σ domains during the entire sampling period. Figure 2A shows that the Cl^- concentrations increased abruptly on September 19, October 1, October 31, 1999 and November 1, 2000, reaching their maximum values of 2970 ppm, 2988 ppm, 3107 ppm and 2987 ppm, which are, respectively, about 34.9%, 35.8%, 41.2% and 35.7% above the mean value. It is important to note that these variations are very sudden. The sulfate contents during the same period seem to show no variations, but they do show high fluctuations from March to June 2000 and December 2000 to February 2001, when the respective concentrations reached their peaks at 126.6 ppm and 111.8 ppm, or about 276.8% and 232.7% above the mean. Figures 3A and 3B show the temporal variations in the hydrogen and oxygen isotopic ratios of the water samples from the Kuantzeling hot springs from September 1999 to September 2000, respectively. The $\delta^{18}O$ and δD ratios of all samples remain fairly constant during the sampling period, firmly indicating that the hot spring waters have come from a stable homogeneous subsurface water body.

The temporal variations in the Cl^- and SO_4^{2-} concentrations of the water samples from the Kuantzeling artesian spring from January 2000 to August 2001 are shown in Figures 4A and 4B, respectively. The sulfate ion with an average concentration of about 25.4 ppm is the major anion in the spring waters, if we compare with an average of about 2.70 ppm for the chloride ion. The chloride and sulfate concentrations of all samples are fairly constant during the sampling period. Two-sigma relative standard deviations (2σ) were calculated for those samples and are 30.4% (Cl^-) and 17.9% (SO_4^{2-}). These 2σ domains can be assumed as representative of the spring water background values. They may be attributed to annual fluctuations in groundwater chemistry, which are themselves mainly as a result of rainfall and other superficial phenomena, such as heterogeneity in the sampling and analytical uncertainties (TOUTAIN et al., 1997). Except for the chloride and sulfate ions, all cations and anions vary within the 2σ domains during the entire sampling period. Figure 4A shows that the Cl^- concentrations increase sharply on April 12, June 13 and July 16, 2000 reaching their maximum values (6.43 ppm), (5.55 ppm) and (4.63 ppm), which are, respectively, about 138.1%, 105.6% and 71.5% above the mean value. Again, of significance is these variations are very abrupt. Except for July 15, 2000 when little change is found, the variations in the sulfate content (Fig. 4B) are the same as those for chloride, with concentrations reaching their maximum values of 50.8 ppm and 64.8 ppm, which are, respectively, about 100.0% and 155.1% above the mean value.

2. Mechanism for the Chemical Changes

Changes in the chemical compositions of hot and artesian springs have previously been attributed to several factors. Different compositions of groundwater recharge,

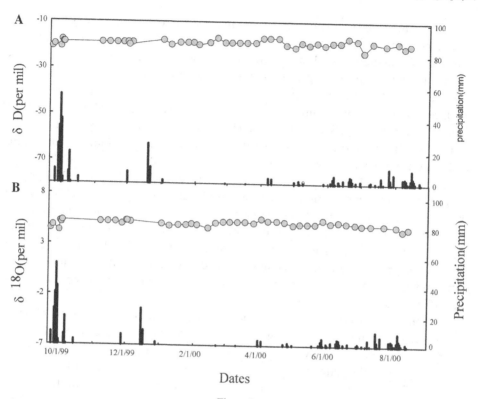

Figure 3
Temporal variations in (A) δD and (B) $\delta^{18}O$ in the Kuantzeling hot springs. Heavy bars are daily amounts of precipitation obtained at the Kuantzeling area (Data from Central Weather Bureau of Taiwan).

petrologic and mineralogical compositions of subsurface rocks, water-rock interactions (DOMENICO and SCHWARTZ, 1990; LANGMUIR, 1997), mixing of different water compositions and artificial pollutants, etc. To evaluate which mechanisms are responsible for the observed chemical changes, two facts must be kept in mind. Firstly, chloride ion is considered chemically stable, and the concentration level is high enough to measure reliably. The second, in contrast to the chloride ion, sulfate is not so stable in groundwater conditions and can be affected by sulfide mineral oxidation, precipitation-dissolution of gypsum in an unsaturated zone, dissolution of anhydrite or gypsum or redox reactions in a saturated zone (DOMENICO and SCHWARTZ, 1990). Such reactions, however, are not quick enough to cause such abrupt changes in SO_4^{2-} concentrations in a stable subsurface water system. Among all those factors that are capable of causing the observed temporal variations in both Cl^- and SO_4^{2-} concentrations in a short duration (Figs. 2 and 4), mixing of different water compositions (KING *et al.*, 1981; THOMAS, 1988) and artificial pollutants are the only two factors that cannot be ruled out.

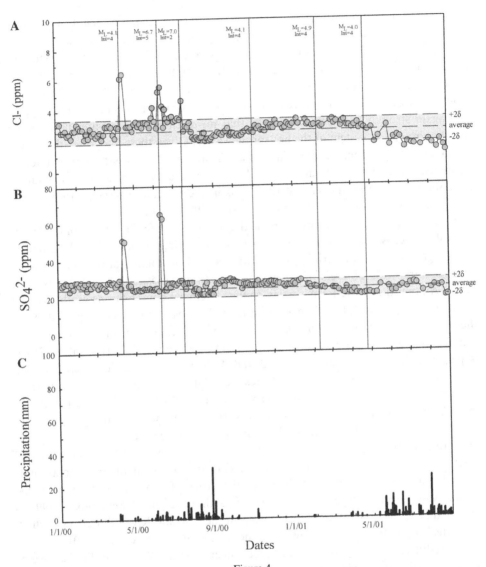

Figure 4
Temporal variations in (A) Cl⁻ and (B) SO₄²⁻ concentrations in the Kuantzeling artesian springs. The average concentration (solid lines) and 2σ variation range (dashed lines) are also shown. The vertical lines represent the earthquakes with magnitudes and intensities greater than 4 occurred in this area. (C) Daily amounts of precipitation obtained at the Kuantzeling area (Data from Central Weather Bureau of Taiwan).

Kuantzeling is located in an industry-free, sparsely populated mountainous area. Accordingly, pollutant solutes require media, i.e., meteoric water, in order to be transported down into the groundwater system. However, no correlated relationships among the anomalies of the ions in the temporal variations and daily amounts of precipitation are found during the sampling period (Figs. 2 and 4), highly suggesting that the observed temporal variations in the Cl⁻ and SO_4^{2-} concentrations are not, in fact, induced by recent meteoric water flowing down into the circulation system of the subsurface water system. Meanwhile, the analysis results of Tritium (^3H) concentration in the hot springs are less than 0.2 (TU < 0.2) (CHEN et al., 2004), and almost no variations in the oxygen and hydrogen isotopic ratios are found (Fig. 3). This again strongly supports the notion that no recent meteoric water flows down into the circulation system of the hot spring waters. Given these lines of evidence, the abrupt change in the temporal variations of Cl⁻ and SO_4^{2-} concentrations cannot be attributed to artifact pollutants. It follows then that the most probable factor controlling the temporal chemical variations in the hot and artesian springs of the Chiayi area is the mixing of water bodies with different compositions.

Several mechanisms can bring about the mixing of different water compositions in a subsurface water body, and these include the mixing of meteoric and formation waters, groundwater and brines or pore waters, and the mixing of different aquifers or reservoirs with different chemical compositions (DOMENICO and SCHWARTZ, 1990). Among these, the formers may change chemical compositions gradually and eventually lead to complete changes or at least changes that last a long period, while the mixing of different aquifers or reservoirs with different chemical compositions can quickly occur and the chemical changes can disappear in a short duration. What is particularly salient is that Figures 2 and 4 clearly illustrate that such chemical changes are abrupt and that the dates correlate well with the occurrence of earthquakes, near the hot and artesian springs. Thus, it is reasonable to attribute the mechanism of the rapid temporal chemical variations in the hot and artesian springs in the Chiayi area to the mixing of different aquifers or reservoirs. It may be equally justifiable to make the claim that the mechanism of the rapid temporal chemical variations may have been induced by an earthquake. Such an interpretation is strongly supported by the observation of the large 1.0- to 11.1-m changes in the groundwater levels induced by the Chi-Chi earthquake on 21 September 1999, as recorded at 157 out of 179 monitoring wells in the Choshui River alluvial fan, which is located about 10–20 km northwest of the Chiayi area (CHIA et al., 2001).

Two different earthquake-induced mechanisms for the mixing of different water systems of different compositions have been proposed in previous studies. The first involves permeability enhancement due to a breaking in the crust (KING et al., 1981; THOMAS, 1988; ROJSTACZER and WOLF, 1992; Rojstaczer et al., 1995; KING et al., 1999), while the second encompasses the changing of pressure in aquifer systems because of elastic compression (MUIR-WOOD and KING, 1993; TOUTAIN et al., 1997), before and after a major earthquake. The former necessitates cracking water barriers, i.e., fault gauge zones or aquicludes and, subsequently, inducing the mixing of initially isolated aquifers. Chemical changes induced by such a mechanism, however, would require large-scale mixing processes (TSUNOGAI and WAKITA, 1996) and would not be compatible with short-term reversible anomalies, such as those observed in the

Table 1

The Cl⁻ and SO₄²⁻ concentrations (ppm) of the Kuantzeling hot spring

Date	Cl^-	SO_4^{2-}	Date	Cl^-	SO_4^{2-}	Date	Cl^-	SO_4^{2-}
7/15/1999	2093	28.1	11/8/1999	2200	11.8	2/19/2000	2132	46.6
8/8/1999	1983	21.7	11/9/1999	2233	20.5	2/22/2000	2272	72.8
8/18/1999	1875	23.5	11/10/1999	2193	44.5	2/25/2000	2102	45.7
8/20/1999	2123	25.4	11/11/1999	2237	40.3	2/28/2000	2123	31.1
8/27/1999	2178	25.8	11/12/1999	2043	41.0	3/1/2000	2385	80.3
9/3/1999	2219	23.7	11/13/1999	2242	40.5	3/2/2000	2272	78.6
9/12/1999	2153	14.1	11/14/1999	2054	27.5	3/5/2000	2055	63.6
9/14/1999	2215	24.7	11/15/1999	2263	41.0	3/8/2000	2141	48.2
9/19/1999	2970	11.7	11/16/1999	2388	39.9	3/11/2000	2057	59.0
9/26/1999	2268	34.1	11/17/1999	2406	37.7	3/17/2000	1994	40.3
10/1/1999	2988	13.5	11/18/1999	2260	31.9	3/20/2000	2022	38.1
10/11/1999	2284	34.4	11/19/1999	2259	38.0	3/23/2000	2089	84.6
10/12/1999	2306	34.6	11/20/1999	2425	33.1	3/26/2000	2133	38.3
10/13/1999	2287	35.5	11/21/1999	2237	33.2	3/29/2000	2220	41.3
10/14/1999	2285	33.9	11/22/1999	2382	35.6	4/1/2000	2258	83.8
10/15/1999	2136	33.8	11/23/1999	2145	30.7	4/4/2000	2359	126.6
10/16/1999	2234	29.3	11/24/1999	2226	32.5	4/7/2000	2398	119.7
10/17/1999	2184	28.3	11/25/1999	2000	32.9	4/10/2000	2320	38.8
10/18/1999	2236	34.8	11/26/1999	2112	33.7	4/13/2000	2266	38.1
10/19/1999	2211	35.4	11/27/1999	2263	27.5	4/16/2000	1764	88.4
10/20/1999	2593	29.8	1/6/2000	2388	41.0	4/19/2000	2168	67.2
10/21/1999	2635	39.4	1/9/2000	2406	39.9	4/25/2000	2196	44.8
10/22/1999	2646	39.9	1/12/2000	2260	44.5	4/28/2000	2164	36.5
10/24/1999	2399	38.6	1/15/2000	2259	27.4	5/1/2000	2340	61.7
10/25/1999	2627	47.4	1/18/2000	2085	38.0	5/4/2000	1714	65.3
10/26/1999	2489	43.8	1/21/2000	2177	33.1	5/7/2000	2299	55.0
10/27/1999	2585	40.8	1/24/2000	2123	20.3	5/10/2000	2534	48.6
10/28/1999	2361	37.5	1/27/2000	2200	35.6	5/13/2000	2485	17.9
10/29/1999	2497	40.9	1/29/2000	2150	30.7	5/16/2000	2413	43.0
10/30/1999	2553	39.1	2/1/2000	2150	32.5	5/19/2000	2426	20.6
10/31/1999	3107	32.4	2/4/2000	2204	32.9	5/25/2000	2464	36.1
11/4/1999	2617	36.5	2/7/2000	2177	33.7	5/28/2000	2183	33.8
11/5/1999	1851	49.6	2/10/2000	2166	40.6	5/31/2000	2317	37.0
11/6/1999	2623	36.5	2/13/2000	2188	32.5	6/3/2000	2290	34.3
11/7/1999	2558	36.4	2/17/2000	2080	48.1	6/6/2000	2306	21.8
6/9/2000	1886	49.7	10/20/2000	2334	40.7	3/4/2001	2243	30.6
6/12/2000	2299	33.9	10/23/2000	2266	62.4	3/7/2001	2262	57.6
6/15/2000	2306	34.5	10/26/2000	2380	29.3	3/10/2001	2314	41.0
6/18/2000	2313	36.0	10/29/2000	2987	18.0	3/13/2001	2300	24.7
6/20/2000	2311	36.3	11/1/2000	2437	42.9	3/21/2001	2376	36.6
6/21/2000	2312	44.1	11/4/2000	2734	20.7	3/27/2001	2620	35.7
6/24/2000	2321	36.5	11/7/2000	2461	35.7	3/30/2001	2295	54.2
6/30/2000	2329	34.9	11/10/2000	2291	15.9	4/2/2001	2022	36.1
7/6/2000	2345	35.7	11/13/2000	2304	37.4	4/8/2001	2173	42.4
7/9/2000	2339	35.5	11/16/2000	2373	20.7	4/14/2001	2289	39.9
7/12/2000	2361	36.9	11/19/2000	2342	26.2	4/17/2001	2385	30.5
7/18/2000	2366	36.5	11/22/2000	2444	26.2	4/20/2001	2259	38.6
7/21/2000	2386	35.3	11/25/2000	2388	79.3	4/23/2001	2108	16.9

Table 1

(Contd.)

Date	Cl^-	SO_4^{2-}	Date	Cl^-	SO_4^{2-}	Date	Cl^-	SO_4^{2-}
7/24/2000	2418	36.9	11/28/2000	2259	53.9	4/26/2001	2171	34.8
8/2/2000	2365	36.2	12/1/2000	2449	39.2	4/29/2001	2189	43.8
8/5/2000	2305	37.5	12/4/2000	2361	73.0	5/2/2001	2363	39.6
8/8/2000	2382	39.1	12/7/2000	2365	28.2	5/9/2001	1997	24.4
8/11/2000	2299	38.7	12/10/2000	2249	59.6	5/18/2001	2216	6.5
8/14/2000	2188	40.3	12/13/2000	2269	33.3	6/2/2001	1786	5.4
8/17/2000	2307	38.3	12/16/2000	2310	14.0	6/11/2001	2001	24.2
8/20/2000	2542	37.9	12/19/2000	2196	54.5	6/17/2001	2006	22.4
8/23/2000	2508	34.9	12/22/2000	2219	32.3	6/23/2001	2140	42.2
8/26/2000	2538	35.6	12/25/2000	2310	59.3	6/29/2001	2034	67.5
8/30/2000	2406	35.2	12/31/2000	2091	74.4	7/2/2001	2256	20.7
9/4/2000	2329	37.2	1/3/2001	2123	54.1	7/11/2001	2367	52.3
9/7/2000	2320	37.7	1/6/2001	2365	58.1	7/17/2001	2372	54.9
9/10/2000	2267	34.4	1/9/2001	2051	51.4	7/23/2001	1974	24.6
9/13/2000	2222	33.3	1/12/2001	2184	50.9	7/29/2001	2491	17.7
9/17/2000	2182	35.5	1/15/2001	2323	59.9	8/1/2001	2522	22.4
9/20/2000	2212	34.4	1/18/2001	2466	73.1			
9/26/2000	2231	33.4	1/21/2001	2421	111.8			
9/29/2000	2225	33.7	1/27/2001	2518	102.9			
10/2/2000	2178	34.9	2/2/2001	2319	55.9			
10/5/2000	1904	32.3	2/4/2001	1900	28.0			
10/8/2000	2486	31.8	2/5/2001	2331	38.4			
10/11/2000	2346	32.4	2/17/2001	2324	63.6			
10/14/2000	2460	30.4	2/20/2001	2451	31.5			
10/17/2000	2023	40.4	2/26/2001	2297	29.9			

Kuantzeling hot and artesian springs, where the chloride and sulfate contents returned to pre-seismic levels within a few days after the onset of the anomaly. The latter mechanism, on the other hand, is induced by elastic compression on aquifers, which generates pressure variations among them, enough to generate reversible changes in hydraulic levels (MUIR-WOOD and KING, 1993) and, therefore, lead to subsequent limited geochemical discharge effects (THOMAS, 1988). It is clear that a limited mixing of aquifers with different compositions, not unlike that in the case of the rapid geochemical changes in the Kuantzeling hot and artesian springs (Figs. 2 and 4), is the most likely model for the generation of the drastic ion concentration changes. Aside from this, the mechanism is supported by the significant changes in groundwater levels induced by the Chi-Chi earthquake (CHIA *et al.*, 2001). It is highly expected, therefore, that the chloride- and sulfate-rich spring waters were introduced into the Kuantzeling subsurface water systems. Although it is difficult to precisely identify the water introduced in the subsurface system, a hot spring with high chloride and sulfate concentrations, located nearby the Kuantzeling hot springs (KU, 2001) is a potential source.

Table 2

The Cl$^-$ and SO$_4^{2-}$ concentrations (ppm) of the Kuantzelingartesian spring

Date	Cl$^-$	SO$_4^{2-}$	Date	Cl$^-$	SO$_4^{2-}$	Date	Cl$^-$	SO$_4^{2-}$
1/7/2000	3.17	26.6	4/24/2000	2.96	26.6	7/31/2000	2.57	23.4
1/10/2000	2.58	27.4	4/27/2000	2.97	24.2	8/2/2000	2.20	24.3
1/13/2000	2.60	28.4	4/30/2000	2.66	23.7	8/4/2000	2.09	24.6
1/16/2000	2.71	27.8	5/3/2000	3.14	24.1	8/6/2000	2.08	21.1
1/19/2000	2.40	23.8	5/6/2000	3.27	24.4	8/8/2000	2.20	24.1
1/22/2000	2.48	27.8	5/9/2000	2.99	23.9	8/10/2000	2.07	21.4
1/25/2000	2.19	25.0	5/12/2000	3.15	24.1	8/12/2000	2.03	21.5
1/28/2000	2.71	28.4	5/15/2000	2.91	24.9	8/14/2000	2.17	21.3
1/31/2000	2.72	28.4	5/18/2000	3.23	24.5	8/16/2000	2.21	21.6
2/3/2000	3.11	26.9	5/21/2000	3.23	24.6	8/18/2000	2.48	26.0
2/6/2000	2.94	28.4	5/24/2000	3.15	24.9	8/20/2000	2.50	27.6
2/9/2000	2.79	26.5	5/27/2000	2.91	24.4	8/22/2000	1.96	21.2
2/12/2000	2.78	27.9	5/30/2000	3.52	24.2	8/24/2000	2.11	22.4
2/15/2000	2.14	24.2	6/1/2000	4.21	24.8	8/26/2000	2.11	22.2
2/18/2000	2.54	28.3	6/4/2000	3.30	25.0	8/28/2000	1.98	21.6
2/21/2000	2.20	25.6	6/7/2000	2.87	23.7	8/30/2000	2.08	21.3
2/24/2000	2.84	27.8	6/10/2000	5.24	64.8	9/1/2000	2.20	24.6
2/27/2000	2.44	26.6	6/13/2000	5.55	62.4	9/4/2000	2.33	27.3
3/1/2000	2.60	27.7	6/16/2000	4.29	24.1	9/7/2000	2.28	27.1
3/4/2000	2.28	25.2	6/18/2000	2.91	24.7	9/10/2000	2.54	28.3
3/7/2000	2.43	27.0	6/19/2000	4.12	24.1	9/13/2000	2.44	28.8
3/10/2000	2.57	25.8	6/20/2000	4.09	25.2	9/16/2000	2.33	25.8
3/13/2000	2.11	25.2	6/22/2000	3.48	27.0	9/19/2000	2.51	28.6
3/16/2000	2.46	24.2	6/25/2000	3.27	25.5	9/22/2000	2.61	28.9
3/19/2000	2.98	28.1	6/28/2000	3.53	27.6	9/25/2000	2.78	29.5
3/22/2000	2.98	28.8	7/1/2000	3.59	27.2	9/28/2000	2.29	28.1
3/25/2000	3.05	27.7	7/4/2000	3.27	26.6	9/30/2000	2.38	28.9
3/28/2000	2.95	28.3	7/7/2000	3.35	27.7	10/1/2000	2.48	28.2
3/31/2000	2.44	26.2	7/10/2000	3.48	28.8	10/4/2000	2.45	27.8
4/3/2000	2.20	27.6	7/13/2000	3.38	28.8	10/7/2000	2.27	27.5
4/6/2000	2.93	29.2	7/16/2000	4.63	26.9	10/10/2000	2.39	27.0
4/9/2000	2.90	25.1	7/19/2000	2.62	24.6	10/13/2000	2.26	24.3
4/12/2000	6.14	50.8	7/22/2000	2.95	27.7	10/16/2000	2.32	26.7
4/15/2000	6.43	50.1	7/25/2000	2.90	27.5	10/19/2000	2.50	26.7
4/21/2000	2.96	26.6	7/28/2000	3.18	27.8	10/22/2000	2.59	26.9
10/25/2000	2.71	26.0	2/5/2001	2.95	24.7	7/7/2001	1.63	26.6
10/28/2000	2.53	26.0	2/13/2001	2.96	24.2	7/11/2001	1.60	26.2
10/31/2000	2.63	26.3	2/15/2001	2.80	24.5	7/21/2001	1.53	21.9
11/3/2000	2.49	26.0	2/18/2001	2.90	24.1	8/3/2001	1.87	23.4
11/6/2000	2.92	28.2	3/3/2001	3.23	24.2	8/8/2001	1.65	24.9
11/9/2000	2.73	25.9	3/6/2001	3.33	24.1	8/15/2001	1.34	24.4
11/12/2000	2.86	26.2	3/8/2001	3.04	24.1	8/20/2001	1.89	24.9
11/15/2000	2.62	26.9	3/12/2001	3.17	23.0	8/26/2001	1.50	19.2
11/18/2000	2.64	27.4	3/19/2001	2.88	21.8	8/29/2001	1.16	19.6
11/21/2000	2.58	26.5	3/21/2001	3.16	21.4			
11/24/2000	3.08	27.7	3/24/2001	3.26	24.6			
11/27/2000	2.79	28.0	3/29/2001	2.90	21.9			
11/30/2000	2.95	26.3	4/1/2001	2.90	21.6			

Table 2

(Contd.)

Date	Cl⁻	SO₄²⁻	Date	Cl⁻	SO₄²⁻	Date	Cl⁻	SO₄²⁻
12/1/2000	2.78	27.8	4/5/2001	2.76	23.1			
12/4/2000	2.78	27.5	4/9/2001	2.92	21.1			
12/7/2000	3.25	26.9	4/14/2001	3.07	21.0			
12/10/2000	2.99	26.8	4/16/2001	2.70	20.7			
12/16/2000	2.95	26.0	4/21/2001	2.84	20.9			
12/19/2000	2.71	25.0	4/24/2001	2.73	21.3			
12/22/2000	2.80	26.1	4/29/2001	2.59	21.5			
12/25/2000	3.06	25.9	5/5/2001	2.72	20.8			
12/28/2000	2.95	26.3	5/10/2001	1.76	21.1			
12/31/2000	2.88	27.1	5/16/2001	2.15	25.6			
1/3/2001	3.02	26.3	5/29/2001	2.88	24.5			
1/9/2001	3.35	28.0	6/3/2001	1.56	22.0			
1/12/2001	3.03	28.6	6/9/2001	2.06	22.3			
1/23/2001	3.29	26.0	6/15/2001	2.19	24.1			
1/25/2001	2.97	25.8	6/19/2001	2.08	25.1			
1/27/2001	3.05	24.4	6/26/2001	1.41	24.1			
2/2/2001	2.90	25.0	7/1/2001	1.74	26.1			

The Cl^- and SO_4^{2-}, especially the SO_4^{2-} variations, which responded to the earthquakes in the Kuantzeling hot and artesian springs, seem to be so different. The variations in the hot springs were more fluctuant and lasted longer than those in the artesian springs (Figs. 2 and 4). This may have been a result of the different depths of the aquifers or reservoirs. In other words, the shallower the artesian springs are, the faster are the responses from the earthquake-induced stresses.

Although the chemical anomalies that occurred in the Kuantzeling hot and artesian springs can be explained by the earthquakes, there still remain several open questions like why some earthquakes do not cause chemical anomalies in the same subsurface water systems (Figs. 2 and 4). The answer surely must lie in the fact that such anomalies are likely related to the unknown characteristics of some subsurface structures and water systems, the complexities of the processes of earthquake-induced stresses on the crust and aquifers or reservoirs, the wide-ranging chemical compositions of subsurface waters, varying water-rock interactions, and so on. Obviously, more data from geochemical monitoring and further investigations into earthquake precursors are required to enhance our understanding vis-à-vis the origin of earthquakes in the future.

5. Conclusions

Located in an orogenic belt with highly active seismicity, Taiwan has often been struck by major devastating earthquakes, which have caused huge numbers of fatalities and casualties as well as the destruction of countless buildings. Nonetheless, in spite of this, it has given rise to numerous opportunities to investigate the potentially hydrological geochemical precursors of the earthquakes. Here, short-term, reversible precursory geochemical anomalies have been recorded in hot and artesian springs prior and subsequent to the major earthquakes occurred September 1999 in the Kuantzeling area of west-central Taiwan. The anomalies were sharp sudden increases in chloride and sulfate ions. These are interpreted here as stress/strain-induced pressure changes in the subsurface water system, followed by limited precursory geochemical discharges generated by limited changes in the levels of the subsurface reservoirs, finally leading to the mixing of previously isolated subsurface water bodies. This strongly suggests that both the hot and artesian springs in the Kuantzeling area may be ideal sites for recording strain changes and that therefore, they should serve well in earthquake precursor research.

Acknowledgements

The authors appreciate the assistance of Mr. Yu, W.Y. for the field samplings and partial IC and ICP-AES analyses. This research was supported by the Central Geological Survey, Ministry of Economic Affairs and partly by the National Science Council, Republic of China under grants NSC 89-2116-M-002-046.

REFERENCES

BARSUKOV, V.L., VARSHAL, G.M., and ZAMOKINA, N.S. (1984/1985), *Recent results of hydrogeochemical studies for earthquake prediction in the USSR*, Pure Appl. Geophys. *122*, 143–156.

BIQ, C. (1990), *Another Coastal Range on Taiwan*, Ti-Chih *12*, 1–14 (in Chinese).

CENTRAL WEATHER BUREAU (CWB) (Taiwan) (1999), *CWB earthquake report*, http:// www.cwb.gov.tw.

CHANG, H.C., LIN, C.W., CHEN, M.M., and LU, S.T. (1998), *An Introduction to the Active Faults of Taiwan: Explanatory Text for the Active Fault of Taiwan, Scale 1:500,000*, Spec. Publ. Central Geol. Survey, Taiwan *10*, 103 pp. (in Chinese with English abstract).

CHEN, C.H., LIU, T.K., SONG, S.R., YANG, T.Y., and LEE, C.Y. (2004), *Environmental geochemistry with respect to the fault activity during 2000–2002 in Chiayi–Tainan and Hsinchu-Miaoli Area, Western Taiwan*, Bull. Centl. Geol. Survey *17*, 129–174 (in Chinese).

CHEN, Y.G., CHEN, W.S., LEE, J.C., LEE, Y.H., LEE, C.T., CHANG, H.C., and LO, C.H. (2001), *Surface rupture of the 1999 Chi-Chi earthquake yields insights on the active tectonics of central Taiwan*, Bull. Seismol. Soc. Amer. *91*, 977–985.

CHENG, S.N., YEH, Y.T., HSU, M.T., and SHIN, T.C., *Photo Album of Ten Disastrous Earthquakes in Taiwan* (CWB, Taiwan, 1999).

CHIA, Y.-P., WANG, Y.S., CHIU, J.J., and LIU, C.W. (2001), *Changes of groundwater level due to 1999 Chi-Chi earthquake in the Choushui River alluvial fan in Taiwan*, Bull. Seismol. Soc. Am. *91*, 1062–1068.

CHUNG, J.K. and SHIN, T.C. (1999), *Implication of the rupture process from the displacement distribution of strong ground motions recorded during the 21 September 1999 Chi-Chi Taiwan earthquake*, Terr. Atmos. Oceanic Sci. *10*, 777–786.

DOMENICO, P.A. and SCHWARTZ, F.W., *Physical and Chemical Hydrogeology* (John Wiley and Sons, Singapore, 1990).

GUIRU, L., FONGLIANG, W., JIHUA, W., and PEIREN, Z. (1984/1985), *Preliminary results of seismogeochemical research in China*, Pure Appl. Geophys. *122*, 218–230.

HAUKSSON, E. (1981), *Radon content of groundwater as an earthquake precursor: Evaluation of worldwide data and physical basis*, J. Geophys. Res. *86*, 9397–9410.

HO, C.S. (1986), *An introduction to the Geology of Taiwan: Explanatory Text of the Geologic Map of Taiwan*, Cent. Geol. Survey, Taiwan, 163 pp.

HSU, C.Y. and WEY, S.K. (1983), *Structural geology in the Chiayi foothills, Taiwan*, Petroleum Geol. Taiwan *19*, 17–28.

KAO, H. and CHEN, W.P. (2000), *The Chi-Chi earthquake sequence: Active out-of-sequence thrust faulting in Taiwan*, Science *288*, 2346–2349.

KING, C.Y. (1986), *Gas geochemistry applied to earthquake prediction. An overview*, J. Geophys. Res. *91*, 12,269–12,281.

KING, C.Y., AZUMA, S., IGARASHI, G., OHNO, M., SAITO, H., and WAKITA, H. (1999), *Earthquake-related water-level changes at 16 closely clustered wells in Tono, central Japan*, J. Geophys. Res. *104*, 13,073–13,082.

KING, C.Y., EVANS, W.C., PRESSER, T., and HUSK, R. (1981), *Anomalous chemical changes in well waters and possible relation to earthquakes*, Geophys. Res. Lett. *8*, 425–428.

KOIZUMI, N., YOSHIOKA, R., and KISHIMOTO, Y. (1985), *Earthquake prediction by means of change of chemical composition in mineral spring water*, Geophys. Res. Lett. *12*, 510–513.

LANGMUIR, D., *Aqueous Environmental Geochemistry* (Prentice-Hall, New Jersey 1997).

LIN, C.W., CHANG, H.C., LU, S.T., SHIH, T.S., and HUANG, W.C. (2000), *An Introduction to the Active Faults of Taiwan, second edition: Explanatory Text of the Active Fault Map of Taiwan, Scale: 500,000*, Special Publication of Central Geological Survey, 13, 122 pp. (in Chinese with English abstract).

LU, C.Y. (1994), *Neotectonics in the foreland thrust belt of Taiwan*, Petroleum Geol. Taiwan *29*, 1–26.

LU, C.Y. and MALAVIELLE, J. (1994), *Oblique convergence, indentation and rotation tectonics in the Taiwan mountain belt: Insights from experimental modeling*, Earth Planet. Sci. Lett. *121*, 477–494.

MA, K.F., LEE, C.T., TSAI, Y.B., SHIN, T.C., and MORI, J. (1999), *The Chi-Chi Taiwan earthquake: Large surface displacement on an island thrust fault*, EOS, *80*, 605–611.

MUIR-WOOD, R. and KING, G.C.P. (1993), *Hydrological signatures of earthquake strain*, J. Geophys. Res. *98*, 22,035–22,068.

ROJSTACZER, S. and WOLF, S. (1992), *Permeability changes associated with large earthquakes: an example from Loma Prieta, California*, Geology 20, 211–214.

ROJSTACZER, S., WOLF, S., and MICHEL, R. (1995), *Permeability enhancement in the shallow crust as a cause of earthquake-induced hydrological changes*, Nature *373*, 237–239.

SENO, T. and MARUYAMA, S. (1984), *Paleogeographic reconstruction and origin of the Philippine Sea*, Tectonophys. *102*, 53–84.

SHIN, T. and CHANG, Z. (1992), *Earthquakes in 1992*, Rep. Metro. *38*, 218–232.

SONG, S.R., CHEN, Y.L., LIU, C.M., KU, W.Y., CHEN, H.F., LIU, Y.J., KUO, L.W., YANG, T.F., CHEN, C.H., LIU, T.K., LEE, M. (2005), *Hydrochemical changes in spring waters in Taiwan: Implications for evaluating sites for earthquake precursory monitoring*, Terr. Atmos. Oceanic Sci. *16*, 745–762.

SUGISAKI, R., ITO, K., NAGAMINE, K., and KAWABE, I. (1996), *Gas geochemical changes at mineral springs associated with the 1995 southern Hyogo earthquake (M = 7.2), Japan*, Earth Planet. Sci. Lett. *139*, 239–249.

TENG, L.S. (1987), *Stratigraphic records of the late Cenozoic Penglai Orogeny of Taiwan*, Acta Geologica Taiwanica 25, 205–224.

TENG, L.S. (1990), *Geotectonic evolution of late Cenozoic arc-continent collision in Taiwan*, Tectonophys. *183*, 57–76.

THOMAS, D. (1988), *Geochemical precursors to seismic activity*, Pure Appl. Geophys. *126*, 241–265.

TOUTAIN, J.P., MUNOZ, M., POITRASSON, F., and LIENARD, A.C. (1997), *Springwater chloride ion anomaly prior to a M_L = 5.2 Pyrenean earthquake*, Earth Planet. Sci. Lett. *149*, 113–119.

TSAI, Y.B., LIAW, Z.S., LEE, T.Q., LIN, M.T., and YEH, Z.H. (1981), *Seismological evidence of an active plate boundary in the Taiwan area*, Mem. Geol. Soc. China 4, 143–154.

TSUNOGAI, U. and WAKITA, H. (1995), *Precursory chemical changes in ground water: Kobe earthquake, Japan*, Science *269*, 61–63.

TSUNOGAI, U. and WAKITA, H. (1996), *Anomalous change in groundwater chemistry- possible precursors of the 1995 Hyogo-ken Nanbu Earthquake, Japan*, J. Phys. Earth *44*, 381–390.

YANG, K.M., CHI, W.R., WU, J.C., and TING, H.H. (1994), *Tectonic evolution and mechanisms for the formations of Neogene extensional basins in southwestern Taiwan: Implications for hydrocarbon exploration*, Extended abstract of Ann. Meeting, Geol. Soc. China, 392–395.

YU, S.B. and CHEN, H.Y. (1994), *Global positioning system measurement of crustal deformation in the Taiwan arc-continent collision zone*, Terr. Atmos. Oceanic Sci. *5*, 477–498.

YU, S.B., CHEN, H.Y., and KUO, L.C. (1997), *Velocity fields of GPS stations in the Taiwan area*, Tectonophys. *274*, 41–59.

YU, S.B., KUO, L.C., PUNONGBAYAN, R.S., and RAMOS, E.G. (1999), *GPS observations of crystal deformation in the Taiwan-Luzon region*, Geophys. Res. Lett. *26*, 923–926.

(Received: January 30, 2003, revised: November 28, 2005, accepted: November 30, 2005)

To access this journal online:
http://www.birkhauser.ch

Pure appl. geophys. 163 (2006) 693–709
0033–4553/06/040693–17
DOI 10.1007/s00024-006-0040-3

© Birkhäuser Verlag, Basel, 2006

┌ Pure and Applied Geophysics

Seismo-Geochemical Variations in SW Taiwan: Multi-Parameter Automatic Gas Monitoring Results

T.F. Yang,[1,2] C.-C. Fu,[1] V. Walia,[2] C.-H. Chen,[1,3] L.L. Chyi,[4] T.-K. Liu,[1] S.-R. Song,[1] M. Lee,[5] C.-W. Lin,[5] and C.-C. Lin[5]

Abstract—Gas variations of many mud volcanoes and hot springs distributed along the tectonic sutures in southwestern Taiwan are considered to be sensitive to the earthquake activity. Therefore, a multi-parameter automatic gas station was built on the bank of one of the largest mud-pools at an active fault zone of southwestern Taiwan, for continuous monitoring of CO_2, CH_4, N_2 and H_2O, the major constituents of its bubbling gases. During the year round monitoring from October 2001 to October 2002, the gas composition, especially, CH_4 and CO_2, of the mud pool showed significant variations. Taking the CO_2/CH_4 ratio as the main indicator, anomalous variations can be recognized from a few days to a few weeks before earthquakes and correlated well with those with a local magnitude > 4.0 and local intensities > 2. It is concluded that the gas composition in the area is sensitive to the local crustal stress/strain and is worthy to conduct real-time monitoring for the seismo-geochemical precursors.

Key words: Active faults, gas geochemical precursor, monitoring station, Taiwan.

1. Introduction

Since the 1960s, geochemical signals preceding significant earthquakes have been used for earthquake prediction, especially in China, Japan, the former Soviet Union, and the United States (KING, 1986; THOMAS, 1988). Gas monitoring on hydrothermal systems and fault zones has been prevailing because gas moves at a relatively high velocity within these areas (TAKAHATA et al., 1997; TOUTAIN and BAUBRON, 1999; WHITEHEAD and LYON, 1999; HEINICKE and KOCH, 2000). Changes in gas compositions (H_2, CH_4, CO_2, and noble gases, especially Rn and He), and He/Ar, CH_4/Ar, N_2/Ar, and $^3He/^4He$ ratios, were reported to be potential

[1]Department of Geosciences, National Taiwan University, Taipei 106, Taiwan, R.O.C.
E-mail: tyyang@ntu.edu.tw
[2]National Center for Research on Earthquake Engineering, NARL, Taipei 106, Taiwan, R.O.C.
[3]National Applied Research Laboratories, Taipei 106, Taiwan, R.O.C.
[4]Department of Geology, University of Akron, Akron, OH 44325-4101, U.S.A.
[5]Central Geological Survey, MOEA, Taipei 235, Taiwan, R.O.C.

precursors (KING, 1978; SUGISAKI, 1978; KAWABE, 1984; SUGISAKI and SUGIURA, 1986; IGARASHI et al., 1995; SANO et al., 1998; FYTIKAS et al., 1999; TEDESCO and SCARSI, 1999; ITALIANO et al., 2001; VIRK et al., 2001; CHYI et al., 2001, 2005; YANG et al., 2005, 2006; WALIA et al., 2005; 2006). Among the potential geochemical precursors, radon changes are probably the most frequently used for earthquake monitoring/predicting purposes, although radon may be significantly affected by nontectonic changes in the environment (KING, 1986). Recently, multi-parameter monitoring stations with simultaneous recording of the main external parameters (e.g., atmospheric pressure, water and air temperature, soil moisture) were strongly recommended because this approach can screen out possible nontectonic factors (TOUTAIN and BAUBRON, 1999).

Systematic radon monitoring, aiming at earthquake prediction, was performed continuously in northern and northeastern Taiwan in 1980s (LIU et al., 1985). However, this project was terminated in 1990 due to a shortage of funds. After the largest and most disastrous earthquake (1999 Chi-chi earthquake, $M_W = 7.6$) of the 20th century in Taiwan, a new phase of a geochemical monitoring project in southwestern Taiwan was started in 2000. Representative sites, such as mud volcanoes and hot springs, with natural gases bubbling from deep sources along tectonic suture zones were chosen for weekly sampling. Gases were collected with pre-evacuated low permeability glass bottles, and analyzed for composition and $^3He/^4He$ isotopic ratios. Results for gases from Chung-lun (CL) (Fig. 1) for two years (from July, 1999 to September, 2001) showed significant temporal variations of $^3He/^4He$ ratio before and after the earthquake (YANG et al., 2006). Hence, it was considered a candidate for further earthquake surveillance. Consequently, a multi-parameter, continuous gas monitoring station, mainly equipped with a radon detector and a quadrupole mass spectrometer (QMS), was installed at the CL site since 2001. Continuous operation for more than one year demonstrates that the gas composition in this area showed significant variations. Possible effects of meteorological factors and earthquakes on the gas composition and flux are major concerns in this paper.

2. Geological Background of the Monitoring Site

Taiwan is situated at the boundary between the Eurasian Plate and the Philippine Sea Plate and has been subjected to an oblique collision with the northern Luzon arc in the Philippine Sea Plate since mid-Miocene (Fig. 1A). As a result of the collision by the northwestward movement of the Philippine Sea Plate, a series of thrust faults were developed in western Taiwan (Fig. 1B). The 1999 Chi-chi earthquake ($M_W = 7.6$) occurred along the Che-lung-pu (CLP) fault (Fig. 1C) and resulted in a surface rupture extending for 100-km in a nearly north-south direction in front of the Western Foothills, central Taiwan (CHEN, Y.G. et al., 2001; CHEN, W.S. et al., 2003).

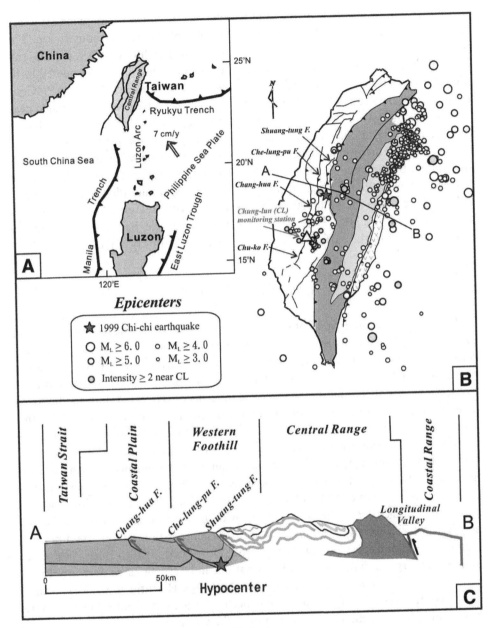

Figure 1

Tectonic settings and the earthquakes that occurred around Taiwan during the period of gas monitoring in this study. (A) Simplified tectonic features around Taiwan; (B) Distributions of the epicenters of sensible earthquakes ($M_L \geq 3.0$) recorded around Taiwan from October 1, 2001 to November 30, 2002. The 1999 Chi-chi disastrous earthquake ($M_W = 7.6$) is also marked for reference. Those earthquakes with local intensity greater than 2 (gravity acceleration velocity ≥ 2.5 gal) near the Chung-lun (CL) monitoring station, which is located on the Chu-ko Fault and is indicated by a triangle symbol, were marked with yellowish-color. Earthquake data are taken from the website of Central Weather Bureau, R.O.C. (http://www.cwb.gor.tw). (C) Geologic profile along A-B cross-section shown in Figure 1B.

The CL site is located on the culmination of an elongated anticline structure trending approximately north-northeast for a traceable length of about 15 km. The east limb of the anticline is cut by the well-known Chu-ko (CK) thrust fault (Fig. 1B), which connects the CLP fault at the northern end and is considered to be one of the most potentially reactive faults in southwestern Taiwan. Previous geological survey showed that abundant oil or gas might have been accumulated near the CL area (CHANG, 1962). However, several test wells drilled by the Chinese Petroleum Corporation indicated that the gas composition is predominantly CO_2. Reservoirs of future potential were identified in the Miocene sandstone strata, which are overlain by Miocene shales. The porosity and permeability of the sandstones can be classified as of the fair to good category (WU, 1968). In addition to the CK fault, at least three faults merge together and form a heavily fractured fault zone in this area (CHYI *et al.*, 2001). Thus, gases of deep lithosphere origin may be able to migrate upward to the surface through the fractured pathways (YANG *et al.*, 2003).

3. The Monitoring Station

Many mud volcanoes and hot springs are distributed along the tectonic sutures in southwestern Taiwan (YANG *et al.*, 2004). The composition of gas and fluid are considered to be sensitive to the earthquake activity (CHYI *et al.*, 2001, 2005; SONG *et al.*, 2005, 2006; YANG *et al.*, 2005). Hot springs (~70°C) and mud-pools (ambient temperature) are also found in the CL area. In terms of $^3He/^4He$ ratios, the bubbling gases are suggested to be originated from a mantle-derived component mixed with a crustal component (YANG *et al.*, 2003). A striking feature is the large variation of gas composition associated with seismic activity (CHEN, C-H. *et al.*, 2004; YANG *et al.*, 2005; 2006). Meanwhile, some local people claimed that vigorous bubbling of gases in the largest mud-pool at CL often occurred several days before big earthquakes, deserving its nickname of "earthquake pond". Thus, the composition and flux variations of gases from this area may provide useful information on earthquake activity. After continuous monitoring the gas composition of CL mud-pool by manually sampling every one-two weeks from December, 2000 to June, 2001, CHEN *et al.* (2004) argued that the CO_2/CH_4 variations in bubbling gases were sensitive to the earthquake events (Fig. 2). Therefore, the CO_2/CH_4 ratio is used as a target parameter for further earthquake surveillance in this area. A monitoring station was built on the bank of the largest mud-pool in the CL area.

Figure 3 shows the configuration of the automated monitoring system. A stainless steel funnel, which is 1 m × 1 m in cross section and covers about one third of the gas bubbling area, was installed inversely and fixed at the bottom of the mud pool. Bubbling gases were then introduced via a PVC pipe into the monitoring station. Gas flow rates, which range from 20 to 1200 liter/hour, are measured with a drum-type gas flow meter. After passing through a condensing water trap, the gases

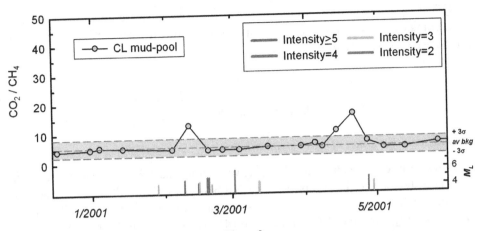

Figure 2

Variations in CO_2/CH_4 ratios of bubbling gases from the CL mud-pool. Earthquake data source is the same as shown in Figure 1.

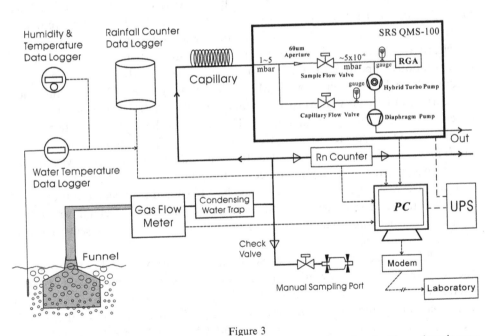

Figure 3

Configuration of the automated gas monitoring system. This system is equipped with quadrupole mass spectrometer, radon counter, drum-type gas flow meter, meteorological sensors, and an uninterrupted power supplier (UPS). Functions of each component are described in the text.

are split to (1) QMS; (2) radon counter; and (3) manual sampling port, respectively. Gas composition can be automatically analyzed by remote control. A manual sampling port, normally closed, is used for weekly sampling in order to calibrate the analytical systems in the laboratory. All the analytical results, including the meteorological data (humidity, air temperature, water temperature, and rainfall), can be accessed via internet.

4. Results and Discussions

4.1 Temporal Variation of Gas Compositions

After testing and calibrating for a few months, the QMS system was able to continuously analyze the composition of bubbling gases from the CL mud-pool at the interval of every two minutes. The system was set to measure H_2, He, CH_4, H_2O, N_2, C_2H_6, O_2, Ar, and CO_2 simultaneously. Of these, argon and helium were measured by the channel electron multiplier (CEM) detector; and others were analyzed by the Faraday Cup (FC) detector. Figure 4 shows the hourly average of continuous recorded raw data before interference correction for the period from October 2001 to October 2002. Some abrupt changes of the time-series are due to manual factor calibration and system servicing. Significant diurnal variations of H_2O are clearly related to variations in ambient temperature. Note that helium and argon concentrations appeared to decrease significantly late in the monitoring period because of the degradation of the CEM detector, although the CEM gain calibration was carried out frequently. Mass discrimination of the CEM detector can also be observed for both the helium and argon. This indicates that the CEM detector may not be as good as expected for long-term monitoring purpose, although it exhibits a much higher sensitivity and a lower detection limit than the FC detector. Meanwhile, hydrogen sensitivity is strongly affected by the temperature of the filament in the RGA analyzer. Whenever the filament was shut down and reactivated, it took a few hours for the hydrogen sensitivity to reach its original state. Helium, argon and hydrogen partial pressures may therefore not be able to represent the true composition of the sample.

To eliminate the effects of H_2O variations and the abrupt change in the spectrum due to manual factor calibration of the system (Fig. 4), each component (after interference correction) was normalized to the anhydrous total pressure to represent its dry gas composition as volume percent in Figure 5. This figure shows that CO_2 is the major component of the gas sample, and CH_4 and N_2 are the second most abundant gases. In contrast to the relative constant percentage of N_2, CO_2, CH_4, and C_2H_6 displayed significant variations. Variations of CO_2 and CH_4 are negatively correlated; whereas C_2H_6 and CH_4 show a positive correlation. Oxygen concentrations are relatively low ($< 0.05\%$), consistent with the gas originated from deep sources because oxygen is a highly active gas which is easily consumed below the depth of a few meters.

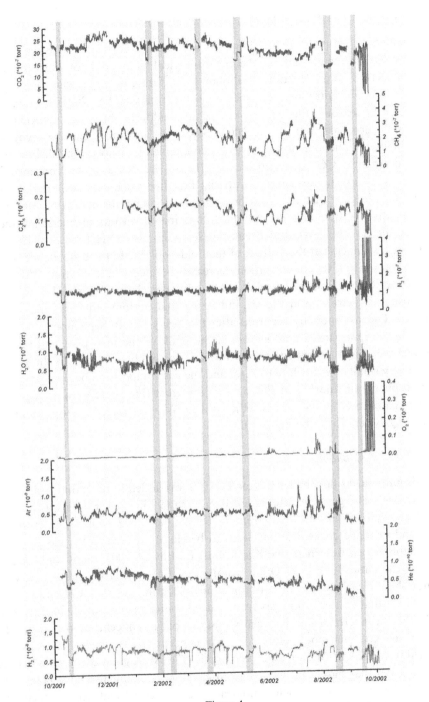

Figure 4

Continuous monitoring results of bubbling gases of the CL mud-pool. Some abrupt changes of the spectrum due to manual factor calibration and system service are marked with a grey-bar.

4.2 Gas Composition Affected by the Meteorological Factors

Although air temperature and humidity are strongly correlated with the concentration of H_2O in the gas sample, the meteorological factors did not affect the major components (CO_2 and CH_4) of the bubbling gases during the dry season (Fig. 6).

N_2 and O_2 concentrations remained constant from October, 2001 to May, 2002, i.e., the dry season (Fig. 5). However, anomalously high O_2 concentration occasionally occurred after June, 2002. Seven O_2 anomalies can be recognized. Interestingly, the phenomenon is always accompanied by heavy rainfall. In contrast to the decreasing CO_2 concentration, CH_4, C_2H_6, N_2 and Ar concentrations increased suddenly in association with the first five O_2 anomalies (marks a-e in Fig. 5). This implies that the local heavy rain may have disturbed the mud-pool. Consequently, the bubbling gas may be contaminated with atmospheric air and/or dissolved air (O_2, N_2, Ar), and organic gases (CH_4 and C_2H_6) released from the organic-rich sediments at the bottom of the mud-pool due to the disturbance by rain water. Meanwhile, CO_2 concentration decreased because of the dilution effect.

The last two O_2 anomalies did not exhibit correlation with CH_4 and C_2H_6 concentrations (marks f-g in Fig. 5), indicating that bottom sediments may not be disturbed. The last anomaly occurred after the heavy rain in September, 2002. This sudden heavy rain caused land sliding at the bank of the mud-pool and seriously disturbed the pool and changed the pathway of the bubbling gases. As the gas flow became very small or even close to zero, and the gas composition became dominated by atmospheric air (mark g in Fig. 5), it is apparent that there were no significant bubbling gases entering the gas collecting system. We were forced to stop running the monitoring system for bubbling gas then.

4.3 Flow Rates of Bubbling Gas

A drum-type gas flow meter (Ritter TG10®) was added to the monitoring system in May, 2002. The gas flow rate varied from 7 to 699 liter/hour, and was even less than 1 liter/hour after the heavy rain in September, 2002 (Fig. 5). Interestingly, radon and methane concentrations seem to be closely related to the gas flow rate. When bubbling gas flow increased significantly on July 24, 2002, radon and CH_4 also increased dramatically (Fig. 7). Other major gases (CO_2, N_2, H_2O, and O_2), however, did not show clear changes in concentrations. This implies that CH_4 may be the major phase that carries radon gas from a deep source to the surface in this area.

To check the relationship between variations of gas concentration and gas flow rate for the CL mud-pool, plots for CH_4 concentrations versus gas flow data recorded from May to August, 2002 were made (Fig. 8). A good correlation between them indicates that significant variations in gas flow rate may also have taken place during the dry season when meteorological effects were minimal. After a reconnaissance for possible air contamination, variations in both the gas composition and flow rate may be useful parameters for earthquake monitoring.

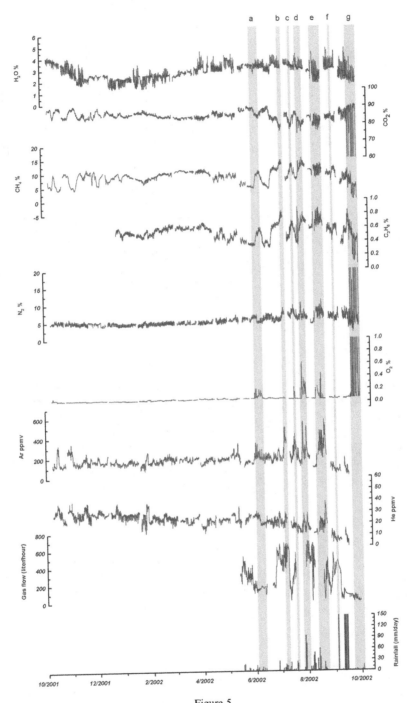

Figure 5

Temporal variations of gas compositions (volume percent) of the CL mud-pool. Anomalous high O_2 contents, indicating contamination of air and/or water dissolved air, are marked.

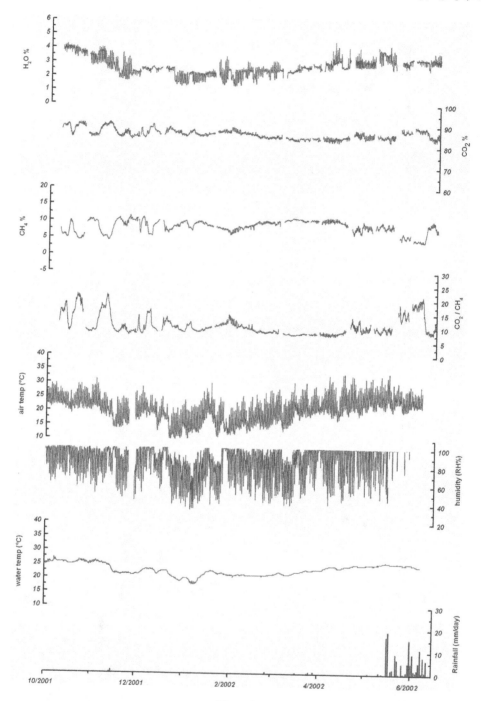

Figure 6
Variations of major compositions (H_2O, CH_4, and CO_2) of bubbling gases and meteorological factors (ambient temperature, humidity, water temperature and rainfall) with time. There is no clear correlation between gas compositions and meteorological factors during the dry season.

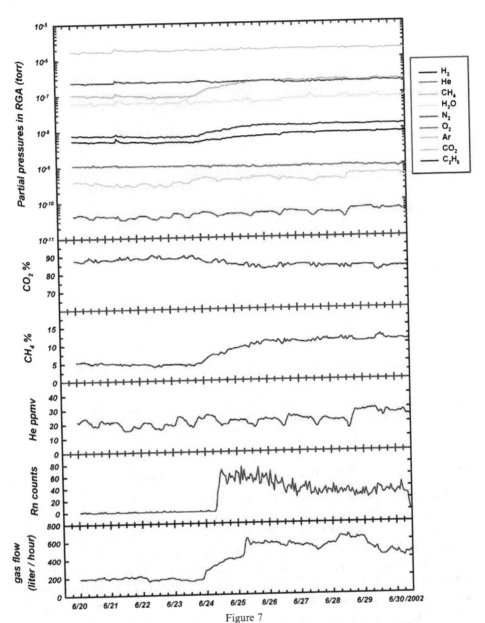

Figure 7
Variations of gas compositions and the gas flow rate of bubbling gases with time.

As discussed previously, the bubbling gases exhaled around the mud-pool in varying positions eventually. Hence, the measured gas flow in this study, especially during the heavy rain season (September to October, 2002), may not represent the variation of the total flux of bubbling gases in this area.

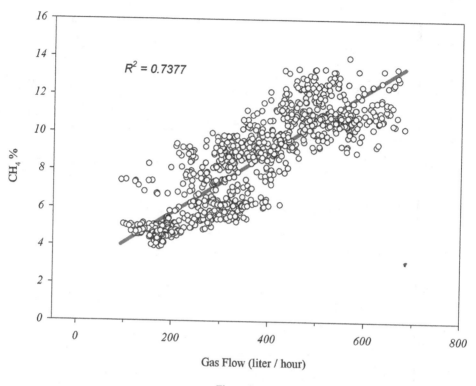

Figure 8

CH_4 concentration versus gas flow plots for the CL mud-pool during the period of May 2002 to August 2002. A positive correlation between these two parameters is observed.

4.4 Variations in Gas Compositions Associated with Earthquakes

Disregarding sample data that may have been contaminated by the atmospheric air and/or water dissolved air due to heavy rain (those marked in Fig. 5), we are able to discuss the relationship between the gas composition and associated earthquake events (Fig. 9). Considering the variation in gas composition and the sensitivity to earthquakes, the CO_2/CH_4 ratio can amplify the individual CO_2 and CH_4 variations, and minimize the influence of potential air contamination (both are in trace amounts in atmospheric air). With reference to the background values shown in Figure 2, indicate a total number of 23 anomalous variation peaks can be identified for further correlation with relevant earthquakes (Fig. 9).

A total of 444 earthquakes have been recorded around Taiwan during the period, from October 1, 2001 to November 30, 2002 (Fig. 1). Twenty-five earthquake events ($M_L \geq 3.0$ and Intensity ≥ 2) were considered to be effective with respect to the monitoring site. We divide the earthquakes into three groups based on the distance between epicentres and the monitoring site (Table 1). Nineteen "effective" earthquakes

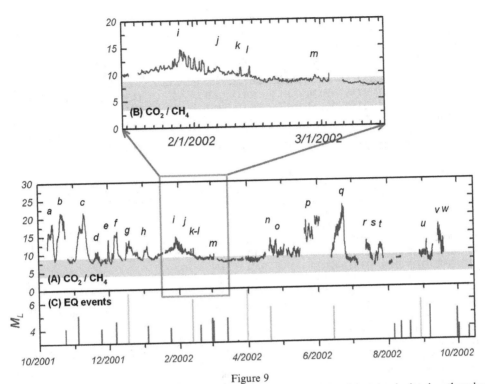

Figure 9

Variations of CO_2/CH_4 ratios of bubbling gases from the CL mud-pool and associated related earthquake events. (A) Twenty-three anomalous peaks (marked with letters) can be recognized; the background range was taken from Figure 2; (B) Enlarged blocked for better resolving the anomalous peaks i-m; (C) Earthquake events (data source from Fig. 1); red: group I; green: group II; blue: group III (see Table 1 for grouping).

belong to Group I (epicentres ≤ 80 km); three earthquakes to Group II (80–150 km), and another three to Group III (≥150 km), respectively. This indicates that only those earthquakes that occurred close to the monitoring site can be effective; unless the magnitude exceeded 5.5.

Aside from four earthquake events (numbers 13, 14, 20, 24), each earthquake can be correlated to one or two precursory anomalous peaks a few days to a few weeks earlier (Table 1). Earthquake 13 failed to show an associated anomaly probably due to its small magnitude ($M_L = 3.1$). On the other hand, epicenters may be too far to reflect anomalous signals for earthquakes 14 and 20 (Group III). No data were available on earthquake 24 because the system had stopped running.

In general, the CO_2/CH_4 anomalies occurred two to four weeks ahead of the earthquake and could last for a few hours to two weeks. Many studies showed that there may be good correlations between the precursory anomalies (time and duration) and the earthquakes (magnitude, depth, distance) (see compilation and

Table 1

Catalog of related earthquakes[1] occurred from October 1, 2001 to December 30, 2002 in Taiwan

No.	Local time	Lat. (°N)	Long. (°E)	Mag. (M_L)	Depth (km)	Dist. (km)	Intensity	Δt^2 (day)	ΔD^3 (day)	Peak[4]	Group[5]
1	2001/10/24 01:47	23.48	120.30	4.1	5.0	28.0	2	14.3	5.0	a	I
2	2001/11/04 16:45	23.93	121.04	5.1	5.9	79.3	2	14.1	6.2	b	I
3	2001/11/24 16:15	23.45	120.35	4.1	14.9	22.0	2	15.1	10.3	c	I
4	2001/12/07 16:07	23.09	120.79	4.6	3.0	39.3	2	17.0	2.2	d	I
5	2001/12/18 12:02	23.89	123.04	6.7	32.2	257.5	3	17.6	0.9	e	III
								11.3	5.8	f	
6	2002/01/04 05:31	23.56	120.58	4.3	13.4	21.2	2	28.1	5.8	f	I
7	2002/01/24 18:20	23.01	120.61	4.1	19.5	40.3	2	20.4	3.3	h	I
8	2002/02/12 11:27	23.77	121.66	6.2	25.1	120.3	3	14.1	14.8	i	II
9	2002/02/19 22:52	23.45	120.73	4.3	9.4	20.2	2	13.7	1.6	j	I
10	2002/03/01 16:47	23.11	120.81	4.8	5.7	38.9	2	18.4	0.3	k	I
11	2002/03/02 16:48	23.09	120.80	4.6	5.0	39.9	2	17.5	0.3	l	I
12	2002/03/14 19:25	23.39	120.68	4.8	9.5	13.3	2	15.2	0.3	m	I
13	2002/03/23 01:29	23.41	120.50	3.1	4.4	6.7	2	—	—	-	I
14	2002/03/31 14:52	24.24	122.17	6.8	9.6	189.6	3	—	—	-	III
15	2002/04/20 16:47	22.80	121.69	5.6	7.0	131.1	2	0.1	5.9	n	II
16	2002/06/14 15:22	23.51	121.63	5.5	24.9	110.0	2	16.1	13.1	p	II
17	2002/08/06 15:20	23.41	120.67	4.1	11.4	12.9	2	24.4	3.5	r	I
18	2002/08/12 04:57	23.34	120.45	4.3	15.1	10.6	3	21.2	0.7	s	I
19	2002/08/20 18:01	23.33	120.76	4.3	15.0	21.6	2	26.8	3.2	t	I
20	2002/08/29 01:05	22.20	121.35	6.0	13.2	152.6	3	—	—	-	III
21	2002/09/06 19:02	23.89	120.73	5.5	26.0	60.4	2	3.2	0.7	u	I
22	2002/09/30 16:35	23.34	120.63	5.3	5.2	8.7	3	16.4	7.9	v	I
23	2002/10/01 07:50	23.35	120.62	4.1	10.8	7.4	3	13.2	0.9	w	I
24	2002/10/09 23:49	23.31	120.62	3.9	10.9	9.7	3	—	—	-	I
25	2002/11/08 11:41	23.31	120.61	4.1	11.2	9.0	3	6.4	9.0	x	I

[1] There were in total 444 earthquakes recorded in Taiwan during this period (Fig. 1). Only those earthquakes with local intensity ≥ 2 are considered significant to the monitoring site.
[2] Time difference between peak of the anomaly and the earthquake event.
[3] Duration of the anomaly.
[4] The relevant anomalous peak marked in Fig. 9.
[5] The earthquakes are arbitrarily divided into three groups according to the distance between epicenter and the CL monitoring station. Group I: ≤80 km; Group II: 80—150 km; Group III: ≥150 km.

discussion by TOUTAIN and BAUBRON, 1999). With only one year observation, however, the present study does not show a clear relationship between them. Continuous surveillance for a longer time is necessary to better understand the characteristics of earthquakes that truly affect the CL area.

5. Conclusions

(1) An automatic multi-parameter monitoring system was set-up for long-term monitoring of gas composition at Chung-lun mud-pool, SW Taiwan. Although

oxygen is not measured for the geochemical monitoring purpose, we successfully used it as a good indicator for monitoring air-contamination in this study. Accordingly we are able to resolve possible effects of the environmental factor on gas samples.

(3) Bubbling gases from CL mud-pool is mainly composed of CO_2, CH_4, N_2, H_2O, and small amount of C_2H_6, O_2, Ar and He. After one year of continuous monitoring from October, 2001 to October, 2002, the gas composition showed systematic variations and was independent of the meteorological factors.

(4) Gas flow rate of the bubbling gases ranged from 7 to 699 liter/hour, and showed positive correlation with CH_4 and radon concentrations, indicating that CH_4 may be the major carrier gas for radon migrating from deep sources in this area. Unfortunately, the flow rate has been down to zero since October 2002 when bubbling gases were disturbed by the land sliding due to localy heavy rain.

(5) Compared to other gases, CH_4 and CO_2 showed significant variations during the monitoring period, and were considered to be related to earthquake events in this area. Taking the variations of CO_2/CH_4 ratio as the main indicator, precursory anomalies can be recognized from a few days to a few weeks before an earthquake.

Acknowledgement

We thank J.H. Jiang, Y.J. Lu, D.R. Hsiao, K.W. Wu for helping in setting up and calibrating the system; L.C. Huang and W.J. Lin for providing local logistical help; Dr. W.S. Chen for providing local geological information and profile. The senior author also benefited from discussions with Dr. J. Heinicke regarding gas flow measurements. Drs. N. Perez, H. Armannsson, W.C. Evans, P. Theodorsson and one anonymous reviewer gave critical comments and improve the manuscript. This research was supported by the Central Geological Survey of Taiwan, ROC (90EC2A380103, 5226902000-01-92-01, 5226902000-02-93-01).

References

CHANG, S.S.L. (1962), *Subsurface geology of the CL-1 Wildcat, Chunglun Structure, Chiayi, and the TK-1 Wildcat, Tishuikan structure, Kaohsiung, Taiwan*, Petrol. Geol. Taiwan *1*, 51–65.

CHEN, C.-H., YANG, T.F., SONG, S.R., LIU, T.K., and Lee, C.Y. (2004), *Environmental geochemistry with respect to the fault activities during 2000–2002 in Chiayi-Tainan and Hsinchu-Miaoli areas, Western Taiwan*, Bull. Cent. Geol. Survey *17*, 129–174. (in Chinese with English abstract)

CHEN, W.S., Chen, Y.G., SHIH, R.C., LIU, T.K., HUANG, N.W., LIN, C.C., SUNG, S.H., and LEE, K.J. (2003), *Thrust-related river terrace development in relation to the 1999 Chi-Chi earthquake rupture, Western Foothills, Central Taiwan*, Jour. Asian Earth Sci. *21*(5), 473–480.

CHEN, Y.G., CHEN, W.S., LEE, J.C., LEE, Y.H., LEE, C.T., CHANG, H.C., and LO, C. H. (2001), *Surface rupture of the 1999 Chi-Chi earthquake yields insights on the active tectonics of Central Taiwan*, Bull. Seismol. Soc. Am. *91*, 977–985.

CHYI, L.L., CHOU, C.Y., YANG, T.F., and CHEN, C-H. (2001), *Continuous Radon measurements in faults and earthquake precursor pattern recognition*, West. Pacific Earth Sci. *1*(2), 227–246.

CHYI, L.L., QUICK, T.J., YANG, T.F., and CHEN, C-H. (2005), *Soil gas Radon spectra and earthquakes*, Terr. Atmos. Oceanic Sci. *16*, 763–774.

FYTIKAS, M., LOMBARDI, S., PAPACHRISTOU, M., PAVLIDES, S., ZOUROS, N., and SOULAKELLIS, N. (1999), *Investigation of the 1967 Lesbos (NE Aegean) earthquake fault pattern based on soil-gas geochemical data*, Tectonophysics *308*, 249–261.

Heinicke, J. and Koch, U. (2000), *Slug flow — A possible explanation for hydrogeochemical earthquake precursors at Bad Brambach, Germany*, Pure Appl. Geophys. *157*, 1621–1641.

IGARASHI, G., SAEKI, S., TAKAHATA, N., SUMIKAWA, K., TASAKA, S., SASAKI, Y., TAKAHASHI, M., and SANO, Y. (1995), *Groundwater Radon anomaly before the Kobe earthquake in Japan*, Science *269*, 60–61.

ITALIANO, F., MARTINELLI, G. and NUCCIO, P.M. (2001), *Anomalies of mantle-derived helium during the 1997–1998 seismic swarm of Umbria-Marche, Italy*, Geophys. Res. Lett. *28*, 839–842.

KAWABE, I. (1984), *Anomalous changes of CH_4/Ar ratio in subsurface gas bubbles as seismo-geochemical precursors at Matsuyama, Japan*, Pure Appl. Geophys. *122*, 194–214.

KING, C.Y. (1978), *Radon emanation on San Andreas Fault*, Nature *271*, 516–519.

KING, C.Y. (1986), *Gas geochemistry applied to earthquake prediction: An overview*, J. Geophys. Res. *91*, 12269–12281.

LIU, K.K., YUI, T.F., YEH, Y.H., TSAI, Y.B., and TENG, T. (1985), *Variations of Radon content in ground waters and possible correlation with seismic activities in Northern Taiwan*, Pure Appl. Geophys. *122*, 231–244.

SANO, Y., TAKAHATA, N., IGARASHI, G., KOIZUMI, N., and STURCHIO, N.C. (1998), *Helium degassing related to the Kobe earthquake*, Chem. Geol. *150*, 171–179.

SONG, S.R., CHEN, Y.L., LIU, C.M., KU, W.Y., CHEN, H.F., LIU, Y.J., KUO, L.W., YANG, T.F., CHEN, C-H., LIU, T.K., and LEE, M. (2005), *Hydrochemical changes in spring waters in Taiwan: Implications for evaluating sites for earthquake precursory monitoring*, Terr. Atmos. Oceanic Sci. *16*, 745–762.

SONG, S.R., KU, W.Y., CHEN, Y.L., LIU, C.M., CHEN, H.F., CHAN, P.S., CHEN, Y.G., YANG, T.F., CHEN, C-H., LIU, T.K., and LEE, M. (2006), *Hydrogeochemical anomalies in the springs of the Chiayi Area in West-central Taiwan as possible precursors to earthquakes*, Pure Appl. Geophys. *163*, DOI 10.1007/s00024-006-0046-x.

SUGISAKI, R. (1978), *Changing He/Ar and N_2/Ar ratio of fault air may be earthquake precursors*, Nature *275*, 209–211.

SUGISAKI, R. and SUGIURA, T. (1986), *Gas anomalies at three mineral springs and a fumarole before an inland earthquake, Central Japan*, J. Geophys. Res. *91*, 12296–12304.

TAKAHATA, N., IGARASHI, G., and SANO, Y. (1997), *Continuous monitoring of dissolved gas concentrations in groundwater using a quadrupole mass spectrometer*, Appl. Geochem. *12*, 377–382.

TEDESCO, D. and SCARSI, P. (1999), *Chemical (He, H_2, CH_4, Ne, Ar, N_2) and isotopic (He, Ne, Ar, C) variations at the Solfatara Crater (Southern Italy): Mixing of different sources in relation to seismic activity*, Earth Planet. Sci. Lett. *171*, 465–480.

THOMAS, D. (1988), *Geochemical precursors to seismic activity*, Pure Appl. Geophys. *126*, 241–265.

TOUTAIN, J P. and BAUBRON, J.C. (1999), *Gas geochemistry and seismotectonics: A review*, Tectonophysics *304*, 1–24.

VIRK, H.S., WALIA, V., and KUMAR, N. (2001), *Helium/Radon precursory anomalies of Chamoli earthquake, Garhwal Himalaya, India*, J. Geodynam. *31*, 210–210.

WALIA, V., VIRK, H.S., and BAJWA, B.S. (2006), *Radon precursory signals for some earthqaukes of magnitude > 5 occurred in N-W Himalaya*, Pure Appl. Geophys. *163*, DOI 10.1007/s00024-006-0044-z.

WALIA, V., VIRK, H.S., YANG, T.F., MAHAJAN, S., WALIA, M., and BAJWA, B.S. (2005), *Earthquake prediction studies using Radon as a precursor in N-W Himalayas, India: A case study*, Terr. Atmos. Oceanic Sci. *16*, 775–804.

WHITEHEAD, N.E. and LYON, G.L. (1999), *Application of a new method of searching for geochemical changes related to seismic activity*, Appl. Radia. Isotop. *51*, 461–474.

WU, F.T. (1968), *Petrographic study of oil sands of the Chunglun Structure, Chiayi, Taiwan*, Petrol. Geol. Taiwan *6*, 183–195.

YANG, T.F., CHOU, C.Y., CHEN, C-H., CHYI, L.L., and JIANG, J.H. (2003), *Exhalation of Radon and its carrier gases in SW Taiwan*. Radiat. Meas. *36*, 425–429.

YANG, T.F., HSIEH, P.S., CHEN, Y.G., and CHEN, C-H. (2006), *Anomalous helium isotopic variations of fluid samples before and after the 1999 Chi-Chi earthquake in SW Taiwan*, Geophy. Res. Lett. (submitted)

YANG, T.F., WALIA, V., CHYI, L.L., FU, C.C., CHEN, C-H., LIU, T.K., SONG, S.R., LEE, C.Y., and LEE, M. (2005), *Variations of soil Radon and Thoron concentrations in a fault zone and prospective earthquakes in SW Taiwan*, Radiat. Meas. 40, 496–502.

YANG, T.F., YEH, G.H., FU, C.C., WANG, C.C., LAN, T.F., LEE, H.F., CHEN, C-H., WALIA, V., and SUNG, Q.C. (2004), *Composition and exhalation flux of gases from mud volcanoes in Taiwan*, Environ. Geol. *46*, 1003–1011.

(Received: June 2, 2003; revised: November 23, 2005; accepted: November 30, 2005)
Published Online First: March 28, 2006

 To access this journal online:
http://www.birkhauser.ch

Pure appl. geophys. 163 (2006) 711–721
0033–4553/06/040711–11
DOI 10.1007/s00024-006-0044-z

© Birkhäuser Verlag, Basel, 2006

Pure and Applied Geophysics

Radon Precursory Signals for Some Earthquakes of Magnitude > 5 Occurred in N-W Himalaya: An Overview

VIVEK WALIA,[1,2] HARDEV SINGH VIRK,[3] and BIKRAMJIT SINGH BAJWA[4]

Abstract—The N-W Himalaya was rocked by a few major and many minor earthquakes. Two major earthquakes in Garhwal Himalaya: Uttarkashi earthquake of magnitude $M_s = 7.0$ ($m_b = 6.6$) on October 20, 1991 in Bhagirthi valley and Chamoli earthquake of $M_s = 6.5$ ($m_b = 6.8$) on March 29, 1999 in the Alaknanda valley and one in Himachal Himalaya: Chamba earthquake of magnitude 5.1 on March 24, 1995 in Chamba region, were recorded during the last decade and correlated with radon anomalies. The helium anomaly for Chamoli earthquake was also recorded and the Helium/Radon ratio model was tested on it. The precursory nature of radon and helium anomalies is a strong indicator in favor of geochemical precursors for earthquake prediction and a preliminary test for the Helium/Radon ratio model.

Key words: Radon, helium, precursor, uttarkashi, chamba, chamoli.

1. Introduction

The Himalayan orogeny is believed to be a product of the on-going collision of the Indian plate with the Eurasian plate. Some of the largest earthquakes in history occurred in the vicinity of Himalaya as a consequence of underthrusting of the Indian plate. The seismicity and tectonics of the Himalaya have been subjects of special investigation. SRIVASTAVA *et al.* (1979) studied earthquake occurrence during the years 1965–75 in the epicentral region of the great Kangra earthquake of 1905 with the help of a closely spaced network of seismic stations. Most of the seismic activity of the region was found to be related to the main boundary fault (MBF) of the N-W Himalaya. They also identified a seismic gap in the eastern part of Himachal Pradesh and suggested that this gap may be the locale of a future major earthquake in the region. BHATTACHARYA *et al.* (1986) carried out a micro-earthquake survey around the Thein Dam during 1983 in the vicinity of the Kangra valley.

[1]National Center for Research on Earthquake Engineering, Taipei-106, Taiwan
[2]Department of Geosciences, National Taiwan University, Taipei-106, Taiwan
[3]360, Sector 71, SAS Nagar, (Mohali)-160071, India
[4]Department of Physics, Guru Nanak Dev University, Amritsar-143005, India

The Kangra valley of Himachal Himalaya and Bhagirthi & Alaknanda valleys of Garhwal Himalaya are good examples of tectonically active areas in N-W Himalaya. The region has suffered several major and minor earthquakes. Each of these earthquakes have created their own unforgettable histories (OLDHAM, 1883). During the last decade two major earthquakes have occurred in the Garhwal Himalaya; Uttarkashi earthquake of magnitude Ms = 7.0 occurred on October 19, 1991 and the Chamoli earthquake of Ms = 6.5 on March 29, 1999. In the Chamba region of Kangra valley, Chamba earthquake of magnitude 5.1 occurred on March 24, 1995. The epicentral zones of major earthquakes are located within the lesser Himalaya close to the Main Central Thrust (MCT) (Fig. 1), which is dominated by north-south compressive tectonics resulting from the continent-continent collision between Indian and Eurasian plates (LE FORT, 1975).

Radon is established as a useful geochemical precursor. Studies of geochemical and hydrological anomalies preceding significant earthquakes had been reported in China, Japan, Uzbekistan, Mexico, Italy, Taiwan, India and Germany (LIU *et al.*, 1984/85; IGARASHI and WAKITA, 1990; ULOMOV and MAVASHEV, 1967; SEGOVIA *et al.*, 1995; HEINICKE *et al.*, 1995; CHYI *et al.*, 2005; YANG *et al.*,

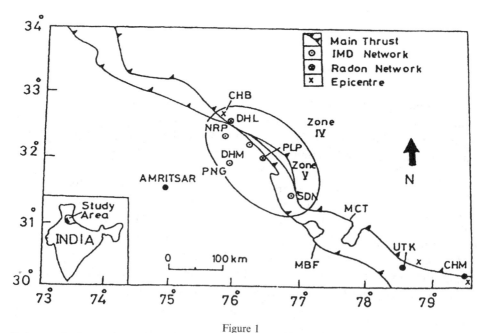

Figure 1
Radon monitoring station Palampur (PLP) and Dalhousie (DHL) together with IMD network stations, viz. Nurpur (NRP), Dharamsala (DHM), Pong Dam (PNG) and Sundernagar (SDN) and Epicenters of Uttarkashi (UTK), Chamba (CHB) and Chamoli (CHM) earthquakes.

2005,06; VIRK, 1996, 1999). However, studies of these preseismic phenomena have been controversial for several reasons (SILVER and WAKITA, 1996; WAKITA, 1996). During the last decade some highly useful data on the correlation of radon anomalies with seismic events which occurred in N-W Himalaya have been reported (VIRK, 1986, 1990, 1995; VIRK and SINGH, 1992, 1993, 1994; VIRK et al., 1995,97; VIRK and SHARMA, 1997; WALIA et al., 2005; RAMOLA et al., 1990). Radon monitoring started in 1989 at Palampur in Kangra valley (Fig. 1) using plastic track detectors and emanometry. Radon anomalies in soil-gas and groundwater were correlated with seismic events in N-W Himalaya. The postdiction of Uttarkashi earthquake at radon network in Kangra valley encouraged us to set up 5 more stations using alpha-logger probes for continuous monitoring of radon in real time.

Helium monitoring was started at Palampur during 1997. In general helium emanates from deeper layers of the crust than radon. Hence helium is a better precursor than radon and the He/Rn ratio model is proposed to be tested in N-W Himalaya for the Chamoli earthquake.

2. Experimental Techniques

2.1 Radon Emanometry

Radon is monitored in soil-gas and groundwater by using the emanometry technique. An emanometer (Model RMS-10) manufactured by the Atomic Minerals Division, Department of Atomic Energy, India is used to measure the alpha emanation rate from radon in the gas fraction of a soil or water sample by pumping the gas into a scintillation chamber using a closed circuit technique (GHOSH and BHALLA, 1966). This technique gives instant values of radon concentration and is highly suitable for a quick radon survey.

In this method, the auger holes, each 60 cm in depth and 6 cm in dia., are left covered for 24 hours so that soil-gas radon and thoron become stable. The soil-gas probe is fixed in the auger hole forming an air-tight compartment. The rubber pump, soil-gas probe and alpha detector are connected in a closed-circuit. The soil-gas is circulated through a ZnS coated chamber (110 ml) for a period of 15 minutes, allowing the radon to form a uniform mixture with air. The detector is then isolated by clamping both the ends, and observations are recorded after four hours when equilibrium is established between radon and its daughters. Alpha particles emitted by radon and its daughters are recorded by the scintillation assembly consisting of a photomultiplier tube and a scaler-counter unit.

Radon monitoring in water is also carried out by using the closed-circuit technique. Groundwater samples are collected daily from a 'bauli' (natural spring) in a sample bottle (250 ml). The air is circulated in the closed-circuit containing a hand-operated rubber pump, the water sample bottle, a drying chamber and a ZnS(Ag) detector cell for 10 minutes. The alpha counts are recorded after four hours during which the equilibrium between radon and its daughters is established. Meteorological effects on radon emanation at Palampur have been reported elsewhere (SHARMA *et al.*, 2000). It is observed that radon emanation in water is not influenced by meteorological effects, whereas the net effect is less than 1σ in the case of soil-gas.

2.2 Helium Monitoring

A helium detector ASM 100 HDS (Alcatel, France) is a complete leak detection system. It uses a sniffing technique and comprises a helium gas analyzer with a pumping system: Molecular Drag Pump (MDP) and a set of dry pumps connected to the MDP exhaust. The main component of a helium leak detector is a spectro-cell (with sensitivity 3.10^{-4} A/mbar) which acts as a mass spectrometer. The helium ion analysis is based on the partial pressure of helium in the system; its magnitude indicates the flow rate of the helium that has been detected. It is calibrated and the logarithmic scale used to display the helium concentration in ppm. The whole operation is fully automatic and helium values from 0.1 to 10^6 ppm (100%) can be measured (Walia *et al.*, 2005). This sniffing probe technique is used for helium analysis in soil-gas from an auger hole at Palampur, using discrete sampling at a fixed site daily.

3. Helium/Radon Ratio Model

It is felt that radon alone may not be relied upon as an earthquake precursor but should be correlated with a deep origin gas-like helium. Radon coming from deeper layers of the crust may not be detectable at the surface due to its half life of 3.8 days and displays a poor intrinsic mobility with diffusion coefficient of 0.12 cm^2/s. Helium, on the other hand, is a highly mobile gas with diffusion coefficient of 1.68 cm^2/s and originates at deeper layers of the crust. It is highly stable and diffuses through interstitial spaces in rocks during strain build up prior to an earthquake. A hypothesis based on the mobility of radon and helium gases in crustal layers may prove to be more useful in earthquake prediction studies.

The various stages of the conceptual model (SHARMA, 1997) shown in Figure 2, are as follows:

(i) Under normal conditions, helium to radon ratio may have some constant value depending on the geology and meteorological conditions at the monitoring site (Segment AB).

(ii) The stresses causing an earthquake buildup around the hypocenter. During this phase, first helium is affected at deeper layers and its emanation rate increases, hence He/Rn ratio rises sharply (Segment BC.)

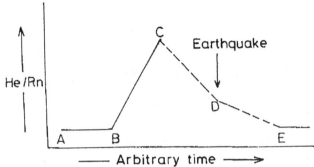

A-B : He/Rn ratio under normal condition
B-C : Rise in He/Rn ratio as stresses accumulate at depth
C-D : Drop in He/Rn ratio prior to triggering of the shock
D-E : Drop back in He/Rn ratio after the shock

Figure 2
A conceptual model of He/Rn ratio as a predictive tool of earthquake prediction in a seismic area.

(iii) When the stress reaches upper crustal layers, radon emanation is enhanced from rocks under excessive strain and hence He/Rn ratio falls suddenly (Segment CD); this is an alarm signal for the impending earthquake.

(iv) After the quake, both Rn and He drop down to normal values (Segment DE) as the ground conditions stabilise.

After plotting helium and radon gas concentrations in soil-gas during March, 1999, a plot of helium/radon ratio (Fig. 7) is prepared after normalizing both helium and radon for the sake of comparison. As predicted, the anomaly in gas ratio, He/Rn, precedes anomalies represented by radon and helium plots (Figs. 5,6).

4. Results and Discussion

The Uttarkashi earthquake of 7.0 M_s (m_b = 6.6) with epicenter at 30.78°N, 78.80°E occurred on 20th October, 1991. A radon anomaly was recorded simultaneously in both the media on October 15 at Palampur (31.10°N, 76.51°E) which is about 293 km from the Uttarkashi earthquake epicenter, with radon activity crossing the 2σ level above the average value (VIRK and SINGH, 1994). Temporal variations of radon in soil-gas and groundwater recorded from September to October 1991 at Palampur are shown in Figure 3. The average radon values recorded by emanometry were 27.55 Bq/L and 48.86 Bq/L with standard deviations of 11.49 Bq/L and 14.89 Bq/L in soil-gas and groundwater, respectively.

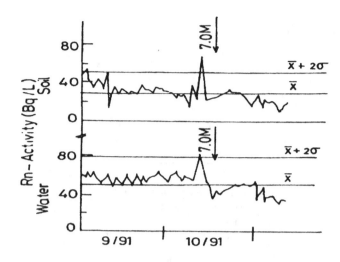

Figure 3
Radon anomalies in soil-gas and groundwater earthquake at Palampur as a precursor to the Uttarkashi earthquake.

The Chamba region was rocked by an earthquake of magnitude 5.1 M on 24 March, 1995 with its epicenter at Pliure (32.60°N, 75.91°E) nearly 7 km away from Chamba Town. The radon anomaly was recoded simultaneously in both media on 21 March, three days before the event at Dalhousie station (32.60°N, 76.00°E) which is only about 10 km from the Chamba earthquake epicenter, with peak values crossing the 2σ level above the average value of 4.52 Bq/L in soil-gas and 4.73 Bq/L in groundwater respectively (Fig. 4) (VIRK *et al.*, 1995). There was another anomaly peak on March 24 followed by a sudden fall in the emanation rate after the strain was released. The simultaneous recording of radon peaks in both soil-gas and groundwater at the same site and under similar meteorological conditions before the occurrence of Chamba earthquake on 24 March established the efficacy of radon as an earthquake precursor. Further, a total of 36 microseismic events were correlated with radon anomalies at the Dalhousie station in both the media during the time window of June 1996 to September 1999 (WALIA, 2001).

The Chamoli earthquake of magnitude $M_s = 6.5$ (6.8 m_b) occurred at 00:50 (IST) on March 29, 1999, with epicenter at 30.2°N, 79.5°E. The epicenters of both the Uttarkashi and Chamoli earthquakes lie along MCT (Fig. 1). The radon and helium anomalies were recorded at Palampur which is about 393 km from the Chamoli earthquake epicenter (VIRK *et al.*, 2001). The average radon values recorded during 1999 at Palampur in soil-gas and groundwater were 24.31 Bq/L and 56.69 Bq/L with a standard deviation of 10.4 Bq/L and 4.66 Bq/L, respectively. Although the radon

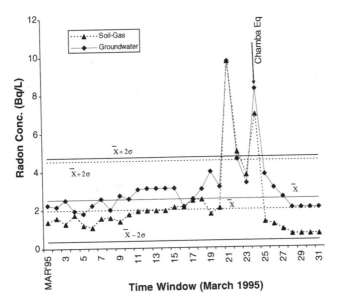

Figure 4
Radon anomalies in soil-gas and groundwater earthquake at Dalhousie as a precursor to the Chamba earthquake.

monitoring station operating during 1989–1991 is about 1 km from the present monitoring station nevertheless the average values and standard deviation do not show sizable difference i.e., only about 12% and 14%, respectively, which is reasonably good keeping in mind the considerable time difference, seasonal and

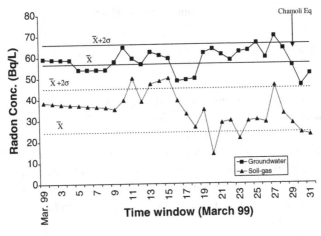

Figure 5
Radon anomaly in soil-gas and groundwater at Palampur as a precursor to the Chamoli earthquake.

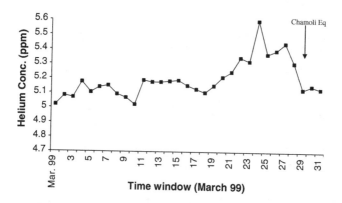

Figure 6
Helium anomaly in soil-gas at Palampur as a precursor to the Chamoli earthquake.

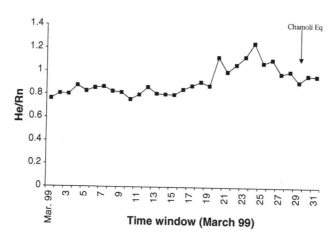

Figure 7
Helium/radon ratio anomaly in soil-gas at Palampur as a precursor to the Chamoli earthquake.

annual variations and variations due to other geodynamic activities in the area. The radon anomalies were recorded in both the media on 27 March, 1999 with the peak values of 46.63 Bq/L and 69.66 Bq/L, crossing $\bar{X} + 2\sigma$ level as shown in Figure 5, respectively. In fact, radon fluctuations in soil-gas started on 10th March with some highs and lows (Fig. 5) attaining the minimum value on 20th March and the final peak on 27th March. There are two positive and one negative radon anomalies before the main shock anomaly. These may be reflecting the stress behavior of crustal rocks before the main event.

Helium variations in soil-gas during March, 1999 are shown in Figure 6. The helium anomaly was recorded on March 24, i.e., three days before the radon anomaly and five days before the Chamoli earthquake. It clearly shows that helium is influenced by strain buildup prior to radon and due to its high mobility reaches the surface earlier than radon. The same trend is observed in He/Rn ratio (Fig. 7) which is plotted after normalizing both radon and helium data. Normalization is needed to nullify the big fluctuations i.e., about 200% in radon as compared to helium which is about 12%. As helium varies in ppm, a sizable percentage variation could not be expected from it. On March 20th there was a sharp rise in He/Rn ratio, with a peak value on March 24th, followed by a fall with minima recorded on March 29 on the day of the event. This sudden rise and then fall in the He/Rn ratio is a precursory signal for the impending earthquake which occurred near Chamoli, and a meteorological variation has not been recorded during that period. The observed trend follows the He/Rn ratio model and may be considered as a preliminary test of this time predictive tool.

In all the above three events, anomalous values were recorded in both soil-gas and groundwater almost simultaneously, which clearly indicate that the degassing system is only disturbed by the ongoing stress before the major event. Further, during the time window of June 1992 to August 1995 and June 1996 to September 1999, about 142 microseismic events with magnitudes ranging between 2.1 to 4.8 were correlated with radon anomalies at both Dalhousie and Palampur stations in both the media (WALIA et al., 2003). The above study related to major events clearly indicates the efficiency of both radon and helium as precursors.

Acknowledgement

The authors are grateful to Professor Wakita for his assistance in the promotion of research activity of our group. The senior author, H.S. Virk, enjoyed his hospitality twice in 1993 and 1996. Two anonymous reviewers' comments and suggestions are beneficial.

REFERENCES

BHATTACHARYA, S.N., PRAKASH, C., and SRIVASTAVA, H.N. (1986), Microearthquake observations around Thein Dam in N-W Himalayas, Phys. Earth Planet. Inter. 44, 169–178.

CHYI, L.L., QUICK, T.J., YANG, T.F., and CHEN, C-H. (2005), Soil gas radon spectra and earthquakes, Terr. Atmos. Oceanic Sci. 16(4), 763–774.

GHOSH , P.C. and BHALLA, N.S., A closed-circuit technique for radon measurement in water and soil, with some of its applications, Proc. All India Symp. on Radioactivity and Meterology of Radionuclides (A.E.E.T., Bombay 1966) pp. 226–239.

HEINICKE, J., KOCH, U., and MARTINELLI, G. (1995), CO_2 and radon measurements in the Vogtland area (Germany)— A contribution to earthquake prediction research, Geophys. Res. Lett. 22, 771–774.

IGARASHI, G. and WAKITA, H. (1990), *Groundwater radon anomalies associated with earthquakes*, Tectonophysics *180*, 237–254.

LE FORT, P. (1975), *Himalayas: The collided range. Present knowledge of the continental arc*, Am. J. Sci. *275*(A), 1–44.

LIU, K.K., YUI, T.F., TASI, Y.B., and TENG, T.L. (1984/85), *Variation of radon content in groundwater and possible correlation with seismic activities in northern Taiwan*, Pure Appl. Geophys. *122*, 231–244.

OLDHAM, T.A. (1883), *Catalogue of Indian Earthquakes*, Mem. Geol. Surv. India *19*, 163–215.

RAMOLA, R.C., SINGH, M., SANDHU, A.S., SINGH, S., and VIRK, H.S. (1990), *The use of radon as an earthquake precursor*, Nucl. Geophys. *4*, 275–287.

SEGOVIA, N., MENA, M., SEIDEL, J.L., MONNIN, M., TAMEZ, E., and PENA, P. (1995), *Short and long term radon-in-soil monitoring for geophysical purpose*, Rad. Meas. *25*, 547–552.

SHARMA *et al.* (2000)

SHARMA, S.C. (1997), *Thermal springs as gas monitoring sites for earthquake prediction*, Proc. 3rd Int. Conf. on Rare Gas Geochemistry (ed. H.S.Virk) Amritsar, India , Dec.10–14, 1995, 193–199.

SILVER , P.G. and WAKITA, H. (1996), *A search for earthquake precursors*, Science *273*, 77–78.

SRIVASTAVA, H.N., DUBE, R.K., and CHAUDHURY, H.M., *Precursory seismic observation in the Himalayan foothills region*, Proc. Internat. Symp. Earthquake Prediction (UNESCO, Paris 1979), pp. 101–110.

ULOMOV, V.I. and MAVASHEV, B.Z. (1967), *A Precursor of a strong tectonic earthquake*, Akad. Sci. USSR Earth Sci. Sect. *176*, 9–11.

VIRK, H.S., *Radon monitoring and earthquake prediction*, Proc. Internat. Symp. Earthq. Prediction-Present Status, (University of Poona Pune, India 1986) pp. 157–162.

VIRK, H.S. (1990), *Radon studies for earthquake prediction, uranium exploration and environmental pollution: A review*, Ind. J of Phys. *64*A, 182–191.

VIRK, H.S. (1995), *Radon monitoring of microseismicity in the Kangra and Chamba valleys of Himachal Pradesh, India*, Nucl. Geophys *9*, 141–146.

VIRK, H.S. (1996), *Radon studies for earthquake prediction*, Himalayan Geology *17*, 91–103.

VIRK, H.S. (1999), *Radon/helium studies for earthquake prediction in N-W Himalaya*, IL Nuovo Cimento *22*C, 423–429.

VIRK, H.S. and SHARMA, A.K. (1997), *Microseismicity trends in N-W Himalaya using radon signals*, Proc. 3rd Int. Conf. on Rare Gas Geochemistry (ed. H.S.Virk) , Amritsar, India , Dec.10–14, 1995, 117–135.

VIRK, H.S., SHARMA, A.K., and WALIA, V. (1997), *Correlation of alpha-logger radon data with microseismicity in N-W Himalaya*, Curr. Sci. *72*(9), 656–663.

VIRK, H.S. and SINGH, B. (1992), *Correlation of radon anomalies with earthquake in Kangra Valley*, Nucl. Geophys. *6*, 293–300.

VIRK, H.S. and Singh, B. (1993), *Radon anomalies in soil-gas and groundwater as earthquake precursor phenomenon*, Tectonophysics *227*, 215–224.

VIRK, H.S. and SINGH, B. (1994), *Radon recording of Uttarkashi earthquake*, Geophys. Res. Lett. *21*, 737–740.

VIRK, H.S., WALIA, V., and KUMAR, N. (2001), *Helium/radon precursory anomalies of Chamoli earthquake, Garhwal Himalaya, India*, J. Geodyn. *31*, 201–210.

VIRK, H.S., WALIA, V., and SHARMA, A.K. (1995), *Radon precursory signals of Chamba earthquake*, Curr. Sci. *69*(5), 452–454.

WAKITA, H. (1996), *Geochemical challenge to earthquake prediction*, Proc. Natl. Acad. Sci. (USA) *93*, 3781–3786.

WALIA, V. (2001), *Critical evaluation of radon data as a predictive tool of earthquakes in Kangra and Chamba Valleys of Himachal Pradesh, India*, Guru Nanak Dev University, Amritsar, India, Ph.D. Thesis (unpublished).

WALIA, V., BAJWA, B.S., VIRK, H.S., and SHARMA, N. (2003), *Relationships between seismic parameters and amplitudes of radon anomalies in N-W Himalaya, India*, Radiat. Meas. *36*, 393–396.

WALIA, V., QUATTROCCHI, F., VIRK, H.S., YANG, T.F., PIZZINO, L., and BAJWA, B.S. (2005), *Radon, helium and uranium survey in some thermal springs located in NW Himalayas, India: Mobilization by tectonic features or by geochemical barriers?*, J. Environ. Monitoring *7*(9), 850–855.

WALIA, V., VIRK, H.S., YANG, T.F., MAHAJAN, S., WALIA, M., and BAJWA, B.S. (2005), *Earthquake prediction studies using radon as a precursor in N-W Himalayas, India: a case study*, Terr. Atmos. Oceanic Sci. *16*(4), 775–804.

YANG, T.F., FU, C.C., WALIA, V., CHEN, C-H., CHYI, L.L., LIU, T.K., SONG, S.R., LEE, M., LIN, C.W., and LIN, C.C. (2006), *Seismo-geochemical variations in SW Taiwan: multi-parameter automatic gas monitoring results*, Pure Appl. Geophy. in this issue.

YANG, T.F., WALIA, V., CHYI, L.L., FU, C.C., CHEN, C-H., LIU, T.K., SONG, S.R., LEE, C.Y., and LEE, M. (2005), *Variations of soil radon and thoron concentrations in a fault zone and prospective earthquakes in SW Taiwan*, Radiat. Meas. *40*, 496–502.

(Received: May 23, 2003; revised: May 15, 2005; accepted: May 30, 2005)

To access this journal online:
http://www.birkhauser.ch

© Birkhäuser Verlag, Basel, 2006

Pure appl. geophys. 163 (2006) 723–744
0033–4553/06/040723–22
DOI 10.1007/s00024-006-0047-9

❘Pure and Applied Geophysics

Modelling the Mixing Function to Constrain Coseismic Hydrochemical Effects: An Example from the French Pyrénées

JEAN-PAUL TOUTAIN,[1] MARGOT MUNOZ,[1] JEAN-LOUIS PINAUD,[2] STÉPHANIE LEVET,[1] MATTHIEU SYLVANDER,[3] ALEXIS RIGO,[3] and JOCELYNE ESCALIER[1]

Abstract—Groundwater coseismic transient anomalies are evidenced and characterized by modelling the mixing function F characteristic of the groundwater dynamics in the Ogeu (western French Pyrénées) seismic context. Investigations of water-rock interactions at Ogeu indicate that these mineral waters from sedimentary environments result from the mixing of deep waters with evaporitic signature with surficial karstic waters. A 3-year hydrochemical monitoring of Ogeu springwater evidences that using arbitrary thresholds constituted by the mean ± 1 or 2σ, as often performed in such studies, is not a suitable approach to characterize transient anomalies. Instead, we have used a mixing function F calculated with chemical elements, which display a conservative behavior not controlled by the precipitation of a mineral phase. F is processed with seismic energy release (E_s) and effective rainfalls (R). Linear impulse responses of F to E_s and R have been calculated. Rapid responses (10 days) to rainwater inputs are evidenced, consisting in the recharge of the shallow karstic reservoir by fresh water. Complex impulse response of F to microseismic activity is also evidenced. It consists in a 2-phase hydrologic signal, with an inflow of saline water in the shallow reservoir with a response delay of 10 days, followed by an inflow of karstic water with a response delay of 70 days, the amount being higher than the saline inflow. Such a process probably results from changes in volumetric strain with subsequent microfracturation transient episodes allowing short inflow of deep salted water in the aquifer. This study demonstrates that groundwater systems in such environments are unstable systems that are highly sensitive to both rainfall inputs and microseismic activity. Impulse responses calculation of F to E_s is shown to be a powerful tool to identify transient anomalies. Similar processing is suggested to be potentially efficient to detect precursors of earthquakes when long time-series (5 years at least) are available in areas with high seismicity.

Key words: Hydrochemical time-series, mixing function, transient anomalies, volumetric strain.

1. Introduction

Earthquakes can trigger or be preceded by various types of geochemical and geophysical transient anomalies, the significance of which is still in debate (GELLER,

---

[1]Observatoire Midi-Pyrénées, LMTG, UMR 5563, 14 Avenue Edouard Belin, 31400, Toulouse, France
[2]BRGM, Orléans, France
[3]Observatoire Midi-Pyrénées, DTP, UMR 5562, 14 Avenue Edouard-Belin, 31400, Toulouse, France

1997; GELLER *et al.*, 1997; SYKES *et al.*, 1999; WYSS, 1997, 2001; BERNARD, 2001). Many hydrological (water temperature and level and stream flow) and geochemical (gas contents or ratios) anomalies have been recognized as related to seismic processes, either as precursors or coseismic signals (KING, 1986; THOMAS, 1988; WAKITA *et al.*, 1988; ROELOFFS, 1988; MOGI *et al.*, 1989; MUIR-WOOD and KING, 1993). Most of these anomalies are interpreted as the result of crustal processes occurring at depth.

However, surficial hydrochemical processes may also occur in the sedimentary pile. Aquifers often result from different groundwater sources which continuously mix in various proportions, and earthquake-related processes may change the relative mixing proportions. Abrupt changes in some ion concentrations in groundwaters have been interpreted as short-term precursory anomalies at weak epicentral distances (D < 50 kms) (TSUNOGAI and WAKITA, 1995; NISHIZAWA *et al.*, 1998; TOUTAIN *et al.*, 1997; BELLA *et al.*, 1998; BIAGI *et al.*, 2000; FAVARA *et al.*, 2001; ITALIANO *et al.*, 2001). These anomalies are interpreted as resulting from preseismic strain changes, with subsequent variations of hydraulic head levels and/or infiltration of surface waters. Both processes lead to mixing of geochemically contrasted aquifers, and may both occur as the result of a single strain event (TOUTAIN *et al.*, 1998; Poitrasson *et al.*, 1999).

Mixing of aquifers appears therefore as a key process associated with earthquakes. However, unlike physical anomalies (level, temperature) where a large set of data is available, very few hydrochemical anomalies are published and those available concern mainly isolated events characterized by a single couple earthquake-geochemical anomaly. Therefore, the assessment of systematic relationships for a large set of data at a single site is still lacking, as well as the method to evidence such relationships. Such an assessment requires a long-term recording of both seismic and hydrochemical data within a limited and well constrained area, that is very rarely performed.

In this study we have evidenced that the mixing function F characteristic of aquifer dynamics in mixed sedimentary contexts is a key parameter to constraint coseismic hydrologic effects. With this aim, we performed a three-year hydrochemical monitoring of major elements on springwater from the western Pyrénées, which are characterized by a moderate but recurrent seismicity (RIGO *et al.*, 1997; SOURIAU and PAUCHET, 1998; SOURIAU *et al.*, 2001).

2. Seismotectonic and Geochemical Background

2.1 Seismotectonics and Thermal Springs Distribution in the French Pyrénées

The Pyrenean range results from the convergence of the Iberian and Eurasian plates (CHOUKROUNE, 1992; OLIVET, 1996). The main structural and geological units of the belt are the Mesozoic sedimentary North Pyrenean zone (NPZ), the granitic

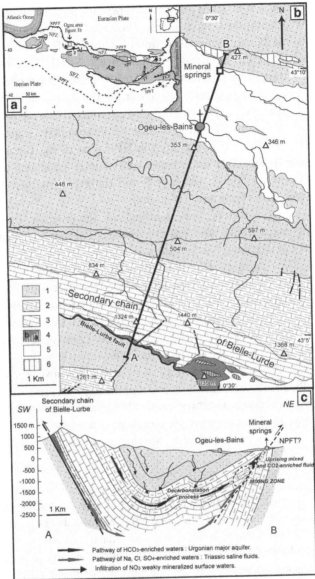

Figure 1

A : Structural sketch map of Pyrénées showing the main faults: NPFT (North Pyrenean Frontal Thrust), BAF (Bigorre-Adour faul), TF (Têt Fault), TCF (Tech Fault) and the main structural domains: NPZ (North Pyrenean Zone), AZ (Axial Zone), SPZ (South Pyreenan Zone). The three main earthquakes of the 20th century are displayed: 1: Arette, M = 5.7, 1981; 2: Arudy, M = 5.6, 1986; 3: St Paul de Fenouillet, M = 5.2, 1996). B: Geological sketch map of the Ogeu area (modified from CASTERAS *et al.*, 1970). C: Interpretative crosssection showing the main structures and potential circulation paths.

and metamorphic axial zone (AZ) and the sedimentary South Pyrenean zone (SPZ), showing an E-W orientation (Fig. 1a), being bounded by three major faults, which are from N to S: the North Pyrenean Frontal Thrust (NPFT), the North Pyrenean Fault (NPF) and the South Pyrenean Frontal Thrust (SPFT). The range displays a permanent, low to moderate seismicity (most of the quakes having magnitudes M ≤ 5, and three quakes of magnitude > 5 during the last century (Arette, 1967, M = 5.7; Arudy, 1980, M = 5.3; St Paul-de-Fenouillet, 1996, M = 5.2) (SOURIAU and PAUCHET, 1998, Fig. 1a). The distribution of the seismicity is not homogeneous all over the belt: it is concentrated along the NPF west of the Bigorre-Adour fault (BAF), whereas in the central and eastern parts, it is associated with structural features such as granitic bodies, the NPFT and the great fractures relative to the Mediterranean tectonics (Tech and Têt faults).

About 120 mineral waters are identified in the French Pyrénées, mainly distributed along major faults and granitic bodies. Three main chemical patterns can be outlined: 1) alkaline Na-S waters discharge in the AZ, 2) waters with carbonated-evaporitic features ($Ca-HCO_3$, $Ca-SO_4$, Na-Cl and intermediate types) rise in the NPZ, and 3) some CO_2-rich springs are recognized along the Tech fault ($Na-HCO_3$ type) and in the NPZ ($Ca-HCO_3$ type). Na-S hot waters and CO_2-rich springs of the eastern Pyrénées are well-documented (MICHARD and FOUILLAC, 1980; MICHARD *et al.*, 1980; YERRIAH, 1986; MICHARD, 1990; ALAUX-NEGREL, 1991; KRIMISSA, 1995), but typical Pyrenean carbonate-evaporite waters are still poorly studied (VANARA, 1997). Recently, LEVET *et al.* (2002) investigated water-rock interactions in Pyrenean tectonically active sedimentary terrains and outlined the extend of mixing processes. TOUTAIN *et al.* (1997) and POITRASSON *et al.* (1999) showed in the eastern Pyrénées that such multiple mixing processes vary continuously with time and may be disturbed by seismic processes.

2.2 Hydrogeological Background of the Ogeu Area

The Ogeu area is made of Upper Triassic detrital and evaporitic (anhydrite and halite) deposits covered by less than 3000 meters thick Jurassic and Cretaceous limestone-dolomite formations associated with marl, sandstone and clay layers (CASTERAS *et al.*, 1970; CASTERAS, 1971). Quaternary deposits (alluvial formations less than 25 m thickness) are crossed by the uprising of Ogeu waters. BERARD and MAZURIER (2000) identified the Urgonian limestones as the main karstic aquifer feeding all the local springs. The local structure is composed of a synclinal followed by an anticline; the bend of which is straight below the Ogeu-les-Bains discharges (Fig. 1c), probably acting as a drain for underground circulations. 'Black Flysch' mainly composed of slaty limestones and marls fill the depression-shaped valley in the study area (Fig. 1b).

The Bielle-Lurbe secondary chain (Fig. 1) is the local recharge area (BERARD and MAZURIER, 2000). The seepage to depth of meteoric waters is probably fast, through the subvertical bedding of the Mesozoic layers and a well-developed network of fractures in karstic terrains (CASTERAS et al., 1970).

Three natural springs ('A', 'B', 'H'), three wells ('AEP1', 'AEP2', 'Labourie') and two drillholes ('C', 'Bel Air') rise in the Ogeu area. Mineral waters are of HCO_3-Ca-(Na-Cl) type (Table 1). Temperatures are in the range $21 \pm 2°C$, pH in the range 5.6–7.7 and TDS in the range 300–848 mg/l. CO_2 degassing occurs at 'A' and 'H' springs and 'Bel Air' drillhole. The lack of CO_2 in C waters may be due to early demixion of the gas owing to the specific geometry of the hydrologic circuit. The 'H' and 'C' waters are bottled for commercialization. We used C water for hydrochemical monitoring.

3. Methods

3.1 Chemical Analyses

326 commercial bottles of C drillhole water have been collected from March, 1997 to June, 2000. Water remained stored in these bottles during a mean period of 1 year prior to processing. TSUNOGAI and WAKITA (1995) and TOUTAIN et al. (1997) showed that a two-year storage in similar conditions does not significantly affect major anion and cation concentrations, except limited changes of HCO_3^- resulting from slow degassing. This makes such bottled waters suitable for chemical monitoring.

Waters were filtered at 0,2 μm with an acetate cellulose membrane. Aliquots for major cations were acidified to pH ≤ 2 with ultrapure 14N HNO_3. Ca^{2+}, Mg^{2+}, Na^+ and K^+ were measured by atomic adsorption spectrophotometry. Anions (Cl^-, SO_4^{2-} and NO_3^-) were analyzed by ion chromatography. The international geo-standard SLRS-4 is used to check the accuracy and reproducibility of the cations analyses. Analytical errors are below 4% for all elements, excepting for calcium ($< 7\%$) probably because of interferences. To correct long-term instrumental drifts, a running sample is analyzed with every set of samples. Concentrations of cations and anions are calculated using an external home-made standard. Twenty-five bottled waters distributed over the whole time-series have been analyzed for total alkalinity (HCO_3^-) and SiO_2 concentrations by using Gran's method and molybdate colorimetry, respectively. The mean concentration of HCO_3^- (3.0 meq/l) has been used to calculate the charge mass balance for all the samples. The charge imbalance is less than 8% for all the samples with a mean value of 3%, which can be considered as an acceptable value.

3.2 Equilibrium Calculations

Major elements ratios and saturation indexes (SI = Log Ion Activity Product/ formation constant K) are used to highlight water/rock interactions. The SI of pertinent mineral phases with respect to the sedimentary environment (calcite, aragonite, ordered- and disordered-dolomite, anhydrite, halite, chalcedony) have been calculated for outlet conditions with the EQ3/6 software (WOLERY and DAVELER, 1992) and the SUPCRT92 thermodynamic database (JOHNSON et al., 1992).

J.-P. Toutain *et al.*

Table 1

Chemical (mmol/kg) and isotope analyses (‰) of selected waters. Saturation Indexes (SI) are calculated for calcite (calc), aragonite (arag), disordered- (dis-) and ordered-dolomite (ord-), anhydrite (anhy), halite (hali) and chalcedony (chal)

| | | | °C | | | | | mmol/kg | | | | | | | meq/l | | % | | SI | | | | | | |
	Type	Date	T	pH	H2CO3*	HCO3⁻	SiO2	Ca^{2+}	Mg^{2+}	Na^+	K^+	Cl^-	SO_4^{2-}	NO_3^-	Ca+Mg/HCO3+SO4	Na/Cl	$\delta^{18}O$	δD	calc.	arag.	dis-	ord-	anhy.	hali.	chal
"C" water																									
C9901	d	25/01/1999	23	7,5	–	3,06	0,20	1.06	0.47	1.49	0.02	1.34	0.02	–	0.99	1.1	–6.4	–50.0	–0.1	–0.2	–0.9	0.7	–3.7	–7.4	0.1
Mean (n=360)	d	–	–	–	–	n.a.	0,21	1.28	0.51	1.47	0.03	1.51	0.18	0.07	1.04	0.97	–	–	–0.02	–0.16	–0.81	0.75	–2.70	–7.31	0.09
Mean (n=25)	d	–	–	–	–	3,08	–	–	–	–	–	–	–	–	1.04	–	–	–	–	–	–	–	–	–	–
A.U.(±) 2s			–	–	–	0,06	0,004	0.04	0.01	0.03	0.001	0.03	0.004	0.003											
						0,15	–	0.08	0.019	0.15	0.003	0.15	0.012	0.015											
AEP1	w	12/04/1994	–	7,6	–	2,87	0.14	1.28	0.35	0.65	0.03	0.73	0.10	0.08	1.06	0.9	–	–	0.0	–0.1	–0.9	0.7	–3.0	–8.0	0.0
AEP2	w	14/09/1992	19	7,6	–	2,82	0.08	1.24	0.38	0.87	0.03	0.96	0.13	0.08	1.05	0.9	–	–	0.0	–0.1	–0.9	0.7	–2.9	–7.7	–0.3
A	s	12/01/1994	21	7,1	0,14	3,00	0.11	1.23	0.46	1.32	0.04	1.37	0.18	0.05	1.01	1.0	–	–	–0.5	–0.6	–1.8	–0.2	–2.9	–7.3	0.2
B	s	03/12/1985	21	7,7	–	3,05	–	1.20	0.50	1.31	0.03	1.29	0.21	0.05	0.98	1.0	–	–	0.1	0.0	–0.6	1.0	–2.7	–7.4	–0.3
H	s	14/05/1997	20	5,6	7,66	3,15	0.16	1.25	0.48	1.48	0.02	1.54	0.19	0.05	0.98	1.0	–	–	–2.2	–2.3	–5.2	–3.6	–2.7	–7.3	0.0
Bel Air	d	15/10/1986	17,4	7,2	0,43	5,10	0.11	1.90	1.17	0.43	0.02	1.05	0.14	0.02	1.14	0.4	–	–	0.0	–0.2	–1.2	0.4	–2.7	–7.6	–0.1

4. Chemical Composition and Water-rock Interactions

4.1. Chemical composition

Only one isotopic measurement is available ($\delta^{18}O = -6,4‰$ and $\delta D = -50‰$, Table 1), which is typical of a meteoric origin with an Atlantic signature. The very weak shift (1‰) with respect to the Global Meteoric Water Line (GMWL; CRAIG, 1963) indicates that no isotopic exchange between rocks and fluids occurs at depth because of low aquifer temperatures.

The chemical composition of Ogeu waters, together with mean concentrations of drillhole 'C' waters are shown in Table 1. The total mineralization ranges from 8 to 13 meq/l. HCO_3^- and $Ca^{2+} + Mg^{2+}$ are the dominant ions indicating a major carbonate contribution. A lower (2 to 4 meq/l, that is to say 22 to 38% of the total mineralization) saline component ($Na^+ + Cl^- + K^+ + SO_4^{2-} + (Ca^{2+} + Mg^{2+} - HCO_3^-)$) indicates a significant evaporitic contribution. This chemistry is typical of water chemistry in complex sedimentary environments (PASTORELLI et al., 1999; LEVET et al., 2002; MINISSALE et al., 2002).

Time-series of cations and anions are displayed in Figure 2. The mean value plus one or two standard deviations (1 or 2σ) are displayed as empirical thresholds above or below which values are often considered as anomalous. Cations and anions vary mainly within their respective 2σ domains during the three years of monitoring. Very few measurements may be considered as anomalous by considering the mean plus two standard deviation threshold. Considering the mean plus one standard deviation should lead to very contrasted results in terms of anomalous/versus non anomalous periods. Moreover, considering alternatively each element should lead to the identification of different anomalous periods. This highlights that assuming an empirical threshold constituted by the mean plus one or two standard deviations is not a reliable method.

Cl^- and SO_4^{2-} ions display seasonal patterns, with higher concentrations in summer (Fig. 2), whereas Ca^{2+}, and Mg^{2+} show no clear seasonal variations. The contribution of chemically different fluids in Ogeu water may explain this different behavior. Morever, Cl^-, SO_4^{2-} and Na^+ likely behave conservatively (and their contents can be affected by dilution/concentration processes) while Ca^{2+} and Mg^{2+} are likely controlled by a mineral phase (Ca-Mg carbonate) either in diluted or concentrated fluids through the seasons.

4.3 Water-rock Interactions

In a carbonate-dominated environment, pCO_2 controls equilibria with respect to carbonate mineral phases (DREYBRODT et al., 1996). "C" is saturated with respect to calcite and aragonite ($SI_{mean} = -0.1$ and -0.2, respectively). The weak oversaturation with respect to ordered-dolomite ($SI_{mean} = 0.7$) likely results from early demixion of CO_2.

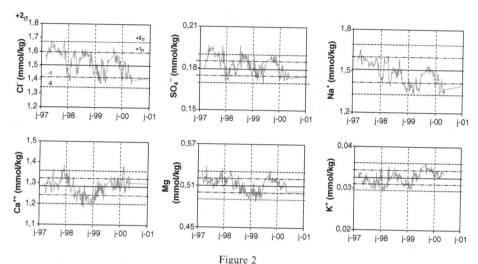

Figure 2
Hydrochemical time-series. Mean and ± 2 sigmas thresholds are displayed.

The calculated $(Ca^{2+} + Mg^{2+})/(HCO_3^- + SO_4^{2-})$ ratios (in meq/l) of the 25 drillhole "C" samples for which alkalinity was measured and for the other emergences vary from 0,98 to 1,14 (Table 1). This suggests that these four ions originate from stoichiometric dissolution of both mixed calcium and magnesium carbonates and sulphates. This confirms Ogeu waters to percolate both Mesozoic carbonates (Urgonian limestones and dolomites) and Triassic rocks (anhydrite- and halite-bearing evaporites).

The Figure 3 displays variations between Cl and Na and SO_4 concentrations (in mmol/kg) for the time-series and the discrete samples. The mean Na/Cl molar ratios of samples from the time-series is 0.97 ± 0.03 (Table 1, Fig. 3a), which confirms that Cl and Na ions originate from stoichiometric dissolution of halite. The linear correlation ($\sigma^2 = 0,80$) shown in Figure 2b between sulphate and chlorine concentrations suggests the Triassic evaporitic layers as a common rock source for these two ions. Ogeu waters are strongly undersaturated with respect to halite and anhydrite ($SI_{mean} = -7,5$ and $-2,9$ for halite and anhydrite respectively) that could result either from a high water/rock ratio (limited occurrence of halite and anhydrite in Triassic evaporites layers) or a mixing between deep saline fluids and weakly mineralized groundwaters, or probably both. Moreover, as anhydrite displays both a high dissolution rate and retrograde solubility, such a strong undersaturation likely indicates that anhydrite is not present along the upflow pathway. This confirms that Ogeu water results from the mixing of two independant reservoirs: a Urgonian carbonated aquifer and a saline fluid resulting from leaching of Triassic evaporitic layers. According to the local tectonics (Fig. 1c) and to strong anhydrite undersaturation (Table 1), the saline body is deeper than the karstic aquifer. Saline fluids

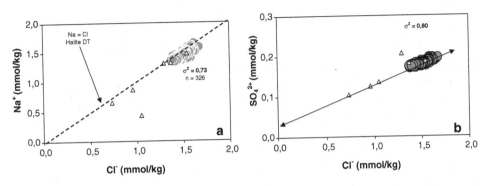

Figure 3

Concentrations of main ions (SO_4^{2-}, Na^+) as a function of chlorine (mmol/kg).

probably migrate along a deep-rooted fracture developed inside the Ogeu's anticlinal, and mixes with the Urgonian water body.

K concentrations are very low (0,02 to 0,04 mmol/kg). In similar sedimentary environments, aqueous potassium commonly originates from interactions (dissolution, ions exchange) with marls and clays interbedded with the dominant limestones and dolomites formations (MARINI et al., 2000; MINISSALE et al., 2000; LEVET et al., 2002). This is confirmed in Ogeu context by the lack of correlation between K and Cl ($\sigma^2 = 0,05$), mainly as the result of ions exchange with clays.

In low-temperature hydrothermal systems hosted in carbonate-evaporite bodies, silica has long been suggested to be controlled by poorly crystalline silica phases, most often chalcedony (ARNÓRSSON, 1975; AZAROUAL et al., 1997; PASTORELLI et al., 1999). This is consistent with the chalcedony SI values (−0,29 to 0,15) calculated for all the waters.

4.4 Quantification of the Mixing Process

Because Cl ions are conservative elements and originate from a single saline deep source in Ogeu context, their concentrations are suitable to quantify the mixing process. The chemistry of the two end-members involved in the mixing process for Ogeu waters (carbonated water and saline water) is unknown, but can be approximated by using hydrochemical data from similar geologic environments.

The Bagnères-de-Bigorre thermal waters which rise in the sedimentary NPZ are typically derived from interactions of fluids with Triassic evaporite deposits at depth since they display a mixed Ca-SO4/NaCl character and an estimated equilibration temperature around 60°C (LEVET et al., 2002). Depending on the relative amounts of choride and sulfate minerals in the Triassic levels, the derived fluids may be Na-Cl or Ca-SO4 dominated. With a mean composition for dominant elements, $Cl^- = 3.1$, $Na^+ = 3.2$, $Ca^{2+} = 13.6$, $Mg^{2+} = 3.1$ and $SO_4^{2-} = 16.1$ mmol/kg, the Bagnères-de-Bigorre waters can be used as the saline end-member for its chloride character.

The chemistry of Mesozoic west Pyrenean karstic aquifers is characterized by the absence of SO_4^{2-}, very low Cl^- contents and displays mean concentrations of about 2.6, 1.3, 0.2 and 0.05 mmol/kg for HCO_3^-, Ca^{2+}, Cl^- and NO_3^-, respectively (VANARA, 1997) which can be used as the carbonate end-member.

The mixing function F corresponds to the percentage of carbonated water in the mixture. It is calculated with a binary mixing equation based on mass balance and Cl contents, Cl_{mix} being a linear function of F:

$$Cl_{mix} = F(Cl_s - Cl_k) + Cl_k, \tag{1}$$

where Cl_{mix} is the measured concentrations in mmol/kg of 'C' drillhole water, Cl_k is the mean concentration of western Pyrenean karstic waters (0.17 mmol/kg) and Cl_s is the concentration of the deep evaporitic end-member (3.1 mmol/kg).

Results (Fig. 4) show that F varies from 49 to 60% within the 2σ interval over the time-series. The absolute values of F obviously depend on the respective concentrations of both saline and carbonate end-members. The mixing function F appears therefore as a good tracer of the dynamics of the Ogeu aquifer.

5. Rainfall and Seismic Time-Series

Surficial karstic aquifers are usually complex hydrological and hydrochemical systems opened to various fluid circulations and therefore may be strongly influenced by rainfall and permanent crustal stresses. Seasonal recharge, particular rainfall events and continuous crustal stress changes may therefore contribute to the control of the chemical composition of the water. In sedimentary areas, strain build-up

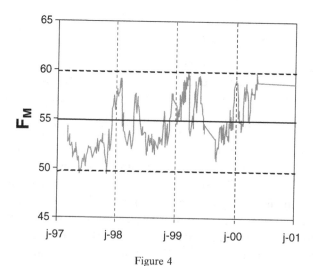

Figure 4
Time-series of the calculated mixing function F. Mean and $\pm$ 2 sigmas thresholds are displayed.

induced by major earthquake preparation process is often suggested to account for the mixing of waters of contrasted chemistry (THOMAS, 1988; WAKITA *et al.*, 1988; TSUNOGAI and WAKITA, 1995; TOUTAIN *et al.*, 1997). Such a basic assumption, however, should be considered together with the dynamic response of the aquifer to current external perturbations, due to rainfall recharge and background seismicity. It is to be noted that up to now, papers dealing with geochemical anomalies related to severe seismic events never took into account such perturbations. In this section, time-series for rainfall, seismic activity and chemical composition of the waters are displayed for a three-year period.

5.1. Rainfall Time-series

Monthly rainfall data recorded from March 1997 to May 2000 at 20 km distance from Ogeu are shown in Figure 5. No clear periodicity of rainfall can be distinguished, probably as the result of both a strong oceanic character and multiple minor climatic influences. However, some periods of heavy rainfall can be identified and may constitute events that potentially trigger impulse responses from the aquifers.

5.2 Seismic Time-series: Energy Released and Epicentral Distributions

The area under consideration is monitored by a homogeneous array of seismological stations from the Observatoire Midi-Pyrénées; five of which lie within 50 km of the spring location (SOURIAU *et al.*, 2001). During the period March 1997

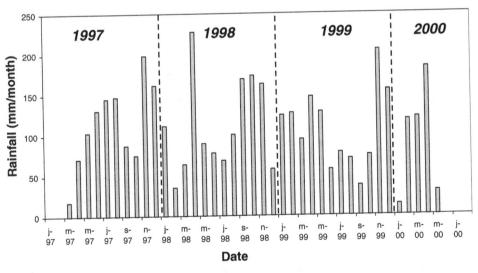

Figure 5

Rainfall time-series. Daily measurements are performed 20 km south of the Ogeu seismic station. Monthly mean variations are displayed.

to June 2000, 448 earthquakes were recorded and accurately located within 30 km distance of the spring, with magnitudes spanning the range 0.3–4.0. 55 of these events, with 0.3 to 3.6 magnitudes, were located closer than 10 km from the spring (Fig. 6). Most of the earthquakes were situated south of the spring, on the north-dipping NPF. Their depths (in the 10-km radius circle) extend from 0 to 16 km, with an average value of about 8 km. The average horizontal and vertical errors on the epicenters are of the order of 1.5 and 3.0 km.

In order to characterize the influence of seismic events on the mixing function, the key parameter is the seismic energy E_S radiated by the earthquakes. This parameter can be connected to the magnitude through empirical laws. Care has to be taken in

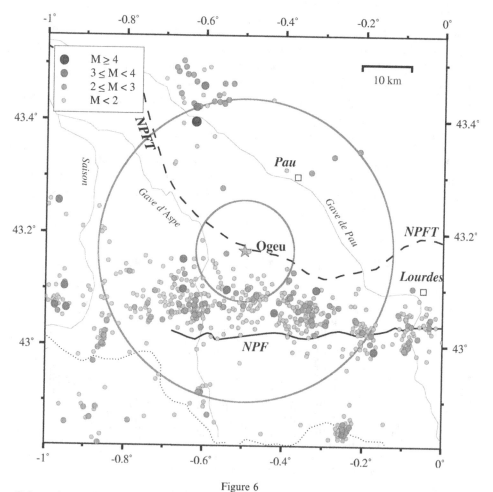

Figure 6

Epicentral map of seimicity in the Ogeu area for the period March 1997 – June 2000. The size of the symbols increases with earthquake magnitude. The star indicates the spring location. Circles with radii 10 km and 30 km centered on the spring are displayed. NPFT : North Pyrenean Frontal Thrust; NPF: North Pyrenean Fault.

the choice of these laws, since they closely depend on the kind of magnitude scale under consideration. In the earthquake catalogue used, the magnitudes of the events are local (or Richter) magnitudes M_L. For this particular kind of magnitude, KANAMORI et al. (1993) obtained a very simple linear relationship $\log_{10}(E_S) = 1.96\ M_L + 9.05$, where the energy E_S is in ergs.

Figure 7 shows the amount of seismic energy radiated during the considered period in the 30 km and 10 km radius circular areas centered on the spring. Events were grouped within time windows of 10 days. The values displayed are lower bound estimates, since the seismological array can fail in the detection of minor events (with magnitudes smaller than about 1.5). However, owing to the exponential dependence of energy on magnitude, the total budget is not greatly affected by the loss of small earthquakes. These plots illustrate well the fact that the seismic activity is not

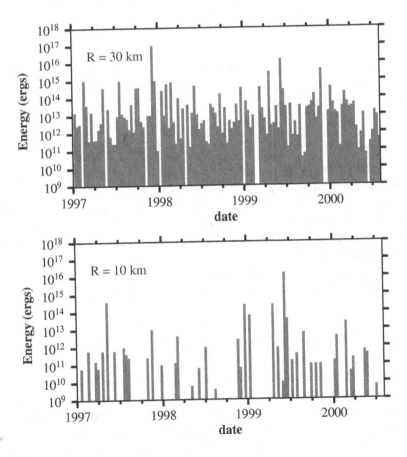

Figure 7

Seismic energy release in a 30 km radius (upper plot) and 10 km radius (lower plot) circular area around the Ogeu sampling site, within 10-day time windows.

continuous, and periods with higher-than-average rates of energy released can be identified, in particular during the year 1999.

6. *The Mixing Function Versus the Energy of Earthquakes and Effective Rainfall*

The purpose of this paragraph is the study of the response of the mixing function F according to both energy E_S of earthquakes whose distance of epicenter to the monitoring site is nearer than 10 kilometers and the effective rainfall R at Ogeu. Variations of the mixing function F represent the amount of fresh carbonated water into the shallow reservoir.

The energy function M is calculated as $\log_{10}(E_S)$ where E_S is the seismic energy of events grouped within 10-day time windows.

The mixing function F can be represented as a linear relationship of M and R such that

$$F = \Gamma_M * M + \Gamma_R * R + C^{st}, \tag{1}$$

where Γ_M and Γ_R are the impulse responses of F to the magnitude M and to the effective rainfall R, respectively; $f * g$ represents the discrete convolution product between time series f and g:

$$[f * g](j) = \sum_{i=0,N} f_i \cdot g_{j-i} = \sum_{i=0,N} f_{j-i} \cdot g_i, \tag{2}$$

N is the order of impulse response f or g.

The calculation of Γ_M, Γ_R and the constant C^{st} from the time series F, M and R uses a regularization method since the inversion of equation (1) is an ill-posed problem. Indeed, an infinity of solutions may fulfill equation (1) and the more likely solution is selected so that its norm is minimum in order to make Γ_M and Γ_R as smooth as possible (PINAULT *et al.*, 2001).

A constant C^{st} is introduced in equation (1) since only the variations of the mixing function F are to be explained from M and R so that:

$$\Delta F = \Gamma_M * \Delta M + \Gamma_R * \Delta R, \tag{3}$$

where $\Delta F = F - \bar{F}$, $\Delta M = M - \bar{M}$ and $\Delta R = R - \bar{R}$, $\bar{F}$, $\bar{M}$ and $\bar{R}$ being the mean of F, M and R.

Equation (2) represents the more general linear relationship between inputs and an output. It is characterized by impulse responses, i.e., the response of the output when inputs are impulses (they last exactly one sampling step, i.e., 10 days). In the present case the inputs are effective rainfall R and the magnitude M. Impulse responses represent the response of the mixing function F after a hypothetical isolated event (a rainfall event or a seismic event depending upon the impulse response). Once the impulse responses are known, the two terms of the second member of equation (3) represent the contribution of both inputs (rainfall and magnitude) to the mixing function F.

Impulse response Γ_M of the mixing function F to the energy function M is equivalent to the variation of F subsequent to a quake for which $M = \log_{10}(E_S)$, i.e., an event of magnitude $M_L = (M - 9.05) / 1.96$. In such a model, the perturbation of F is proportional to the magnitude.

In the same way, impulse response Γ_R of the mixing function F to the effective rainfall R represents the variation of F subsequent to a rainfall event whose height is R and, here again, the perturbation of F is proportional to the height of the effective rainfall event.

Resolution of equation (1) is performed with a 10 day sampling rate. Figure 8 shows the output of the mixing function F versus both inputs M and R. Figure 9(1) exhibits the comparison between the model (equation 2) and the mixing function F. Discrepancies arise during the first year due to the truncation of time series M and R in equation (1): the length of impulse responses is therefore 250 days. Figure 9(2) shows the comparison between the model (equation 1) and the mixing function F when only effective rainfall R is used as input. Discrepancies between the model and F increase significantly between (1) and (2), evidencing the relevancy of the energy M used as input in the model.

The time series representing both inputs and the output are centered in order to remove the constant in equation (2). The calculation of the two impulse responses requires the use of a regularization technique since, as a mathematical point of view, it is an ill-posed problem (many solutions can fulfill the requirements). The regularization technique consists in selecting the 'smoother' solution while the discrepancies between the observed mixing function F and the model are minimal. To accomplish that the Tikhonov regularization technique is used (TIKHONOV and GONCHARSKY, 1987). The solutions that are obtained from this regularization technique are the more realistic according to the parsimony principle, which tries to explain a phenomenon from the minimum number of arbitrary parameters. Moreover, this technique minimizes the error propagation since the norm of the impulse responses is minimal.

Figure 10 (1) shows impulse responses of F to M (Γ_M) and $R(\Gamma_R)$. Impulse response of F to R represents the recharge of the shallow reservoir due to an increase in pressure head after a rainfall event. It increases drastically after a rainfall event, reaching its maximum 10 days after the event and then decreases for 150 days.

Impulse response of F to M may exhibit the intrusion of salted water into the shallow reservoir after a quake that exerts a constraint on the deep feeding system. It decreases very rapidly after a quake, owing to the increase of chloride concentration into the shallow reservoir. This burst is very short since the impulse response increases 30 days after the quake, when fresh-carbonated water replaces salted water, and reaches a maximum 70 days after the quake. The amount of fresh-carbonated water introduced after the quake is higher than the amount of salted water it is

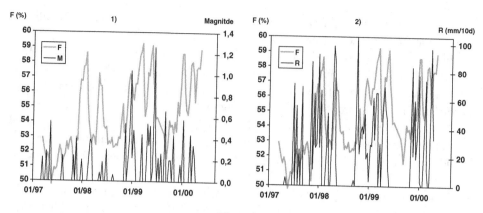

Figure 8
Direct correlation between F (mixing function) and M (energy of events at epicentral distances < 10 km)
and between F and R (effective rainfall).

assumed to replace, as shown in Figure 10 (2) that displays the two components of F related to M and R, namely $\Gamma_M * M$ and $\Gamma_R * R$. The response of F to earthquakes manifests therefore long-term variations and it increases in 1999, due to the high magnitude of earthquakes. The deep reservoir behaves as if the compression that occurs during the quake was followed by decompression a few ten-day periods after the quake. The impulse response of F to M becomes inaccurate when the lag overreaches 100 days because of the shortness of time series in comparison to the length of impulse responses, which makes its interpretation hazardous beyond this lag. Correspondingly, the impulse response is too noised to be interpreted at negative lags to bring to the fore precursor signals.

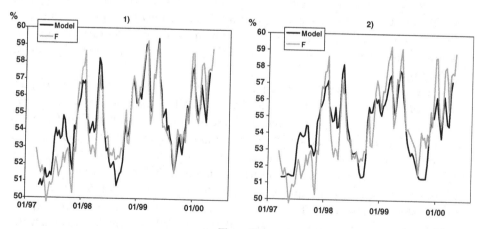

Figure 9
Comparison of the mixing function F and the model with R and M as inputs (equation 1). 2: Comparison
of the mixing function F and the model with only R (rainfall) as input.

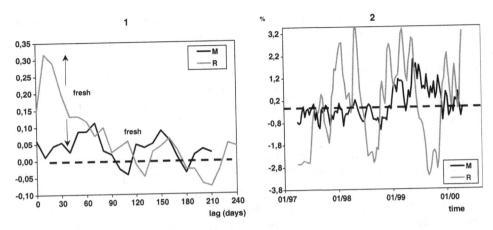

Figure 10

1) Impulse responses of F to M and R. 2) Response of F to earthquakes (M) and rainfall (R).

7. Discussion

Limitations of using arbitrary thresholds. Bulk chemical time-series display variations for which interpretations vary strongly when using either the mean $+ 1\sigma$ or 2σ thresholds, or when considering one or another element. Even if this common practice may be convenient, our results evidence that in many case it is not suitable to detect transient anomalies, at least when the signal/background ratio is limited. Further evidence exists that local water-rock interactions are key constrains for selecting chemical tracers because elements may be controlled either by superficial interactions (e.g., K^+ through ions exchange with clays) or by dissolution/ precipitation reactions (e.g., Ca^{++} controlled by carbonates precipitation resulting from CO_2 degassing). Calculation of Saturation Index and observation of the behavior of elements with respect to conservative elements (mainly Cl) have therefore to be performed prior to selecting elements for geochemical survey.

Methodology. Our results confirm that groundwater in mixed carbonated/ evaporitic environments is the result of complex interactions between chemically contrasted reservoirs. Mixing processes are the most likely to occur and can be described as discontinuous and fast processes which appear to be controlled mainly by rainfall inputs when strain is assumed to be constant. TOUTAIN *et al.* (1997) and POITRASSON *et al.* (1999) evidenced that multiple mixing processes can affect karstic-dominated groundwater under strain built-up. LEVET *et al.* (2002) demonstrated that using a simple binary mixing equation based on mass balance and Cl^- contents enabled to the calculation of mixing proportions between the two main poles (deep saline and surficial carbonated poles). Considering the temporal fluctuations of the mixing function F may therefore supply information pertinent to crustal processes linked to earthquake preparation and/or triggering. Figure 9(1) evidences that in the sedimentary Ogeu area, which undergoes both discontinuous rainfall recharge and seismic activity, the mixing function F is the result of both inputs. The involvement of earthquakes in F is confirmed by Figure 9(2), which demonstrates that F cannot be the result of only rainfall inputs. In the Ogeu context, because chlorine concentrations in the fresh-karstic end-member are close to zero, the mixing function is very similar to the $1/Cl$ concentration of the total fluid. In other contexts, F may be calculated by using any other method based on major or trace elements as well as isotopic ratios.

Crustal processes. As expected, impulse response of F to R constitutes an increase of the contribution of fresh-carbonated water in the discharge. The delay of response in Ogeu conditions is immediate (10 days with a 10-day sampling resolution). On the contrary, the impulse response of F to M is a two-step process with step 1 being a decrease of F, which corresponds to an increase of the contribution of salted water into the shallow reservoir, followed by step 2 that is an increase of F which results from either a decrease of the salted content or an increase of the carbonated contribution. The two-phase process shows response delays of 10 and 70 days for steps 1 and 2, respectively. It is to be noted that the increase of F during step 2 is higher than its decrease during step 1. In other words, the inflow of fresh-carbonated water (step 2) in the reservoir is higher than the inflow of salted water (step 1). This may suggest that step 2 can be due to an inflow of fresh-carbonated water rather than the decrease of salted contribution. Another hypothesis is a simple relaxation process during which normal conditions are reached.

Few crustal processes can be evokated to interpret these observations. The most immediate is the crustal stress variation and/or migration following the earthquakes and allowing the inflow of deeper salted water. Nevertheless, considering the small magnitudes, the epicentral distances and the depths of the seismic events, it is difficult to invoke this phenomenon alone. Preseismic fluid overpressures and variations of the permeability in the fault zones inducing fluid migrations are closely established for earthquakes with magnitude greater than 5 (e.g., MUIR-WOOD and KING, 1993; SIBSON, 1994; HUSEN and KISSLING, 2001; YECHIELI and BEIN, 2002). These fluid migrations highly dependent on the earthquake mechanism are invoked to explain the poro-elastic postseismic deformation (PELTZER *et al.*, 1998) and the triggering of aftershock sequences and the migration of seismic ruptures (NOIR *et al.*, 1997). Regarding micro-seismicity, the focal mechanisms, and therefore, the deformation features are mostly unknown in detail. Consequently, small events frequently distributed in clusters must be considered as volumetric deformation. For example, KOIZUMI *et al.* (1999) showed that earthquake swarms with magnitude of two could be responsible for groundwater level variations in instrumented boreholes associated with changes in volumetric strain. In our Pyrenean case, the two-step feature of the impulse response of F should be associated with the volumetric strain variations linked to the seismic activity. This deformation induced by or inducing micro-seismicity, may allow the opening of microfractures and, then, the inflow of deeper salted water (YECHIELI and BEIN, 2002). Then, the opened fractures may be closed several days later if no other seismic event occurred. Unfortunately, the 10-day resolution of our results does not allow us (i) to ascertain if the inflow salted water is a response of a seismic event (co- and postseismic response) as we hypothesize in our model, or if it is a preseismic response as described by KOIZUMI *et al.* (1999), and (ii) if the hypothesis of a linear response of the groundwater to an event is reasonable. Nevertheless, it is unquestionable that our observations cannot be explained without considering the moderate, but continuous, regional seismic activity.

8. Conclusion

Thermo-mineral waters rising in sedimentary environments such as Ogeu often result from the mixing of chemically contrasted end-members. Temporal fluctuations of the fluid chemistry can be described by using a mixing function (F), which is calculated by using Cl^- contents in the Ogeu case. However, various parameters (isotopic ratios, trace elements concentrations, elemental ratios, ...) may be used to constrain and to calculate the mixing function F according to hydrogeological contexts.

Modelling F over a 3.5-year period by using E_s (seismic energy release in a 10 km radius) and R (effective rainfall) as inputs evidence the involvement of both phenomena in the temporal fluctuations of F. This is the first time that microseismic activity is shown to contribute to the control of the chemistry of water over an extended period. Impulse response of F to E_s, which is calculated with a 10-day sampling rate, indicates volumetric strain episodes leading to successive and transient migration phases of deep saline fluids in the aquifer.

Such modelling confirms that seismically induced fluid transfers can occur episodically and systematically in a given area. Moreover, this modelling appears suitable to precisely assess the role of various parameters such as hypocentral depths, epicentral distances, and earthquake magnitude on the fluid transfers. Unfortunately, such data processing appears unrealistic in this study as the result of the overly short monitoring period (3.5 years) and the exceedingly moderate earthquake activity. However, it is possible and highly desirable to process altogether hydrochemical and seismic time series already available in areas with strong seismic activity.

REFERENCES

ALAUX-NEGREL, G. (1991), *Etude de l'évolution des eaux profondes en milieux granitiques et assimilés. Comportement des éléments traces.* Ph.D. Thesis. University of Paris VII, 216 pp.

ARNÓRSSON, S. (1975), *Application of the silica geothermometer in low temperature hydrothermal areas in Iceland,* Am. J. Sci. *275,* 763–784.

AZAROUAL, M., FOUILLAC, C., and MATRAY, J.M. (1997), *Solubility of silica polymorphs in electrolyte solutions, II. Activity of aqueous silica and solid silica polymorphs in deep solutions from the sedimentary Paris basin,* Chem. Geol. *140,* 167–179.

BELLA, F., BIAGI, P.F., CAPUTO, M., COZZI, E., DELLA MONICA, G., ERMINI, A., GORDEEZ, E.I., KHATKEVICH, Y.M., MARTINELLI, G., PLASTINO, W., SCANDONE, R., SGRIGNA, V., and ZILPIMIANI, D. (1998), *Hydrogeochemical anomalies in Kamchatka (Russia)*, Phys. Chem. Earth. *23*(9–10), 921–925.

BERARD, P. and MAZURIER, C. (2000), *Ressources en eaux thermales et minérales des stations du département des Pyrénées-Atlantiques. Usine d'embouteillage d'Ogeu-les-Bains*, Rapport BRGM/RP 50174-FR, 43 pp.

BERNARD, P. (2001), *From the search of 'precursors' to the research on 'crustal transients'*, Tectonophysics *338*, 225–232.

BIAGI, P.F., ERMINI, A., KINGSLEY, S.P., KHATKEVICH, Y.M., and GORDEEV, E.I. (2000), *Possible precursors in groundwater ions and gases content in Kamchatka (Russia)*, Phys. Chem. Earth. *25* (3), 295–305.

CASTERAS, M., *Notice Explicative de la Carte Géologique d'Oloron-Ste-Marie (1/50000)*, 2d edition (BRGM, France 1971) *19* pp.

CASTERAS, M., CANÉROT, J., PARIS, J.P., TISIN, D., AZAMBRE, M., and ALIMEN, H., *Carte géologique d'Oloron-Ste-Marie (1/50000)* (BRGM, France 1970).

CHOUKROUNE, P. (1992), *Tectonic evolution of the Pyrénées*, Annu. Rev. Earth Planet. Sci. *20*, 143–158.

CRAIG, H. (1963), *Isotopic variations in meteoric waters*, Science *123*, 1702–1703.

DREYBODT, W., LAUCKNER, J., LIU, Z., SVENSSON, U., and BUHMANN, B. (1996), *The kinetics of reaction $CO_2 + H_2O = H^+ + HCO_3^-$ as one of the rate limiting steps for the dissolution of calcite in the system H_2O-CO_2-$CaCO_3$*, Geochim. Cosmochim. Acta. *60* (18), 3375–3381.

FAVARA, R., ITALIANO, F., and MARTINELLI, G. (2001), *Earthquake-induced chemical changes in the thermal waters of the Umbria region during the 1997–1998 seismic swarm*, Terra Nova *13*, 227–233.

GELLER, R.J. (1997), *Earthquake prediction: A critical review*, Geophys. J. Int. *131*, 425–450.

GELLER, R.J., JACKSON, D.D., KAGAN, Y.Y., and MULARGIA, F. (1997), *Earthquakes cannot be predicted.* Science *275*, 1616–1617.

HUSEN, S. and KISSLING, E. (2001), *Postseismic fluid flow after the large subduction earthquake of Antofagasta, Chile*, Geology *29*, 847–850.

ITALIANO, F., MARTELLI, M., MARTINELLI, G., NUCCIO, P.M., and PATERNOSTER, M. (2001), *Significance of earthquake-related anomalies in fluids of Val d'Agri (southern Italy)*, Terra Nova *13*, 249–257.

JOHNSON, J.W., OELKERS, E., and HELGELSON, J.W. (1992), SUPCRIT92: *A software package for calculating the standard molal thermodynamic properties of minerals, gases, aqueous species and reactions from 1 to 5000 bars and 0 to 1000°C*, Computer Geosci. *18*, 899–947.

KANAMORI, H., MORI, J., HAUKSSON, E., HEATON, Th.H., HUTTON, L.K., and JONES, L.M. (1993), *Determination of earthquake energy release and M_L using TERRASCOPE*, Bull. Seismol. Soc. Am. *83*(2), 330–346.

KING, C.Y. (1986), *Gas geochemistry applied to earthquake prediction. An overview*, J. Geophys. Res. *91* (B12), 12269–12281.

KING, C.Y., KOIZUMI, N., and KITAGAWA, Y. (1995), *Hydrochemical anomalies and the 1995 Kobe earthquake*, Science *269*, 38–39.

KOIZUMI, N., TSUKUDA, E., KAMIGAICHI, O., MATSUMOTO, N., TAKAHASHI, M., and SATO, T. (1999), *Preseismic changes in groundwater level and volumetric strain associated with earthquake swarms off the east coast of the Izu Peninsula, Japan*, Geophys. Res. Lett. *26*, 3509–3512.

KRIMISSA, M. (1995), *Application des méthodes isotopiques à l'étude des eaux thermales en milieu granitique (Pyrénées, France)*, Ph.D. Thesis, University d'Orsay, Paris XI, 273 pp.

LEVET, S., TOUTAIN, J.P., MUNOZ, M., NEGREL, P., JENDRZEJEWSKI, N., AGRINIER, P., and SORTINO, F. (2002), *Geochemistry of the Bagnères-de-Bigorre thermal waters from the North Pyrenean Zone (France)*, Geofluids *2*, 25–40.

MARINI, L., BONARIA, V., GUIDI, M., HUNZIKER, J.C., OTTONELLE, G., and VETUSCHI ZUCCOLINI, M. (2000), *Fluid geochemistry of the Acqui Terme-Visone geothermal area (Piemonte, Italy)*, Appl. Geochem. *15*, 917–935.

MICHARD, G. and FOUILLAC, C. *Contrôle de la composition chimique des eaux sulfurées sodiques du Sud de la France. In Géochimie des interactions entre les eaux, les minéraux et les roches* (eds. Tardy) (Tarbes, France 1980) pp. 147–166.

MICHARD, G., FOUILLAC, C., OUZOUNIAN, G., BOULÈGUE, J., and DEMUYNCK, M., *Geothermal applications of the geochemical study of hot springs in eastern Pyrénées*. In *Advances in European Geothermal Research. Proceeding of the second Internal Seminary. On the results of the EC Geothermal Energy Research* (eds. Strub and Ungemach) (Strasbourg, France 1980), pp. 387–395.

MICHARD, G. (1990), *Behaviour of the major elements and some trace elements (Li, Rb, Cs, Sr, Fe, Mn, W, F) in deep hot waters from granitic areas*, Chem. Geol. *89*, 117–134.

MINISSALE, A., MAGRO, G., MARTINELLI, G., VASELLI, O., and TASSI, G.F. (2000), *Fluid geochemical transect in the Northern Apennines (central-northern Italy): Fluid genesis and migration and tectonic implications*, Tectonophysics *319*, 199–222.

MINISSALE, A., VASELLI, O., TASSI, F., MAGRO, G., and GRECHI, G.P. (2002), *Fluid mixing in carbonate aquifers near Rapolano (central Italy): Chemical and isotopic constraints*, Appl. Geochem. *17*, 1329–1342.

MOGI, K., MOCHIZUKI, H., and KUROKAWA, Y. (1989), *Temperature changes in an artesian spring at Usami in the Izu peninsula (Japan) and their relation to earthquakes*, Tectonophysics *159*, 95–108.

MUIR-WOOD, R. and KING, G.C.P. (1993), *Hydrological signatures of earthquake strain*, J. Geophys. Res. *98* (B12), 22035–22068.

NISHIZAWA, S., IGARASHI, G., SANO, Y., SHOTO, E., TASAKA, S., and SASAKI, Y. (1998), *Radon, Cl⁻ and SO₄²⁻ anomalies in hot spring water associated with the 1995 earthquake swarm of the east coast of the Izu peninsula, central Japon*, Appl. Geochem. *13*, 89–94.

NOIR, J., JACQUES, E., BÉKRI, S., ADLER, P.M., TAPPONNIER, P., and KING, G.C.P. (1997), *Fluid flow triggered migration of events in the 1989 Dobi earthquake sequence of Central Afar*, Geophys. Res. Lett. *24*, 2335–2338.

OLIVET, J.L. (1996), *La cinématique de la plaque Ibérique*, Bull. Centres Rech. Explor. Prod. Elf Aquitaine *20* (1), 131–195.

PASTORELLI, S., MARINI, L., and HUNZIKER, J.C. (1999), *Water chemistry and isotope composition of the Acquarossa thermal system, Ticino, Switzerland*, Geothermics *28*, 75–93.

PELTZER, G., ROSEN, P., and ROGEZ, F. (1998), *Poroelastic rebound along the Landers 1992 earthquake surface rupture*, J. Geophys. Res. *103*(30),131–30,145.

PINAULT J-L., PAUWELS, H., and CANN, Ch. (2001), *Inverse modeling of the hydrological and the hydrochemical behavior of hydrosystems: Application to nitrate transport and denitrification*, Water Res. Res. *37* (8), 2179–2190.

POITRASSON, F., DUNDAS, S.H., TOUTAIN, J.P., MUNOZ, M., and RIGO, A., (1999), *Earthquake-related elemental and isotopic lead anomaly in a springwater*, Earth. Planet. Sci. Lett. *169*, 269–276.

RIGO, A., PAUCHET, H., SOURIAU, A., GRÉSILLAUD, A., NICOLAS, M., OLIVERA, C., and FIGUERAS, S., (1997), *The February 1996 earthquake sequence in the eastern Pyrénées: Fisrt results*, J. Seismol. *1*, 3–14.

ROELOFFS, E. (1988), *Hydrologic precursors to earthquakes: A review*, Pure Appl. Geophys. *126*, 177–209.

SIBSON, R.H. (1994), *Crustal stress, faulting and fluid flow*. In *Geofluids: Origin, Migration and Evolution of Fluids in Sedimentary Basins*, Geolog. Soc. Special Pub. *78*, 68–84.

SOURIAU, A. and PAUCHET, H. (1998), *A new synthesis of Pyrenean seismicity and its tectonic implications*, Tectonophysics *290*, 221–244.

SOURIAU, A., SYLVANDER, M., RIGO, A., DOUCHAIN, J.M., and PONSOLLES, C. (2001), *Pyrenean tectonics: Mean seismological constraints*, Bull. Soc. Geol. France. *172* (1), 25–39.

SYKES, L.R., SHAW, B.E., and SCHOLZ, C.H. (1999), *Rethinking earthquake prediction*, Pure Appl. Geophys. *155*, 207–232.

THOMAS, D. (1988), *Geochemical precursors to seismic activity*, Pure Appl. Geophys. *126*, 241–266.

TIKHONOV, A.N. and GONCHARSKY, A.V. (Eds.), *Ill-Posed Problems in the Natural Sciences* (MIR, Moscow, 1987).

TOUTAIN, J.P., MUNOZ, M., POITRASSON, F., and LIENARD, A.C. (1997), *Springwater chloride ion anomaly prior to a ML = 5.2 Pyrenean earthquake*, Earth. Planet. Sci. Lett. *149*, 113–119.

TSUNOGAI, U. and WAKITA, H. (1995), *Precursory chemical changes in groundwater: Kobe earthquake, Japan*, Science *269*, 61–63.

VANARA, N., *Dissolution et spéléogenèse en contexte tectonique actif: Le Massif des Arbailles (Pyrenees-Atlantiques, F)*. In *Proceeding on the 6ᵗʰ Conference on Limestone Hydrology and Fissured Media*. (Ed. Jeannin P.Y.) (La Chaux de Fonds, Switzerland 1997), *vol. 2*, pp. 115–118.

WAKITA, H., NAKAMURA, Y., and SANO, Y., (1988), *Short-term and intermediate-term geochemical precursors*, Pure Appl. Geophys. *126*, 267–278.

WOLERY, T. and DAVELER, S.A. (1992), *EQ3/6, A Software Package for Geochemical modelling of Aqueous Systems, UCRL-MA-110772 PT IV. Lawrence Livermore National Laboratory*, Livermore.

WYSS, M. (1997), *Cannot earthquakes be predicted*? Science *278*, 487.

WYSS, M. (2001), *Why is earthquake prediction research not progressing faster?* Tectonophysics *338*, 217–223.

YECHIELI, Y. and BEIN, A. (2002), *Response of groundwater systems in the Dead Sea Rift Valley to the Nuweiba earthquake: Changes in head, water chemistry and near-surface effects*, J. Geophys. Res. *107*, 2332, doi:10.1029/2001JB001100.

YERRIAH, J. (1986), *Le thermominéralisme carbo-gazeux du Sud-Est de la France (domaine sédimentaire) dans son contexte sismotectonique*, Ph.D. Thesis. University of Montpellier II, France, 108 pp.

(Received: July 25, 2003; revised: November 22, 2005; accepted: November 24, 2005)

To access this journal online:
http://www.birkhauser.ch

Pure appl. geophys. 163 (2006) 745–757
0033–4553/06/040745–13
DOI 10.1007/s00024-006-0035-0

© Birkhäuser Verlag, Basel, 2006

❘Pure and Applied Geophysics

Geographical Distribution of ^{3}He/^{4}He Ratios in the Chugoku District, Southwestern Japan

Yuji Sano,[1] Naoto Takahata,[1] and Tetsuzo Seno[2]

Abstract—We have collected 34 hot spring and mineral spring gases and waters in the Chugoku and Kansai districts, Southwestern Japan and measured the ^{3}He/^{4}He and ^{4}He/^{20}Ne ratios by using a noble gas mass spectrometer. Observed ^{3}He/^{4}He and ^{4}He/^{20}Ne ratios range from 0.054 R_{atm} to 5.04 R_{atm} (where R_{atm} is the atmospheric ^{3}He/^{4}He ratio of 1.39×10^{-6}) and from 0.25 to 36.8, respectively. They are well explained by a mixing of three components, mantle-derived, radiogenic, and atmospheric helium dissolved in water. The ^{3}He/^{4}He ratios corrected for air contamination are low in the frontal arc and high in the volcanic arc regions, which are consistent with data of subduction zones in the literature. The geographical contrast may provide a constraint on the position of the volcanic front in the Chugoku district where it was not well defined by previous works. Taking into account the magma aging effect, we cannot explain the high ^{3}He/^{4}He ratios of the volcanic arc region by the slab melting of the subducting Philippine Sea plate. The other source with pristine mantle material may be required. More precisely, the highest and average ^{3}He/^{4}He ratios of 5.88 R_{atm} and 3.8 ± 1.6 R_{atm}, respectively, in the narrow regions near the volcanic front of the Chugoku district are lower than those in Kyushu and Kinki Spot in Southwestern Japan, but close to those in NE Japan. This suggests that the magma source of the former may be related to the subduction of the Pacific plate, in addition to a slight component of melting of the Philippine Sea slab.

Key words: Helium isotopes, magma source, subduction, Philippine Sea plate.

1. Introduction

It is well documented that helium isotopic ratios can be useful for evaluating a variety of geophysical and geological environments (Mamyrin and Tolstikhin, 1984; Ozima and Podosek, 2002). In the subduction zones, a clear geographical contrast of the ^{3}He/^{4}He ratio; lower value in the frontal arc (forearc) and higher in the volcanic arc (backarc) regions is found in northeastern Japan (Sano and Wakita, 1985), northern New Zealand (Giggenbach et al., 1993), and southern Italy (Sano et al., 1989). The higher ratios with a mantle-derived helium in the

[1]Center for Advanced Marine Research, Ocean Research Institute, The University of Tokyo, Tokyo, 164-8639, Japan. E-mail: ysano@ori.u-tokyo.ac.jp
[2]Division of Geodynamics, Earthquake Research Institute, The University of Tokyo, Tokyo, 113-0032, Japan.

volcanic arc are probably associated with the diapiric uprise of a magma and lower ratios in the frontal arc may be due to radiogenic helium produced by the decay of U and Th in the crustal and sedimentary rocks.

The Japanese Islands are divided by the "Itoigawa-Shizuoka tectonic line" (I-S TL) into major tectonic blocks, northeastern (NE) Japan and southwestern (SW) Japan (Fig. 1).

In NE Japan the old, cold and thick oceanic lithosphere of the Pacific plate subducts beneath the Eurasia plate. A well-defined island arc system feature such as a deep trench, a frontal arc region, a volcanic arc region and a backarc region is developed (MATSUDA, 1964). Geographical contrast of $^3He/^4He$ ratios was found at the volcanic front of NE Japan (SANO and WAKITA, 1985). In SW Japan, in contrast, an exceptional $^3He/^4He$ distribution pattern was observed in the Kansai district. Extraordinarily high $^3He/^4He$ ratios were found in the forearc region. The roughly circular area with the high ratio was called the "Kinki Spot". Taking into account the high ^3He emanation and seismic swarm activity in the region, WAKITA *et al.* (1987) suggested the presence of a shallow magma body beneath the area, and SENO *et al.* (2001) further suggested that partial melting might be induced by dehydration from the serpentinized slab mantle beneath Kii Peninsula.

In the Kyushu district, the most western part of SW Japan and a part of the Kyushu-Ryukyu arc, the contrast of $^3He/^4He$ ratios exists at the volcanic front (SANO and WAKITA, 1985; MARTY *et al.*, 1989; STURCHIO *et al.*, 1996; NOTSU *et al.*, 2001), which is similar to those observed in NE Japan. The Chugoku district is located between the Kinki Spot and the Kyushu district, where the volcanic front is not well defined (SUGIMURA, 1960). The initial purpose of the present work is to identify the front based on the geographical distribution of $^3He/^4He$ ratios. Second is to provide new information on the magma source using the highest and average $^3He/^4He$ ratio of the narrow region near the front together with the $^{87}Sr/^{86}Sr$ ratios of volcanic rocks in literature. The melting of the Philippine Sea slab is not likely to be a major magma source of the basis of the high $^3He/^4He$ ratio. We suggest that the magma source might have a deeper origin comparing the $^3He/^4He$ ratio with those of Kyushu and NE Japan.

2. Experimental

We have collected 32 hot spring and mineral spring gases and waters in the Chugoku district and 2 mineral spring gases in the Kansai district, SW Japan. Gas samples were collected by a displacement method in water using a 50 cm^3 lead glass container with vacuum valves at both ends. Spring waters were sampled in copper tubes (about 20 cm^3) when care was taken to avoid air contamination by air bubbles attaching themselves to the inner wall of the tubes. The tube was sealed at both ends using stainless-steel pinch clamps.

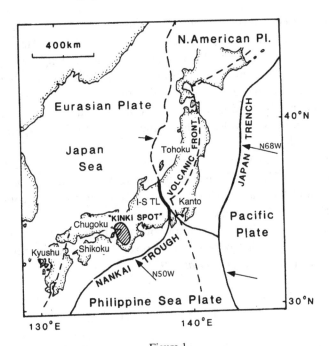

Figure 1
Plate boundaries around the Japanese Islands, trench, and volcanic front and the location of "Kinki Spot".
Arrows show relative direction of the motions between the Pacific and Philippine Sea plates. Volcanic front
was defined by Sugimura (1960).

The ^{3}He/^{4}He and ^{4}He/^{20}Ne ratios of the samples were measured by a noble gas
mass spectrometer (6–60-SGA, Nuclide Co.) installed at the Center for Advanced
Marine Research, Ocean Research Institute, the University of Tokyo, after
purification and separation of noble gases using hot Ti-Zr getters and activated
charcoal traps held at liquid N_2 temperature. Experimental errors of the helium
isotopic ratio and ^{4}He/^{20}Ne ratios are about 3% and 10%, respectively, at 1σ
estimated by repeated measurements of air standard gas (SANO and WAKITA, 1985).
Helium was not separated from Ne in the analysis, which may cause some
uncertainty in absolute ^{3}He/^{4}He ratios (RISON and CRAIG, 1983; SANO and WAKITA,
1988). Accordingly correction was made based on the comparison of ^{3}He/^{4}He ratios
measured by using the VG5400 system with a cryogenic Ne separater and Nuclide
mass spectrometer without the separater (SANO et al., 1998).

3. Results and Discussion

Observed ^{3}He/^{4}He and ^{4}He/^{20}Ne ratios are listed in Table 1 together with the
location and sample type. In the Chugoku district, SW Japan, the ^{3}He/^{4}He and
^{4}He/^{20}Ne ratios vary significantly from 0.054 R_{atm} to 5.04 R_{atm} (where R_{atm} is the

atmospheric ^{3}He/^{4}He ratio of 1.39×10^{-6}) and from 0.25 to 37.7, respectively. Figure 2 shows a correlation diagram between the ^{3}He/^{4}He and ^{4}He/^{20}Ne ratios. The distribution of all samples in the diagram is located in a mixing region of three end-members, primordial helium derived from a mantle beneath the Chugoku district, radiogenic helium produced from uranium and thorium in crustal rocks, and atmospheric helium dissolved in water (air saturated water; ASW) at relatively low temperature compared with volcanic-hydrothermal system. This suggests that helium in the sample is well explained by a simple mixing of those which originated from the three sources (SANO and WAKITA, 1985). If there exists a tritiogenic helium (decay product of tritium, ^{3}H) in the sample, it should be located outside of the mixing between ASW and the mantle. However this is not the case. The contribution of a tritiogenic helium may be significantly small.

Assuming that the ^{4}He/^{20}Ne ratios of mantle and radiogenic helium are significantly larger than that of ASW, it is possible to correct atmospheric helium contamination as follows (CRAIG *et al.*, 1978):

$$R_{cor} = [(^3He/^4He)_{obs} - r]/(1 - r),$$
$$r = (^4He/^{20}Ne)_{ASW}/(^4He/^{20}Ne)_{obs},$$

where R_{cor} and $(^3He/^4He)_{obs}$ denote the corrected and observed ^{3}He/^{4}He ratios, and $(^4He/^{20}Ne)_{ASW}$ and $(^4He/^{20}Ne)_{obs}$ are the ASW and observed ^{4}He/^{20}Ne ratios, respectively. If the observed ^{4}He/^{20}Ne ratio is close to the ASW value (^{4}He/^{20}Ne $= 0.25$ at $0°C$), r becomes about unity and the correction may be significantly erroneous due to an experimental error of ^{4}He/^{20}Ne ratio, that is about 10%. Sano *et al.* (1997) reported the error in the correction for ASW helium contamination as follows:

$$\sigma_{cor} = (R_{0.9} - R_{1.1})/2R_{cor},$$
$$R_{0.9} = [0.9 \times (^3He/^4He)_{obs} - r]/(0.9 - r),$$
$$R_{1.1} = [1.1 \times (^3He/^4He)_{obs} - r]/(1.1 - r),$$

where σ_{cor} denotes the error of the correction. The total error of the corrected ^{3}He/^{4}He ratio is defined as follows:

$$\sigma_{total} = \sqrt{\sigma_{cor}^2 + \sigma_{exp}^2},$$

where σ_{total} and σ_{exp} denote the total error of the corrected ^{3}He/^{4}He ratio and an experimental error of ^{3}He/^{4}He measurement, respectively. The error assigned to the corrected ^{3}He/^{4}He ratio in Table 1 includes all possible ^{3}He/^{4}He errors.

3.1 Volcanic Front in the Chugoku District

Based on the geographical distribution of Quaternary volcanoes in Japan, SUGIMURA (1960) has defined a volcanic front as a strikingly abrupt trenchward limit of volcanoes, which may provide evidence for partial melting in the mantle wedge beneath the volcanic arc behind the volcanic front. It was well identified in the Kurile arc, the NE Japan arc and the Izu-Ogasawara arc of NE Japan and the Kyushu-Ryukyu arc of SW Japan. In contrast the volcanic front was not identified in the Chugoku district (MATSUDA and UYEDA, 1971). Recently KIMURA et al. (2003a) have reported late Cenozoic volcanic activity in the Chugoku district using 108 newly obtained K-Ar ages and they suggested the position of Quaternary volcanic front in the region, which is similar to that indicated by NAKANISHI et al. (2002).

Figure 3 shows the corrected ^{3}He/^{4}He ratios and sampling sites of water and gas samples in this work together with those in Kyushu (STURCHIO et al., 1996; MARTY et al., 1989; NOTSU et al., 2001) and Shikoku (WAKITA et al., 1987). A solid circle has a higher ^{3}He/^{4}He ratio than the open circle, suggesting stronger mantle signature. It is noted that Quaternary volcanic front (QVF) proposed by KIMURA et al. (2003a) cannot match the geographical distribution of ^{3}He/^{4}He ratios. Several samples located in the trench side of QVF such as Tonbara, Tawara, Kakinoki and Kibedani indicate the ^{3}He/^{4}He ratio higher than 2 R_{atm}. Based on the distribution, we draw a helium volcanic front (HVF) in Figure 3. The HVF is located about 20 km from the south side of the QVF in the Chugoku district. This may suggest that the magma source is moving southward.

Figure 4 is the ^{3}He/^{4}He profile in the Chugoku district, showing corrected ^{3}He/^{4}He ratio versus geographic distance from the sampling site to the HVF. There is a clear contrast in ^{3}He/^{4}He ratio between the frontal arc and the volcanic arc regions in the district, which is consistent with those observed in the NE Japan arc and the Izu-Ogasawara arc of NE Japan and the Kyushu-Ryukyu arc of SW Japan (SANO and WAKITA, 1985). The contrast is understood to reflect the absence or presence of magma sources beneath the respective regions, since the high ^{3}He/^{4}He ratio can be of mantle origin and implies the close presence of a rising magma in the volcanic arc.

3.2 Helium and Strontium Isotope Signature of the Chugoku District

In order to discuss the geochemical characteristic of the magma source in NE Japan, SANO and WAKITA (1985) showed the variations in the ^{3}He/^{4}He ratios in narrow areas along the volcanic front, parallel to the trench axis. Data were selected for samples collected in the transition region with a width of 25 km, 5 km on the frontal arc side, and 20 km on the backarc side of the volcanic front. Significant variation of ^{3}He/^{4}He ratio among various arcs was observed, that is, relatively lower ratios ($\sim$3.6 R_{atm}) in the NE Japan arc and higher ($\sim$5.3 R_{atm}) in the Izu-Ogasawara arc. Similar variation in ^{87}Sr/^{86}Sr ratios of volcanic rocks, higher ratios (0.7038–0.7045) in the former arc and lower ratios (0.7032–0.7038) in the latter was reported by NOTSU (1983). Less radiogenic contamination (high ^{3}He/^{4}He and low ^{87}Sr/^{86}Sr ratios) of the magma source in the Izu-Ogasawara arc than the NE Japan arc was attributable to the geotectonic

Table 1

^{3}He/^{4}He, ^{4}He/^{20}Ne, *and corrected* ^{3}He/^{4}He *ratios of hot spring gas and water samples in SW Japan*

No.	Sample	Prefecture	Type	Location (°N)	(°E)	^{3}He/^{4}He (R_{atm})	^{4}He/^{20}Ne	^{3}He/^{4}He$_{cor}$ (R_{atm})	Error	Ref.
	Kansai									
1	Shiota	Hyogo	G	34.95	134.70	0.757	19	0.753	0.023	1
2	Shikano	Hyogo	G	34.92	134.88	1.186	110	1.19	0.04	1
3	Yumura	Hyogo	G	35.55	134.48	3.850	59	3.87	0.12	1
4	Inagawa	Hyogo	G	34.12	134.67	0.830	36	0.829	0.025	
5	Ishimichi	Hyogo	G	34.15	134.58	1.090	251	1.09	0.03	
	Shikoku									
6	Muroto	Kochi	G	33.30	134.13	1.657	7.9	1.69	0.05	2
7	Umaji	Kochi	G	33.57	134.07	1.036	2.2	1.04	0.03	2
8	Okudogo.4	Ehime	G	33.87	132.83	1.264	14	1.27	0.04	2
9	Okudogo.8	Ehime	G	33.87	132.83	1.264	10	1.27	0.04	2
10	Okudogo.K	Ehime	G	33.87	132.83	1.271	16	1.28	0.04	2
11	Yunotani	Ehime	G	33.88	133.17	1.686	17	1.70	0.05	2
12	Bessi	Ehime	G	33.90	133.32	2.214	830	2.22	0.07	2
13	Kamiyama	Tokushima	G	33.97	134.37	0.986	5.9	0.985	0.030	2
	Chugoku									
14	Misasa	Tottori	G	35.40	133.88	5.271	6.3	5.50	0.17	2
15	Sekigane	Tottori	G	35.35	133.77	4.300	30	4.34	0.13	2
16	Koyahara	Shimane	G	35.15	132.58	4.171	59	4.19	0.13	2
17	Tonbara	Shimane	G	35.07	132.80	2.586	18	2.61	0.08	2
18	Tamatsukuri	Shimane	W	35.40	133.02	3.086	29	3.10	0.09	2
19	Yunotsu	Shimane	W	35.08	132.35	4.699	1.01	5.92	0.24	
20	Asahi	Shimane	W	34.87	132.27	3.053	1.39	3.50	0.12	
21	Aribuku	Shimane	W	34.93	132.20	2.338	2.05	2.52	0.08	
22	Mimata	Shimane	W	34.88	132.23	5.038	1.45	5.88	0.24	
23	Kakinoki	Shimane	W	34.43	131.87	4.743	5.92	4.91	0.15	
24	Sanbe	Shimane	W	35.12	132.62	0.847	0.49	0.688	0.040	
25	Mito	Shimane	W	34.68	132.02	1.546	0.82	1.79	0.06	
26	Kibedani	Shimane	W	34.42	131.90	2.331	0.35	5.52	1.24	
27	Yobizuru	Yamaguchi	W	34.03	131.95	1.662	2.48	1.74	0.05	
28	Yumoto	Yamaguchi	W	34.23	131.17	3.150	1.64	3.54	0.12	
29	Kawatana	Yamaguchi	W	34.13	130.93	2.793	2.9	2.96	0.09	
30	Ichinomata	Yamaguchi	G	34.27	131.07	1.594	16.4	1.61	0.05	
31	Ganseiju	Yamaguchi	W	34.43	131.77	4.370	25.8	4.40	0.13	
32	Tawarayama	Yamaguchi	W	34.28	131.10	2.986	3.17	3.16	0.10	
33	Yuno	Yamaguchi	W	34.08	131.68	1.997	37.7	2.00	0.06	
34	Jiseiji	Yamaguchi	W	34.03	131.27	1.818	3.0	1.89	0.06	
35	Yutani	Yamaguchi	W	34.10	131.10	2.169	3.89	2.25	0.07	
36	Suo	Yamaguchi	W	33.95	132.18	0.875	0.30	0.292	0.251	
37	Yumen	Yamaguchi	W	34.35	131.25	1.135	0.27	2.83	0.93	
38	Yuda	Yamaguchi	W	34.15	131.45	0.800	0.48	0.583	0.051	
39	Tawara	Hiroshima	W	34.73	132.43	2.007	2.18	2.14	0.07	
40	Yusaka	Hiroshima	W	34.42	132.83	0.054	12.1	0.034	0.002	
41	Yunoyama	Hiroshima	W	34.48	132.28	0.215	4.21	0.166	0.007	
42	Rakan	Hiroshima	W	34.35	132.07	1.422	4.47	1.45	0.04	
43	Ushiobara	Hiroshima	W	34.47	132.12	1.859	23	1.87	0.06	
44	Yano	Hiroshima	W	34.67	133.10	1.663	8.53	1.68	0.05	

Table 1

(Contd.)

No.	Sample	Prefecture	Type	Location (°N)	(°E)	^{3}He/^{4}He (R_{atm})	^{4}He/^{20}Ne	^{3}He/^{4}He$_{cor}$ (R_{atm})	Error	Ref.
45	Konu	Hiroshima	W	34.70	133.08	0.773	1.03	0.700	0.023	
46	Hiwa	Hiroshima	W	35.00	133.00	1.127	36.8	1.13	0.03	
47	Buttsuji	Hiroshima	W	34.45	133.02	0.975	0.28	0.769	0.435	
48	Yoro	Hiroshima	W	34.43	133.20	1.294	0.39	1.82	0.17	
49	Harada	Hiroshima	W	34.48	133.20	0.975	0.27	0.663	0.513	
50	Iwakura	Hiroshima	W	34.37	132.13	1.033	0.25	–	–	
51	Chiyoda	Hiroshima	W	34.65	132.53	1.504	0.55	1.93	0.10	
	Kyushu									
52	Southern Beppu	Oita	G	33.28	131.38	6.390	4.64	6.79	0.21	3
53	Northern Beppu	Oita	G	33.28	131.38	6.070	4.81	6.43	0.20	3
54	Yufuin	Oita	G	33.27	131.05	6.420	17.1	6.52	0.20	3
55	Kuju	Oita	G	33.08	131.25	5.829	83	5.85	0.18	4
56	Obama	Nagasaki	G	32.72	130.20	4.320	93.6	4.33	0.13	5
57	Unzen	Nagasaki	G	32.73	130.27	5.230	67.8	5.25	0.16	5
58	Shimabara	Nagasaki	G	32.77	130.37	7.060	178	7.07	0.21	5
59	Aso,Yunotani	Kumamoto	G	32.88	131.10	4.529	36	4.56	0.14	4
60	Ebino	Miyazaki	G	31.93	130.85	5.993	37	6.04	0.18	1
61	Shinmoedake	Kagoshima	G	31.90	130.88	6.108	25	6.17	0.19	1
62	Iodani	Kagoshima	G	31.88	130.83	3.683	8.3	3.79	0.11	1
63	Yunotani	Kagoshima	G	31.88	130.80	4.655	2.3	5.24	0.17	1
64	Ramune	Kagoshima	G	31.82	130.73	5.324	4.2	5.68	0.18	1
65	Shikine	Kagoshima	G	31.70	130.78	2.496	27	2.51	0.08	1
66	Sakamoto	Kagoshima	G	30.78	130.28	5.266	3.1	5.75	0.18	1
67	Satsumaiodake	Kagoshima	G	30.78	130.32	6.863	17	6.98	0.21	1

1: SANO and WAKITA (1985); 2: WAKITA et al. (1987); 3: STURCHIO et al. (1996); 4: MARTY et al. (1989); 5: NOTSU et al. (2001). G: gas sample; W: water sample.

setting of the former arc such as the steeper dip angle of the Wadati-Benioff zone and thinner crust over the mantle wedge (SANO and WAKITA, 1985).

Table 2 lists the average and highest ^{3}He/^{4}He ratios of samples collected in the transition region of the NE Japan arc, the Izu-Ogasawara arc, Chugoku district and the Kyushu district together with those of the Kinki spot. The ^{87}Sr/^{86}Sr ratios of volcanic rocks are also referred from the literature (NOTSU, 1983; KIMURA et al., 2005). Both the average and highest ^{3}He/^{4}He ratios are significantly lower in the Chugoku district than the Kyushu district. The ^{3}He/^{4}He signature of the district is consistent with that of the NE Japan arc. In contrast the ^{87}Sr/^{86}Sr ratios of volcanic rocks in the Chugoku district resemble those of the Kyushu district, but are significantly higher than those of the Izu-Ogasawara arc. Therefore the geochemical environment of the magma source in the transition region of the Chugoku district may be similar to that of the NE Japan arc and they are more contaminated by crustal materials than the Kyushu district in terms of helium isotopes. On the other

Table 2

The average and highest ^{3}He/^{4}He ratios and ^{87}Sr/^{86}Sr ratios of NE and SW Japan samples

	Number of samples	Average ^{3}He/^{4}He (R_{atm})	Highest ^{3}He/^{4}He (R_{atm})	^{87}Sr/^{86}Sr
NE-Japan arc[a]	5	3.6 ± 0.7	4.33	0.7038–0.7045
Izu-Ogasawara arc[a]	6	5.3 ± 1.1	6.80	0.7032–0.7038
Kinki spot[b]	11	4.7 ± 1.5	6.97	
Chugoku district[c]	15	3.8 ± 1.6	5.80	0.7035–0.7053
Kyushu district[d]	16	5.6 ± 1.3	6.98	0.7038–0.7054

[a] SANO and WAKITA (1985); NOTSU (1983).

[b] WAKITA et al. (1987).

[c] this work; NOTSU et al. (1990); KIMURA (2005).

[d] MARTY et al. (1989); NOTSU et al. (1990); STURCHIO et al. (1996); NOTSU et al. (2001).

hand, the ^{3}He/^{4}He ratios of backarc region volcanoes were higher than those of the transition region in the NE Japan arc (SANO and WAKITA, 1985), while this is not the case in the Chugoku district. The ^{3}He/^{4}He ratios of the backarc region in the Chugoku district are lower than those of the transition region (see Fig. 4). The geochemical environment of the magma source in the backarc region of the Chugoku district should be discrepant from that of the NE Japan arc.

3.3 Implications for the Magma Source of the Chugoku District

Geotectonic settings of the Chugoku district are different from those of the NE Japan arc. It is well documented that subduction of young plates manifest different geophysical processes than subduction of old plates (e.g., SHIONO, 1988; OKINO et al., 1994). Volcanism in the NE Japan arc is explained by partial melting of the mantle wedge due to dehydration of serpentine, which in turn is hydrated by the dehydration of the slab (IWAMORI, 1998). On the other hand the dehydration of the Philippine Sea slab beneath southwest Japan occurs at a shallow depth (SENO et al., 2001; YAMASAKI and SENO, 2003) and it seems difficult to explain the Quaternary volcanism in this region by dehydration from the subducted crust. Especially beneath Kinki Spot, the dehydration from the serpentinized mantle may have induced incipient partial melting of the mantle wedge (SENO et al., 2001). Melting of the subducted slab is one possibility (NAKANISHI et al., 2002). DEFANT and DRUMMOND (1990) explained the volcanic rocks, Adakites, by partial melting of young subducted slabs. MORRIS (1995) reported that chemical characteristics of volcanic rocks from Sambe and Daisen volcanoes are similar to Adakites, and NAKANISHI et al. (2002) suggested the aseismicity of the Philippine Sea slab at depths exceeding 60 km represents melting of the slab. Although it is tenuous whether the P-T path of the crust of the subducting Philippine Sea slab at present passes through the solidus of amphibolite or not (YAMASAKI and SENO, 2003), it is likely to have passed through the solidus for the geological past (FURUKAWA and TATSUMI, 1999).

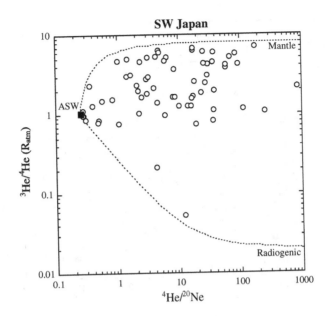

Figure 2
A correlation diagram between the ^{3}He/^{4}He and ^{4}He/^{20}Ne ratios of hot spring gas and water samples in SW Japan. Dotted lines show the mixing lines between mantle derived helium and air saturated water at 0 °C (ASW) and between radiogenic helium and ASW.

We examine here whether the slab melting produces the recent volcanism of the region in terms of the aging effect on the helium isotopes. Formation age of the Shikoku Basin, which is a part of the subducting Philippine Sea plate, is about 20 to 30 Ma (KOBAYASHI and NAKADA, 1978) or about 15 to 25 Ma (SHIH, 1980). Then it is definitely older than 10 Ma. TORGERSEN and JENKINS (1982) reported helium isotope decline due to magma aging. If we assume the holocrystalline Tholeiite as a representative material of the Shikoku Basin, uranium and thorium abundances are 0.1 and 0.18 ppm, respectively (TATSUMOTO, 1966). Again if the magma which consists of the Shikoku Basin evolves as a closed system, the ^{3}He/^{4}He ratio decreases with geological time due to radiogenic production of ^{4}He. Assuming that the initial ^{3}He/^{4}He ratio and ^{3}He content in the holocrystalline tholeiite are 8 R_{atm} and 1.6×10^{-10} cm^{3}STP/g (OZIMA and PODOSEK, 2002), respectively, the estimated ^{3}He/^{4}He ratios of 1 Ma and 10 Ma magma are 0.24 R_{atm} and 0.086 R_{atm}, respectively. The highest (5.8 R_{atm}) and average (3.8 R_{atm}) ^{3}He/^{4}He ratio of the transition region in the Chugoku district are significantly higher than the value of 0.086 R_{atm} estimated in 10 Ma magma. This suggests that the slab melting alone cannot account for the helium isotope data in the region. New and pristine mantle material with a high ^{3}He/^{4}He ratio should be involved in the magma source of the Chugoku district.

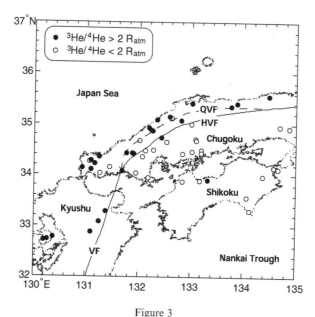

Figure 3

Locations of sampling sites and corrected ^{3}He/^{4}He ratios. Solid circles indicate that the ^{3}He/^{4}He ratio is higher than 2 R_{atm}, and open circles show that the ratio is lower than that. Quaternary volcanic front (QVF) reported by KIMURA *et al.* (2003a) does not agree with Helium Volcanic Front (HVF) defined in this work.

Since the dehydration from the crust occurs at shallow depths, and the dehydration from the serpentinized mantle is limited beneath SW Japan (SENO *et al.*, 2001), typical island-arc volcanism is not expected. Another candidate to provide the mantle helium should be required in addition to the possible slab melting of the Philippine Sea plate with a low ^{3}He/^{4}He ratio. Recently the geometry of the subducting Pacific slab beneath SW Japan has been estimated, by a high density seismic network, to be a continuation of that beneath NE Japan along the same strike (UMINO *et al.*, 2002; SEKINE *et al.*, 2002). The depth of the slab surface is between 400 km and 500 km in the Chugoku district. We suggest that dehydration and fluid migration in the deep mantle associated with the subduction of the Pacific plate beneath SW Japan could be one possible source of pristine magmas beneath the Quaternary volcanic front of SW Japan (IWAMORI, 1991). Although it is not well known whether such deep phenomena may be consistent with the geochemical characteristics of the magma source, it is consistent with IWAMORI's (1992) inference that the source is estimated to be deep in the mantle, based on the incompatible element concentrations. Because the physical and chemical mechanism to link such deep processes to the slab melting and surface volcanism is not understood and beyond the scope of the present paper, further discussion of the problem is needed.

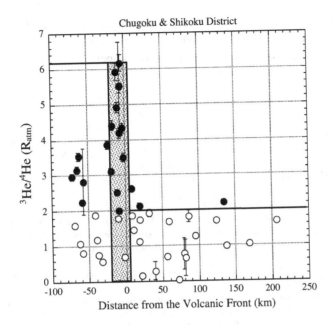

Figure 4

The ^{3}He/^{4}He profile in the Shikoku-Chugoku district, showing corrected ^{3}He/^{4}He ratio versus geographic distance from sampling site to Helium Volcanic Front (HVF in Fig. 3). Dotted region shows a transition region of ^{3}He/^{4}He ratios with a width of 25 km.

4. Conclusion

A clear geographical contrast of ^{3}He/^{4}He ratio was observed in the Chugoku district, SW Japan, which may provide a constraint on the position of volcanic front in the region where it was not well defined by previous works. Higher ^{3}He/^{4}He ratios found in the volcanic arc side cannot be explained by the slab melting of the subducting Philippine Sea plate because of the magma aging effect. New and pristine mantle component is required. The lower ^{3}He/^{4}He ratios in the narrow regions near the volcanic front of the Chugoku district than those ratios of the Kyushu district indicate that the magma source of the former is currently affected by more radiogenic contamination than the latter. The ^{3}He/^{4}He ratios near the volcanic front of the Chugoku district are rather close to those near the volcanic front of NE Japan, which suggests that the magma source may be related to the Pacific plate subduction, even though the mechanism is not well understood at present.

Acknowledgements

We thank J. Kimura for valuable comments and T. Kosugi for their help in field work. This work was partly supported by Grant-in-Aid for Scientific Research

Program No. 08404036 from Ministry of Education, Science and Culture of Japan. We also thank an associate editor and two anonymous reviewers for their valuable comments and suggestions, which enhanced the paper.

REFERENCES

CRAIG, H., LUPTON, J.E., and HORIBE, Y. (1978), *A mantle helium component in Circum-Pacific volcanic gases: Hakone, the Marianas and Mt. Lassen*. In *Terrestrial Rare Gases* (Alexander E.C. and Ozima M. eds.) (Center for Academic Publishing Japan, Tokyo) pp. 3–16.

DEFANT, M.J. and DRUMMOND, M.S. (1990), *Derivation of some modern arc magmas by melting of young subducted lithosphere*, Nature *347*, 662–665.

FURUKAWA, Y. and TATSUMI, Y. (1999), *Melting of subducting slab and production of high-Mg andesite magmas: Unusual magmatism in SW Japan at 13~15 Ma*, Gephys. Res. Lett. *26*, 2271–2274.

GIGGENBACH, W.F., SANO, Y., and WAKITA, H. (1993), *Isotopic composition of helium, and CO_2 and CH_4 contents in gases produced along the New Zealand part of a convergent plate boundary*, Geochim. Cosmochim. Acta *57*, 3427–3455.

IWAMORI, A. (1991), *Zonal structure of Cenozoic basalts related to mantle upwelling in southwest Japan*, J. Geophys. Res. *96*, 6157–6170.

IWAMORI, A. (1992), *Degree of melting and source composition of Cenozoic basalts in southwest Japan: Evidence for mantle upwelling by flux melting*. J. Geophys. Res. *97*, 10983–10995.

IWAMORI, A. (1998), *Transportation of H_2O and melting in subduction zone*, Earth Planet. Sci. Lett. *160*, 65–80.

KIMURA, J., KUNIKIYO, T., OSAKA, I., NAGAO, T., YAMAUCHI, S., KAKUBUCHI, S., OKADA, S., FUJIBAYASHI, N., OKADA, R., MURAKAMI, H., KUSANO, T., UMEDA, K., HAYASHI, S., ISHIMARU, T., NINOMIYA, A., and TANASE, A. (2003), *Late Cenozoic volcanic activity in the Chugoku area, southwest Japan arc during back-arc basin opening and reinitiation of subduction*, The Island Arc *12*, 22–45.

KIMURA, J., STERN, R.J., and YOSHIDA, T. (2005), *Re-initiation of subduction and magmatic responces in SW Japan during Neogene time*, Geol. Soc. Amer. Bull. 117, 969–986.

KOBAYASHI, K. and NAKADA, M. (1978), *Magnetic anomalies and tectonic evolution of the Shikoku inte-arc basin*. J. Phys. Earth *26* (Suppl.), s392–s402.

MAMYRIN, B.A. and TOLSTIKHIN, I.N., *Helium Isotopes in Nature* (Elsevier, Amsterdam 1984) p. 273.

MARTY, B., JAMBON, A., and SANO, Y. (1989), *Helium isotopes and CO_2 in volcanic gases from Japan*, Chem. Geol. *76*, 25–40.

MATSUDA, T. (1964), *Island arc features and the Japanese Islands*, Chigaku Zasshi *73*, 271–280.

MATSUDA, T. and UYEDA, S. (1971), *On the Pacific-type orogeny and its model – Extension of the paired belts concept and origin of marginal seas*, Tectonophysics *11*, 5–27.

MORRIS, P.A. (1995), *Slab melting as an explanation of Quaternary volcanism and aseismicity in southwest Japan*, Geology *23*, 395–398.

NAKANISHI, I., KINOSHITA, Y., and MIURA, K. (2002), *Subduction of young plates: A case of the Philippine Sea plate beneath the Chugoku region, Japan*, Earth Planets Space *54*, 3–8.

NOTSU, K. (1983), *Strontium isotope composition in volcanic rocks from the Northeast Japan arc*, J. Vocanol. Geotherm. Res. *18*, 531–548.

NOTSU, K., ARAKAWA, Y., and KOBAYASHI, T. (1990), *Strontium isotopic characteristics of arc volcanic-rocks at the initial-stage of subduction in western Japan*, J. Volcanol. Geotherm. Res. *40*, 181–196.

NOTSU, K., NAKAI, S., IGARASHI, G., ISHIBASHI, J., MORI, T., SUZUKI, M., and WAKITA, H. (2001), *Spatial distribution and temporal variation of $^3He/^4He$ in hot spring gas released from Unzen volcanic area, Japan*, J. Vocanol. Geotherm. Res. *111*, 89–98.

OKINO, K., SHIMAKAWA, Y., and NAGANO, S. (1994), *Evolution of the Shikoku Basin*, J. Geomag. Geoelectr. *46*, 463–479.

OZIMA, M. and PODOSEK, F.A., *Noble Gas Geochemsitry* (Cambridge University Press, Cambridge 2002) p. 286.

RISON, W. and CRAIG, H. (1983), *Helium isotopes and mantle volatiles in Loihi Seamount and Hawaiian Island basalts and xenoliths*, Earth Planet. Sci. Lett. *66*, 407–426.

SANO, Y. and WAKITA, H. (1985), *Geographical distribution of ^{3}He/^{4}He ratios in Japan: Implications for arc tectonics and incipient magmatism*, J. Geophys. Res. *90*, 8729–8741.

SANO, Y. and WAKITA, H. (1988), *Precise measurement of helium isotopes in terrestrial gases*, Bull. Chem. Soc. Japan *61*, 1153–1157.

SANO, Y., WAKITA, H., ITALIANO, F., and NUCCIO, P.M. (1989), *Helium isotopes and tectonics in southern Italy*, Geophys. Res. Lett. *16*, 511–514.

SANO, Y., GAMO, T., and WILLIAMS, S.N. (1997), *Secular variations of helium and carbon isotopes at Galeras volcano, Colombia*, J. Volcanol. Geotherm. Res. *77*, 255–265.

SANO, Y., NISHIO, Y., SASAKI, S., GAMO, T., and NAGAO, K. (1998), *Helium and carbon isotope systematics at Ontake Volcano, Japan*, J. Geophys. Res. *103*, 23863–23873.

SEKINE, S., OBARA, K., SHIOMI, K., and MATSUBARA, M. (2002), *Three-dimensional attenuation structure beneath the Japan Islands derived from NIED Hi-net data*. Abstract of Fall Meeting, Seismol. Soc. Japan, A49.

SENO, T., ZHAO, D., KOBAYASHI, Y., and NAKAMURA, M. (2001), *Dehydration of serpentinized slab mantle: Seismic evidence from southwest Japan*, Earth Planets Space *53*, 861–871.

SHIH, T.C. (1980), *Magnetic lineations in the Shikoku Basin*. Initial Rep. Deep Sea Drilling Project *58*, 783–788.

SHIONO, K. (1988), *Seismicity of the SW Japan arc – subduction of the young Shikoku Basin*, Modern Geology *12*, 449–464.

STURCHIO, N.C., OSAWA, S., SANO, Y., AREHART, G., KITAOKA, K., and YUSA, Y. (1996), *Outflow plume of the Beppu hydrothermal system at Yufuin, Japan*, Geothermics *25*, 215–230.

SUGIMURA, A. (1960), *Zonal arrangement of some geophysical and petrological features in Japan and its environs*, J. Fac. Sci. Univ. Tokyo, Sect. 2, *12*, 133–153.

TATSUMOTO, M. (1966) *Genetic relation of oceanic basalts as indicated by lead isotopes*, Science *153*, 1094–1095.

TORGERSEN, T. and JENKINS, W.J. (1982), *Helium isotopes in geothermal systems: Iceland, the geysers, raft river and steamboat springs*. Geochim. Cosmochim. Acta *46*, 739–748.

UMINO, N., ASANO, Y., OKADA, T., MATSUZAWA, T., and HASEGAWA, A. (2002), *Geometry of the subducted Pacific slab estimated from ScSp phases observed by the high density seismic network*, Abstract of Fall Meeting, Seismol. Soc. Japan, A50.

WAKITA, H., SANO, Y., and MIZOUE, M. (1987), *High ^{3}He emanation and seismic swarm activities observed in a non-volcanic, frontal arc region*, J. Geophys. Res. *92*, 12539–12546.

YAMASAKI, T. and SENO, T. (2003), *Double seismic zones and dehydration embrittlement of the subducting slab*, J. Geophys. Res. *108*(B4), 2212, doi:10.1029/2002JB001918.

(Received: January 27, 2003; revised: December 7, 2004; accepted: December 15, 2004)
Published Online First: March 28, 2006

 To access this journal online:
http://www.birkhauser.ch

Pure appl. geophys. 163 (2006) 759–780
0033–4553/06/040759–22
DOI 10.1007/s00024-006-0037-y

© Birkhäuser Verlag, Basel, 2006

Pure and Applied Geophysics

Geochemistry of the Submarine Gaseous Emissions of Panarea (Aeolian Islands, Southern Italy): Magmatic vs. Hydrothermal Origin and Implications for Volcanic Surveillance

Giovanni Chiodini, Stefano Caliro, Giorgio Caramanna, Domenico Granieri, Carmine Minopoli, Roberto Moretti, Lavinia Perotta, and Guido Ventura

Abstract—The marine sector surrounding Panarea Island (Aeolian Islands, South Italy) is affected by widespread submarine emissions of CO_2-rich gases and thermal water discharges which have been known since the Roman Age. On November 3rd, 2002 an anomalous degassing event affected the area, probably in response to a submarine explosion. The concentrations of minor reactive gases (CO, CH_4 and H_2) of samples collected in November and December, 2002 show drastic compositional changes when compared to previous samples collected from the same area in the 1980s. In particular the samples collected after the November 3rd phenomenon display relative increases in H_2 and CO and a strong decrease in the CH_4 contents, while other gas species show no significant change. The interaction of the original gas with seawater explains the variable contents of CO_2, H_2S, N_2, Ar and He which characterize the different samples, but cannot explain the large variations of CO, CH_4 and H_2 which are instead compatible with changes in the redox, temperature and pressure conditions of the system. Two models, both implying an increasing input of magmatic fluids are compatible with the observed variations of minor reactive species. In the first one, the input of magmatic fluids drives the hydrothermal system towards atypical (more oxidizing) redox conditions, slowly pressurizing the system up to a critical state. In the second one, the hydrothermal system is flashed by the rising high-T volcanic fluid, suddenly released by a magmatic body at depth. The two models have different implications for volcanic surveillance and risk assessment: In the first case, the November 3rd event may represent both the culmination of a relatively slow process which caused the overpressurization of the hydrothermal system and the beginning of a new phase of quiescence. The possible evolution of the second model is unforeseeable because it is mainly related to the thermal, baric and compositional state of the deep magmatic system that is poorly known.

1. Introduction

Submarine gas emissions occur in the neighborhood of many of the Aeolian Islands, such as Salina, Lipari, Vulcano, Stromboli and Panarea (Figs. 1a,1b). At Panarea, submarine exhalations are located to the east of the island mainly along a NE-SW strike and within a subcircular, submerged depression, with 30 m maximum

INGV "Osservatorio Vesuviano", Operative Unit of Fluids Geochemistry, Via Diocleziano 328, 80124, Naples, Italy.

depth (Fig. 1c, GABBIANELLI *et al.*, 1990). The rim of this structure, which probably represents a submerged crater, is evidenced by the islets of Bottaro, Lisca Bianca, Panarelli and Dattilo. These submerged hydrothermal manifestations have been known since the Roman age, and have been described by DÉODAT DE DOLOMIEU in *Voyage aux îles Lipari* (1783). ITALIANO and NUCCIO (1991) distinguished different hydrothermal systems at temperatures from 170 to 240 °C feeding the submarine gas emissions of Panarea. These authors hypothesized that sea water and a magmatic component as the main sources of fluids circulating in the hydrothermal systems. CALANCHI *et al.* (1995) recognized two different re-equilibration zones: A deeper one characterized by equilibrium temperatures up to 200 °C and an upper one with temperatures around 100 °C.

Since November 3rd 2002 an anomalous release of gas in the area of the submerged subcircular depression of Panarea was first observed by some fishermen who alerted the Italian Civil Defense. The occurrence of such phenomena was promptly confirmed by researchers of the National Institute of Geophysics and Volcanology (INGV) in Catania, which on November 4th, 2002 described three distinct spots aligned along a NE-SW strike (INGV-CATANIA, 2002). They in fact noted: (a) the occurrence of dead fish, (b) whitening of sea water due to the presence of clay and sand, (c) changes in the pH of the sea water (pH = 5.6–5.7). As visible at the sea surface, the largest rising column of bubbles reached approximately 8–10 m in diameter and was located near the Bottaro islet (Fig. 2a). At the bottom emission outlet, a maximum temperature of 35 °C was measured during the sampling campaigns of November and December 2002, whereas 47 °C was measured at #8 (northeastern sector of Bottaro islet). The areas surrounding the vents were characterized by colloidal sulfur deposits. Marine geology observations revealed the occurrence of dislocated blocks of rocks and gravels remobilized around the main gas emissions. Microearthquakes with M_d generally less than 1 (M_d max = 1.8) were recorded between November 3rd and 13th. The accurate hypocentral/epicentral location of these low energy events is unknown but their source area falls in the vent field. All these observations may suggest that on November 3rd, 2002 the seafloor was affected by phreatic explosions which opened the strong vents subsequently observed.

In the proceeding of this study we show that the chemical compositions of the gaseous emissions sampled at Panarea from November to December 2002 do provide evidence of increased magmatic influence since the 1980s, although a main problem in interpreting data from the gas vents is that the original water content of deep gas cannot be determined.

2. Geological, Geodynamic and Volcanic Setting

The Aeolian Islands, Southern Tyrrhenian Sea, consist of seven major islands and several associated seamounts (Fig. 1a). The volcanic rocks belong to the calc-alkaline

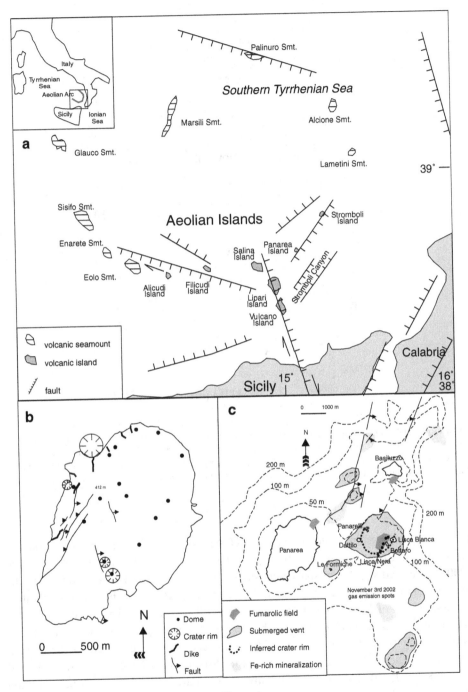

Figure 1

a) Sketch map of the Aeolian arc with main structural alignment; b) Panarea map with details of volcano-related tectonic features; c) simplified bathymetric chart showing location of Panarea islets and submerged emission vent.

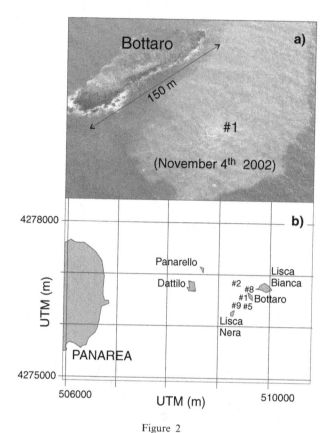

Figure 2

a) November 4th, 2002 airborne image of #1 emission zone, in the proximity of Bottaro islet (modified from http://www.ct.ingv.it/Report/Panareaweb/). b) Location of sampled submarine emission zones indicated by Arabic numbers preceded by symbol '#'. different sampling sites within the same emission zone are labeled with stand-alone Arabic numbers, whereas Roman numbers refer to different samples collected at the same site.

(CA), high-potassium calc-alkaline (HKCA), shoshonitic and potassic-alkaline associations. ELLAM *et al.* (1989), FRANCALANCI *et al.* (1993) and DE ASTIS *et al.* (2000) described significant space and time-related compositional variations from the western islands of Alicudi and Filicudi to the CA, HKCA, shoshonitic and alkaline potassic rocks of Vulcano, Lipari, and Stromboli to the east. The eastern sector of the Aeolian Islands is characterized by deep-focus earthquakes depicting a narrow NW-dipping Benioff-Wadati plane. The deep seismicity is restricted to the eastern branch of the Aeolian arc (Panarea and Stromboli Islands, FALSAPERLA and SPAMPINATO 1999). This sector of the arc is also characterized by a thermal anomaly (heat flow > 100 mW/m^2; WANG *et al.*, 1989) due to the upwelling of the mantle in the Stromboli-Panarea area (NERI *et al.*, 2002).

Panarea (421 m a.s.l.) is a volcanic complex situated between the central sector of the Aeolian arc and Stromboli (Figs. 1a, b, c) and rises about 1700 m above the sea floor, but only a very minor part of the volcano crops out at the surface to form the main island and the scattered neighboring islets. Panarea and the associated islets emplaced on 15–20 km thick continental crust (NERI *et al.*, 2002). Most of the rocks outcropping, which mainly consist of domes and lava flows, show an acid CA composition but less abundant HKCA and shoshonitic products also occur (CALANCHI *et al.*, 2002). The geochemical features of these rocks are consistent with a heterogeneous, possibly, fluid-rich, mantle source that shows characteristics intermediate between the Stromboli magmas and those of Alicudi and Filicudi (CALANCHI *et al.*, 2002). The submerged part of Panarea shows an almost flat surface at a depth of about 100 m b.s.l. (Fig. 1c). Both the Panarea Island and the submerged flat surface are affected by NE-SW to NNE-SSW striking faults along which Fe-mineralizations and sulfide deposits occur (Figs. 1b,c) (GABBIANELLI *et al.*, 1990; GAMBERI *et al.*, 1997). The strike of these faults is the same of the faults affecting the Island of Stromboli and the Stromboli Canyon, a NE-SW striking submarine valley located south of Panarea (Fig. 1a) (VOLPI *et al.*, 1997). These observations suggest that the uprising of magmas at Panarea and Stromboli are strongly controlled by NE-SW striking structures. Available geochronological data indicate that the volcanism in the Panarea area developed between 149 and 54 kyr but 42 to 13 kyr B.P. old ash-flow deposits erupted from unidentified submerged vents outcrop on the island (CALANCHI *et al.*, 2002).

3. Field Operations and Analytical Procedure

Gases were collected from a narrow area in the neighborhood of Bottaro and Lisca Bianca islets (Fig. 2b) from five different emission zones.

Four different sampling campaigns were performed (analytical data are shown in Table 1): November 29 and 30 (emission zones #1, #2, #8), December 4 (#5 and #9), December 10 (#1) and December 17 (#1).

Gases were collected in typical sampling flasks connected through silicone tubes to a funnel opportunely overburdened with lead. At each site gases were collected in flow-through bottles (*dry gases*) and also in pre-evacuated solution-filled bottles (NaOH 4N; GIGGENBACH, 1975), where acid condensable gases are absorbed.

Chemical analyses were performed at the INGV-OV Laboratory of Fluid Geochemistry as follows:

— Determination of CO and H_2 contents (at low concentrations) on dry gas samples by gas-chromatographic separation coupled to a high sensitivity Reduced Gas Detector (HgO);

— gas-chromatographic analyses with Thermal Conductivity detection of the non-absorbed gases (H_2, He, CH_4, Ar, O_2, N_2) in the headspace of soda-filled bottles;

G. Chiodini *et al.*

Table 1

Analytical data (vol%) of the submarine gas emissions of Panarea (November–December 2002)

Sample	Date	CO_2	H_2S	Ar	O2	N_2	CH_4	H_2	He	CO
#1, 1 I	29-Nov.02	98.2	1.17	0.00540	0.0815	0.439	0.0005	0.1196	0.00102	0.00066
#1, 1 II	29-Nov.02	98.2	1.17	0.00498	0.0701	0.416	0.0004	0.1152	0.00098	0.00066
# 1, 1	10-Dec.02	97.8	1.86	0.00090	0.0015	0.201	0.0002	0.1061	0.00078	0.00062
# 1, 1	17-Dec.02	97.8	1.83	0.00115	0.0002	0.213	0.0002	0.1085	0.00075	0.00061
#1, 2	29-Nov.02	98.4	1.17	0.00298	0.0127	0.332	0.0011	0.0514	0.00096	0.00042
#2, 1 I	30-Nov.02	97.3	2.38	0.00098	0.0091	0.232	0.0003	0.0533	0.00080	0.00043
#2,1 II	30-Nov.02	97.4	2.34	0.00082	0.0074	0.218	0.0002	0.0518	0.00078	0.00043
#5, 1	4-Dec.2	98.7	0.98	0.00189	0.0002	0.247	0.0004	0.0532	0.00085	0.00084
#8, 1 I	29-Nov.02	98.0	1.60	0.00474	0.0404	0.395	0.0002	0.0045	0.00101	0.00011
#8, 1 II	29-Nov.02	97.9	1.68	0.00459	0.0368	0.402	0.0002	0.0048	0.00100	0.00011
#8, 2	30-Nov.02	97.9	1.87	0.00160	0.0133	0.254	0.0001	0.0010	0.00088	0.00008
#9, 1	4-Dec.02	98.0	1.61	0.00216	0.0007	0.270	0.0004	0.0691	0.00084	0.00047

— determination of CO_2, S_{tot} (H_2S + SO_2), Cl, F in alkaline solutions through acidimetric titration and ion chromatography.

4. Data Discussion

4.1 Chemical Compositions of Discharged Gases

Gases emitted in the marine sector of Panarea Island are mainly made up of CO_2 (97.3–98.4 vol.%). Minor amounts of H_2S (1–2.4 vol.%), H_2 (0.001–0.12 vol.%), N_2 (0.2–0.5 vol.%), He (0.0008–0.001 vol.%), Ar (0.0008–0.0054 vol.%), CH_4 (0.0004–0.001 vol.%) and CO (0.0001–0.0008 vol.%) are present.

In order to ascertain the origin of these emissions, we must consider that the original gaseous composition has undergone chemical changes in response to the interaction with seawater, which is at equilibrium with the atmosphere. In order to evaluate the extent of possible interpretative biases, it is important to discuss two processes affecting the 'deep' gas: 1) enrichment in atmospheric components dissolved in sea water, such as N_2, O_2, Ar and 2) partial dissolution of the 'deep' gas, thus accounting for the different solubility behavior of its species in water (i.e., H_2S and CO_2 considerable more soluble that the remaining species).

The mixing between 'deep' gaseous and dissolved atmospheric end-members emerges from the ternary He-Ar-N_2 (Fig. 3). Panarea samples plot on a straight line highlighting the mixing between the deep component, characterized by relatively high He contents and N_2/Ar ratios, and the ASW composition (Air-Saturated Water) rich in Ar. Gases released within zone #2 are the least contaminated, while those emitted at #1 (and sampled on November 29) are the most enriched in the dissolved atmospheric component. Figure 3 highlights that the deep component has the typical

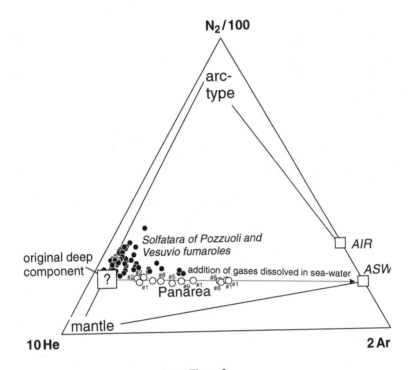

Figure 3

He-Ar-N₂ ternary plot. Note the alignment of Panarea Samples along the mixing line between deep-component and ASW. Since plotted samples do not tend toward the composition of air, such a representation excludes that the enrichment in atmospheric component would be the result of a mere post-collection contamination of samples with air. Compositions from fumaroles of Vesuvius and Solfatara of Pozzuoli have been plotted for comparative purposes.

composition of deep gases associated with other volcanic-hydrothermal systems of Italy. In order to study the effects induced by the interaction between the 'deep' gas and seawater, i.e., partial dissolution of most soluble gases and addition of the dissolved atmospheric component, we first considered the correlations of Ar with respect to CO_2, H_2S, He and N_2. For the sake of clarity, we have divided points in two families, depending on the relative amount of H_2S. In particular the samples collected northwest of Bottaro islet (#2 and #8) show H_2S contents systematically higher than gases from the southwest of the islet (#1, #5 and #9). This allows us to distinguish significant correlations between the considered species and Ar (Fig. 4). Such correlations may be easily employed to infer the composition of the original gas. A hypothesis quite close to reality is that of assuming a nil concentration of Ar in the original gas. Therefore, it is possible to calculate the relative original content of each species by considering its specific regression (Fig. 4).

These compositions are used as original deep gases in order to study the effects induced by interaction of the 'deep' gas in seawater. From field observation

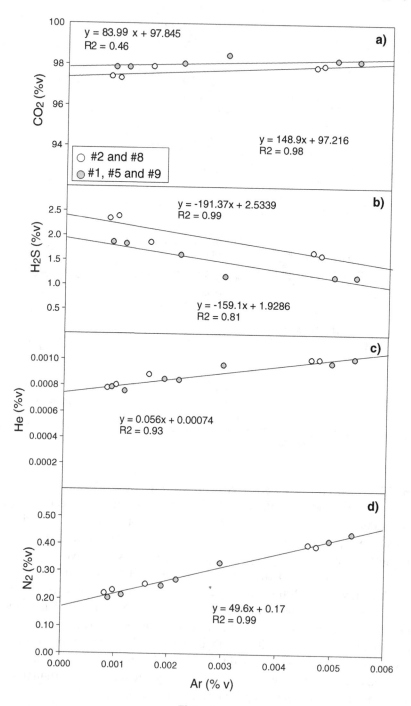

Figure 4

Binary diagrams of CO_2, H_2S, He, N_2 vs. Ar. Samples are divided in two families, depending on the relative amount of H_2S. The concentration of these species in the original gas has been retrieved on the basis of given regressions by assuming a limiting condition of Ar = 0 for the deep component. For N_2 and He regressions all the samples are considered, due to the slight differences in their concentrations.

the strong energy of the phenomenon was evident, appearing as a mixed turbulent gas-water column moving at high speed towards the surface, which conveyed the surrounding seawater from the bottom towards the central part of the rising gas emission. The effect was so evident that during the first survey the scuba divers were attracted by the moving low density central column and were expulsed at the surface at the center of the large bubble (see Fig. 2). In order to model the interaction between deep gases and seawater we considered that the different species present in the original deep gas or dissolved in seawater distribute between the gas and the liquid phases according to their solubilities. The computation was done at seawater temperature (20 °C) and at a pressure of 2.5 bar, corresponding to a depth of 15 m (average depth of submarine emissions). The process was simulated through a multi-step procedure. At each step we computed the compositions of the dissolved gas phase and of the free gas phase. The latter was employed in the subsequent step to model interaction with seawater at a fixed gas/water molar ratio. Figures 5a,b and 6a,b where the simulated and measured values are compared, suggest that the model well reproduced the observed compositions including the increase in the atmospheric gases N_2 and Ar (Figs. 6c,b).

In general, dissolution of a Panarea-like gas implies on one hand the increase of the relative contents in atmospheric species, He and, to a minor extent, of CO_2, on the other hand a decrease of H_2S concentration. In the case of gases sampled at Panarea we estimate the molar fraction of dissolved CO_2 lower than 0.3 (<0.2 in most cases). Assuming an initial content of Ar > 0 would obviously imply lower contents of (estimated) dissolved CO_2, so that we can reasonably conclude that the value of 0.3 represents the maximum fraction of 'deep' gas which may dissolve in seawater.

It is worth remarking that the high degrees of correlation among He, Ar, N_2, CO_2 and H_2S strongly support the proposed model. Nevertheless, the procedure is not sufficient to explain the strong variations of minor reactive species, such as CO, H_2 and CH_4, as clearly evidenced in Figure 7. In fact, the absence of any significant correlation between these three gaseous species and He suggests that dissolution in seawater is not the main process responsible for the chemical speciation governing the relative amounts of CO, H_2 and CH_4. Furthermore, we may here anticipate that the effects of gas dissolution in water are limited (see above) and may be neglected for the geothermometric applications developed in the following.

In Figures 8 and 9 we observe a strong correlation between CO and H_2 contents and, to a minor extent, between CO and CH_4.

Hydrogen, CO and CH_4 are reactive gaseous species very sensitive to changes in temperature, pressure and redox conditions, supporting that sampled gases may originate under different thermochemical conditions among emission-feeding zones.

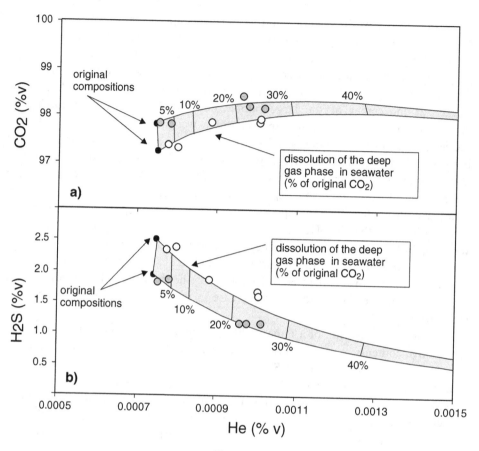

Figure 5

CO$_2$ vs. He (a) and H$_2$S vs He (b) diagrams. Measured compositions are compared with theoretical compositions of residual gas upon dissolution in seawater of different molar fractions of the original gas. Molar fraction of dissolved gas is less than 0.3 in any case.

4.2 Gaseous Geoindicators of Temperature and Pressure (System H$_2$O-H$_2$-CO$_2$-CO-CH$_4$)

Before proceeding with the application of geothermobarometric methods based on the relative contents of hydrogen, methane and carbon monoxide, we first make some general considerations regarding the origin of 'deep' emitted gases.

In Figure 10a (log H$_2$/CH$_4$ Vs. log CO/CH$_4$) we have reported:

1. The theoretical composition of gaseous mixtures at equilibrium in the temperature range 100–373 °C and at P$_{CO_2}$ between 0.1 and 100 bar. The theoretical grid for typical conditions of hydrothermal systems has been calculated:

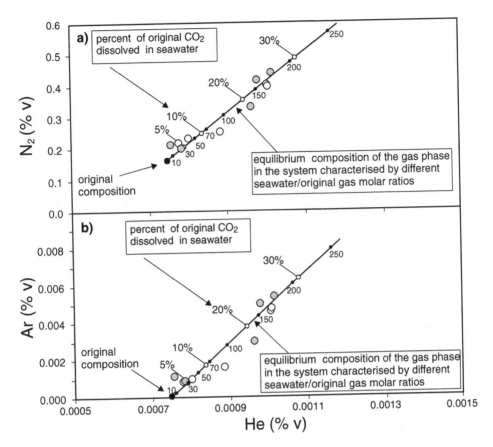

Figure 6

He vs. Ar (a) and N_2 (b) diagrams. Measured compositions are compared with theoretical compositions of residual gas after interaction with seawater. The number refer to the different H_2O/gas molar ratios used in the model.

a) assuming redox buffer of hydrothermal environments, expressed as (D'AMORE and PANICHI, 1980):

$$\log fO_2(bar) = 8.20 - 23643/T(K); \qquad (1)$$

b) assuming the presence of pure liquid water, therefore fixing fH_2O as a function of temperature (GIGGENBACH, 1980) along the liquid-vapor univariant equilibrium:

$$\log fH_2O(bar) = 5.510 - 2048/T(K); \qquad (2)$$

c) considering the formation reactions (i.e., involving oxygen) of H_2, CO and CH_4 and their equilibrium constants (thermodynamic data from GIGGENBACH, 1980):

G. Chiodini *et al.*

Pure appl. geophys.,

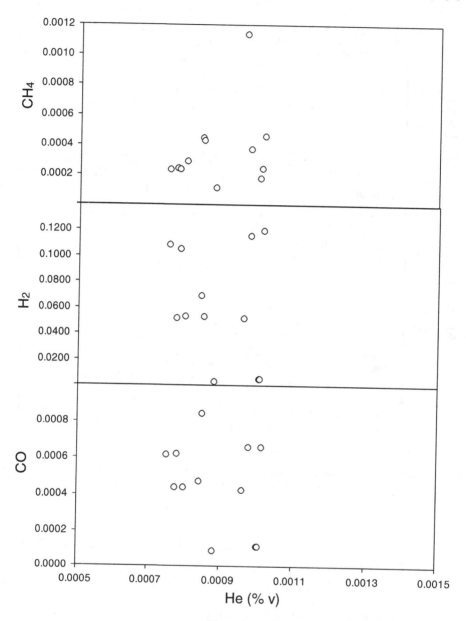

Figure 7
Binary diagrams of CH$_4$ (a), H$_2$ (b) and CO (c) vs. He. Points show no appreciable correlation, thus suggesting that the partial dissolution of gas in seawater is not the key factor controlling the sampled amounts of CO, CH$_4$ and H$_2$.

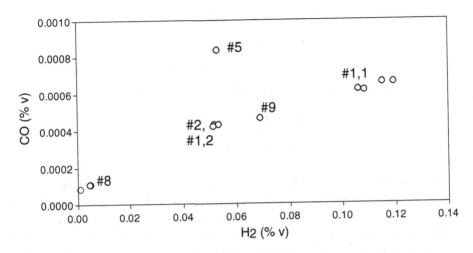

Figure 8
CO vs. H_2 diagram showing the fairly good correlation between these two reactive species, indicative of redox equilibrium.

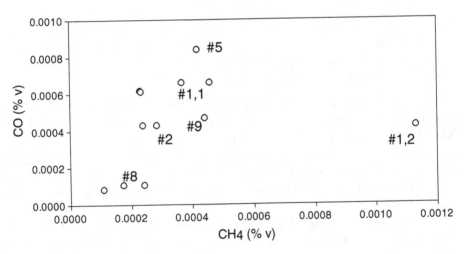

Figure 9
CO vs. CH_4 diagram. The correlation is not as good as CO vs. H_2, but still indicates a redox control on composition of minor reactive species.

$$H_2O \Leftrightarrow H_2 + \frac{1}{2}O_2, \tag{3}$$

$$\log K_5 = 2.548 - 12707/T(K), \tag{3a}$$

$$CO_2 \Leftarrow CO + \frac{1}{2}O_2, \tag{4}$$

$$\log K_6 = 5.033 - 14955/T(K), \tag{4a}$$

$$CO_2 + 2H_2 \Leftrightarrow CH_4 + O_2, \tag{5}$$

$$\log K_7 = -4.569 - 16593/T(K); \tag{5a}$$

d) deriving the following functions of temperature and P_{CO_2}

$$\log(H_2/CH_4) = 8.811 - 4121.5/T(K) - \log P_{CO_2} \tag{6}$$

$$\log(CO/CH_4) = 5.786 - 4326.5/T(K). \tag{7}$$

2. The compositional data of 151 gaseous samples from the following 23 hydrothermal systems: Larderello, Travale, Amiata, Bagnore, Mofete, Vulcano, Campi Flegrei, Vesuvio, Ischia, Pantelleria and Lipari (Italy), Nysiros (Greece), Teide (Canary Islands, Spain), Ahuachapan (El Salvador), Monteserrat (West Indies), Guagua Pichincha (Ecuador), Tambora (Indonesia), Kizildere (Turkey), Cagua, Alto Peak and Mahagnao (Philippines), St. Lucia (West Indies), Ngqu (China). Data and complete analytical characterization are given in CHIODINI and MARINI (1998).

3. The compositional data of high-T (> 280 °C) volcanic fumaroles, from White Islands (New Zealand; GIGGENBACH, 1987), Fossa di Vulcano crater (Italy; CHIODINI et al., 1993) and from volcanoes Klyuchevskoy, Mutnovsky, Mount Usu, Ngauruhoe, Papandayan, St. Helens, Satsuma Iwo J, Sierra Negra, Showa-Shinzan, Tokachi (GIGGENBACH, 1996).

The large compositional difference between the two families of gases emerges from this figure: volcanic gases are characterized by markedly high H_2/CH_4 and CO/CH_4 ratios, having orders of magnitude larger than those pertinent to hydrothermal systems, both calculated and measured. It should be emphasized that all the experimental data on hydrothermal gases plot within the theoretical grid while all the high-T volcanic fumaroles fall outside the hydrothermal compositional field.

The "Panarea" emissions (Fig. 10b) subsequent to the 2002 event of November 3rd lie outside the hydrothermal field and show contents of H_2, CH_4 and CO that are typical of high-T volcanic fumaroles. Data from samples collected in the 1980s and reported by ITALIANO and NUCCIO (1991) show instead compositions closer to those from the hydrothermal environment, but still different because of the very low H_2/CH_4 ratio. This is likely to be due to the H_2 removal operated by surface oxidative processes, which are favored by the interaction of gases with water (CHIODINI, 1994). Figure 10b then evidences both the strong compositional variation between the 1980s and 2002 and the non-hydrothermal features of these emissions in 2002. If now we consider the presence of saline solutions or brines instead of pure water, the field of vapor-liquid equilibrium can be expanded. In Figure 11 the theoretical ''hydrothermal' grid has been computed for a brine NaCl

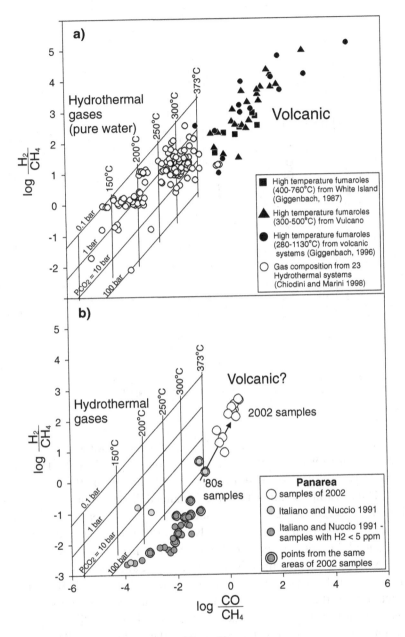

Figure 10

Diagram of log H_2/CH_4 vs log CO/CH_4 for well-known hydrothermal and volcanic systems studied in literature (part a) and for Panarea samples (part b). The theoretical grid for hydrothermal gases was calculated by assuming coexistence with pure liquid water.

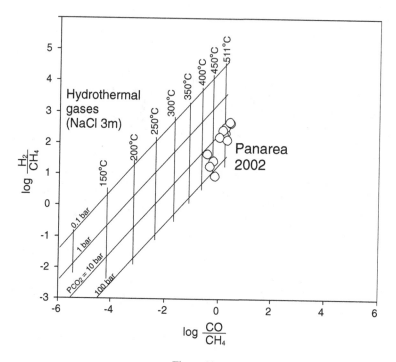

Figure 11

Diagram of log H_2/CH_4 vs log CO/CH_4. In this case, the theoretical grid for hydrothermal gases was calculated by assuming coexistence with NaCl 3 m brine.

3 m, with a critical temperature of 511 °C (KNIGHT and BODNAR, 1989; CHIODINI *et al.*, 2001). Obviously equation (2) is no longer valid and fH_2O must be recomputed as outlined in CHIODINI *et al.* (2001). With these modifications, "Panarea 2002" samples fall within the expanded hydrothermal field, indicating equilibrium temperatures between 400 and 500 °C. However, if we also consider the CO_2 content and then extend our analysis to the gaseous system CO_2-CO-CH_4-H_2, such an observation is not confirmed by the log H_2/CH_4 vs. log CO/CO_2 diagram of Figure 12. In this figure, in fact, "Panarea 2002" gases indicate temperatures which are much lower (180 to 250 °C) and then in evident contradiction with the hypothesis of a vapor in equilibrium with a saline solution. This may be attributed to the fact that sampled gases did not achieve the equilibrium at redox conditions typical of hydrothermal systems. This would imply that equation (1) cannot be applied and that T and P_{CO_2} estimates from Figures 10 and 11 have no physical sense.

We still have two hypotheses to test and evaluate: a) the presence of a hydrothermal system with 'atypical' redox conditions although in any case at equilibrium within the system H_2O-H_2-CO_2-CO-CH_4 and b) the possibility that gases

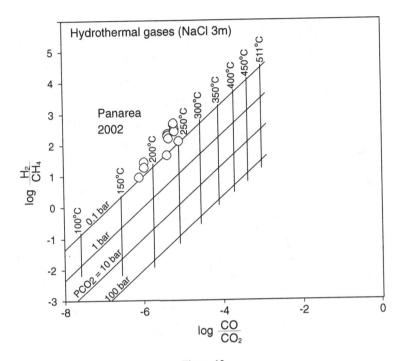

Figure 12

Diagram of log H_2/CH_4 vs log CO/CO_2. The theoretical computation for hydrothermal gases is the same as in Fig. 11. Contrarily to previous representation, Panarea samples exhibit different equilibrium temperatures conforming to the $CO-CO_2$ couple, perhaps suggesting methane disequilibrium (GIGGENBACH, 1987) in the system $H_2-H_2O-CO-CO_2$ or high-T volcanic features (see text).

discharged in 2002 at Panarea represent the residual fraction of a 'volcanic' fluid, like those emitted from high-T fumaroles after steam condensation.

In the first hypothesis, we may follow CHIODINI and CIONI (1989) and combine equations (3, 4, and 5) in order to obtain two reactions without the involvement of gaseous O_2 and then retrieve the following T and P_{CO_2} functions, which are independent of the redox conditions of the system:

$$T(K) = 13606 \Big/ \Big(8.065 - \log(X_{CO}^4 \big/ X_{CO_2}^3 X_{CH_4})\Big), \tag{8}$$

$$\log P_{CO_2}(bar) = 3.573 - 46/T(K) - \log\frac{X_{H_2}}{X_{CO}}. \tag{9}$$

The system derived by these two equations is graphically solved in Figure 13 ($\log(X_{H_2}/X_{CO})$ vs. $\log(X_{CO}^4\big/X_{CO_2}^3 X_{CH_4})$). Collected samples in general indicate equilibrium temperatures of about 300 °C and P_{tot} of about 100 bar. Plotting points corrected for gas dissolution in water does not affect the estimates obtained from analytical data. Let us recall that these estimates make sense only under the hypothesis that H_2, CO, CH_4 and CO_2 are at equilibrium under 'atypical'

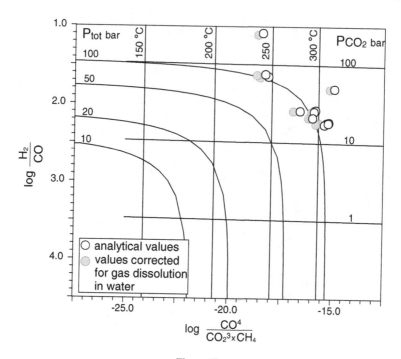

Figure 13

Gas-geoindicators in the system H_2O-H_2-CO_2-CO-CH_4 (Chiodini and Cioni, 1989; Chiodini and Marini, 1998). Theoretical lines refer to vapour-liquid equilibrium.

hydrothermal redox conditions that would be more oxidizing than those previously observed. This could be ascribed to the input of magmatic gases, rich in SO_2, to the hydrothermal system.

In the case of hypothesis b) the above estimates are meaningless. In fact, under the relatively high-fO_2 conditions of volcanic fumaroles, methane, present in very minor amounts, is not in equilibrium with other species, whereas H_2/H_2O and CO/CO_2 ratios quickly re-equilibrate with temperature and pressure at redox conditions which are fixed by the H_2S-SO_2 couple and at specific values of P_{H_2O} (Giggenbach, 1987, 1996; Chiodini *et al.*, 1993). In this case, the only geothermometric indication is given by the CO-CO_2 couple. In Figure 14 (log CO/CO_2 vs. 1000/T) sampled gases show temperature estimates between 350–450 °C along the H_2S-SO_2 buffer and at reasonable values of P_{H_2O}.

5. Conclusions

This study on submarine gaseous emissions in the Aeolian district of Panarea shows that:

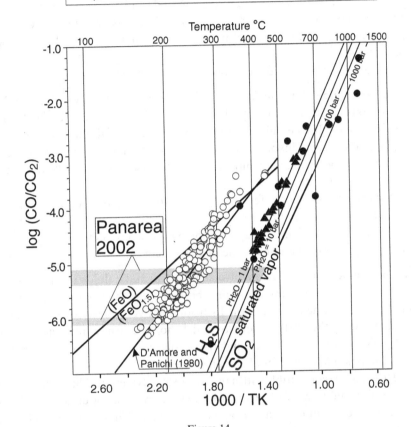

Figure 14

Stability diagram of the CO-CO₂ couple for different redox buffers. FeO-FeO1.5 and D'AMORE PANICHI (1980) buffers refer to the hydrothermal environment. The H₂S-SO₂ buffer applies to volcanic gases (GIGGENBACH 1987, CHIODINI et al., 1993). Under the hypothesis that gases sampled at Panarea would represent residual volcanic gases, temperatures of 350–450 °C would be estimated.

1. The concentrations of nonreactive species (He, Ar, N₂) and CO₂ and H₂S are mainly controlled by the interaction of discharged gas with seawater. The interaction takes place through two different processes: a) Mixing of the 'deep' original gas with the air dissolved in seawater and b) dissolution of a gaseous fraction accompanied by relative enrichment in less soluble components. We estimated that the molar fraction of dissolved gas, although it is variable in time

and space, was always less than 0.3. These low values suggest that the effects induced by the gas dissolution process on the concentration of reactive species H_2, CO and CH_4 are minor and do not affect the geothermometric and geobarometric applications.

2. A strong compositional variation in the relative contents of H_2, CO and CH_4 allows to distinguish samples collected in 2002, after the November 3rd event, from those collected in the 1980s. Such a variation is very probably due to the input of magmatic fluids in the hydrothermal system feeding the exhaling field since the 1980s.

3. Present compositions are compatible with two models:

   a) Submarine emissions are fed by a hydrothermal system with 'atypical' redox conditions, that is more oxidizing than that expected for hydrothermal environments. If true, this could be again explained with the input of magmatic fluids in the system, invoked to explain compositional variations since the 1980s. In this case, we estimated temperatures up to 300 °C and bulk fluid pressures of about 100 bar;

   b) submarine emissions are fed by 'residual' volcanic gases which lost acid gases because of condensation and interaction with seawater. In this case, we estimated temperatures roughly ranging between 350 and 450 °C, which probably refer to the upper zone of the system.

Finally, both models imply the involvement of magmatic fluids in the triggering of the November 2002 degassing event and both suggest that at present temperature and fluid pressure at depth within the hydrothermal-volcanic system of Panarea are higher than during the 1980s. The two models however carry different implications for that which hazard assessment and risk management, especially for that which concerns the possible evolution of the phenomenon.

Model a) may be compatible with a relatively slow dynamics. The explosive event of November 3rd would have been caused by the achievement of critical conditions of overpressurization within the hydrothermal system, which at a certain moment has been unable to release the continuous addition of gas and energy from the magmatic source. In this case, it is probably a new period of "hydrothermal quiescence" of the system, during which the "recharge" from depth may take place similarly to what was observed in the past.

Model b) implies that the November 3rd explosion was triggered by the sudden release from a magmatic body of a gaseous phase separated at depth. Released gases would have flashed the existing hydrothermal system and reached the surface, thus generating the explosion. In this case, the future evolution of the system cannot be foreseen, since it is controlled by the thermal, baric and compositional state of a magmatic body not yet known.

Acknowledgements

We wish to thank the GNV for financial support. The Civil Defense Department of the Italian Government is acknowledged for its logistical help and assistance. We also thank W. C. Evans and J. C. Varekamp for their helpful comments to the first version of manuscript.

REFERENCES

CALANCHI, N., CAPACCIONI, B., MARTINI, M., TASSI, F., and VALENTINI, L. (1995), *Submarine gas-emission from Panarea Island (Aeolian Archipelago); distribution of inorganic and organic compounds and inferences about source conditions*, Acta Vulcanologica 7, 43–48.

CALANCHI, N., PECCERILLO, A., TRANNE, C.A., LUCCHINI, F., ROSSI, P.L., KEMPTON, P., and BARBIERI, M., (2002), *Petrology and geochemistry of volcanic rocks from the Island of Panarea: Implications for mantle evolution beneath the Aeolian Island arc (southern Tyrrhenian Sea)*, J. Volcanol. Geotherm. Res. *115*, 367–395.

CHIODINI, G. (1994), *Temperature, pressure and redox conditions governing the composition of the cold CO_2 gases discharged in north Latium (Central Italy)*, Appl. Geochem. *9*, 287–295.

CHIODINI, G. and CIONI, R. (1989), *Gas geobarometry for hydrothermal systems and its application to some Italian geothermal areas*, Appl. Geochem. *4*, 465–472.

CHIODINI, G. and MARINI, L. (1998), *Hydrothermal gas equilibria: The H_2O-H_2-CO_2-CO-CH_4 system*, Geochim. Cosmochim. Acta. *62*, 2673–2687.

CHIODINI, G., CIONI, R., and MARINI, L. (1993), *Reactions governing the chemistry of crater fumaroles from Vulcano Island, Italy, and implications for volcanic surveillance*, Appl. Geochem. *8*, 357–371.

CHIODINI, G., D'ALESSANDRO, W., and PARELLO, F. (1996), *Geochemistry of gases and waters discharged by the mud volcanoes at Paternò, Mt Etna (Italy)*, Bull. Volcanol. *58*, 51–58.

CHIODINI, G., MARINI, L., and RUSSO, M. (2001), *Geochemical evidence for the existence of high-temperature brines at Vesuvio volcano, Italy*, Geochem. Cosm. Acta *65*, 2129–2147.

D'AMORE, F. and PANICHI, C. (1980), *Evaluation of deep temperature of hydrothermal systems by a new gas-geothermometer*, Geochim. Cosmochim. Acta. *44*, 549–556.

DE ASTIS, G., PECCERILLO, A., KEMPTON, P.D., LA VOLPE, L., and WU TSAI, W. (2000), *Transition from calc-alkaline to potassium-rich magmatism in subduction environments: Geochemical and Sr, Nd, Pb isotopic constraints from the island of Vulcano (Aeolian arc)*, Contrib. Mineral. Petrol. *139*, 684–703.

ELLAM, R.M., HAWKESWORTH, C.J., MENZIES, M.A., and ROGERS, N.W. (1989), *The volcanism of Southern Italy: Role of subduction and relationship between potassic and sodic alkaline magmatism*, J. Geophys. Res. *94*, 4589–4601.

FALSAPERLA, S. and SPAMPINATO, S. (1999), *Tectonic seismicity at Stromboli volcano (Italy) from historical data and seismic records*, Earth Planet. Sci. Lett. *173*, 425–437.

GABBIANELLI, G., GILLOT, G.Y., LANZAFAME, G., ROMAGNOLI, C., and ROSSI, P.L. (1990), *Tectonic and volcanic evolution of Panarea (Aeolian Islands, Italy)*, Marine Geology *92*, 313–326.

GAMBERI, F., MARANI, M., and SAVELLI, C. (1997), *Tectonic, volcanic and hydrothermal features of a submarine portion of the Aeolian arc (Tyrrhenian Sea)*, Marine Geology *140*, 167–181.

GIGGENBACH, W.F. (1975), *A simple method for the collection and analysis of volcanic gas samples*, Bull. Volcanol. *39*, 132–145.

GIGGENBACH, W.F. (1980), *Geothermal gas equilibria*, Geochim. Cosmochim, Acta *44*, 2021–2032.

GIGGENBACH, W.F. (1987), *Redox processes governing the chemistry of fumarolic gas discharges from White Island, New Zeland*, Appl. Geochem. *2*, 143–161.

GIGGENBACH, W.F., *Chemical composition of volcanic gases. In Monitoring and Mitigation of Volcano Hazards* (R. Scarpa, R.I. Tilling, eds.) (Springer, Berlin 1996) pp. 221–256

INGV CATANIA (2002), *Attività scientifiche multidisciplinari iniziate e in corso nell'area di Panarea*, Internal report, 21 November 2002, Catania.

ITALIANO, F. and NUCCIO, P.M. (1991), *Geochemichal investigations of submarine exhalations to the east of Panarea, Aeolian Islands, Italy*, J. Volcanol. Geotherm. Res. *46*, 125–141.

KNIGHT, C.L. and BODNAR, R.J. (1989). *Synthetic fluid inclusions: IX. Critical properties of NaCl-H₂O solutions*, Geochim. Cosmochim. Acta *53*, 3–8.

NERI, G., BARBERI, G., ORECCHIO, B., and ALOISI, M. (2002), *Seismotomography of the crust in the transition zone between the southern Tyrrhenian and Sicilian tectonic domains*, Geophys. Res. Lett. *29*,23 doi:10.1029/2002GL015562.

WANG, C.Y., HWANG, W .T., and SHI, Y. (1989), *Thermal evolution of a rift basin: The Tyrrhenian Sea*, J. Geophys. Res. *94*, 3991–4006.

(Received: February 4, 2003; revised: October 21, 2003; accepted: November 5, 2003)
Published Online First: March 28, 2006

To access this journal online:
http://www.birkhauser.ch

Pure appl. geophys. 163 (2006) 781–807
0033–4553/06/040781–27
DOI 10.1007/s00024-006-0043-0

© Birkhäuser Verlag, Basel, 2006

⌐ **Pure and Applied Geophysics**

Chemical and Isotopic Composition of Waters and Dissolved Gases in Some Thermal Springs of Sicily and Adjacent Volcanic Islands, Italy

Fausto Grassa,[1] Giorgio Capasso,[1] Rocco Favara,[1] and Salvatore Inguaggiato[1]

Abstract—Hydrochemical (major and some minor constituents), stable isotope (δD_{H_2O} and $\delta^{18}O_{H_2O}$, $\delta^{13}C_{TDIC}$ total dissolved inorganic carbon) and dissolved gas composition have been determined on 33 thermal discharges located throughout Sicily (Italy) and its adjacent islands. On the basis of major ion contents, four main water types have been distinguished: (1) a Na-Cl type; (2) a Ca-Mg > Na-SO_4-Cl type; (3) a Ca-Mg-HCO_3 type and (4) a Na-HCO_3 type water. Most waters are meteoric in origin or resulting from mixing between meteoric water and heavy-isotope end members. In some samples, $\delta^{18}O$ values reflect the effects of equilibrium processes between thermal waters and rocks (positive ^{18}O-shift) or thermal waters and CO_2 (negative ^{18}O-shift). Dissolved gas composition indicates the occurrence of gas/water interaction processes in thermal aquifers. N_2/O_2 ratios higher than air-saturated water (ASW), suggest the presence of geochemical processes responsible for dissolved oxygen consumption. High CO_2 contents (more than 3000 cc/litre STP) dissolved in the thermal waters indicate the presence of an external source of carbon dioxide-rich gas. TDIC content and $\delta^{13}C_{TDIC}$ show very large ranges from 4.6 to 145.3 mmol/Kg and from −10.0‰ and 2.8‰, respectively. Calculated values indicate the significant contribution from a deep source of carbon dioxide inorganic in origin. Interaction with Mediterranean magmatic CO_2 characterized by heavier carbon isotope ratios ($\delta^{13}C_{CO_2}$ value from −3 to 0‰ vs V-PDB (Capasso et al., 1997, Giammanco et al., 1998; Inguaggiato et al., 2000) with respect to MORB value and/or input of CO_2-derived from thermal decomposition of marine carbonates have been inferred.

Key words: Thermal waters, chemical and isotope composition, dissolved gases, δ13C, Sicily.

1. Introduction

Thermal springs are quite widespread in Sicily and in the adjacent volcanic islands such as Pantelleria and the Aeolian Archipelago. In this paper we report geochemical data from 33 thermal waters collected from wells and springs (T < 85°C). Thermal reservoirs are essentially linked to the local geological setting as they can be arranged in two main types. One group includes the thermal aquifers hosted within thick carbonate sequences or the metamorphic units. Their location coincides with the presence of main or secondary tectonic faults and thrust zones

[1]Istituto Nazionale di Geofisica e Vulcanologia, Sezione di Palermo, Via Ugo La Malfa 153, 90146, Palermo, Italy. E-mail: f.grassa@pa.ingv.it

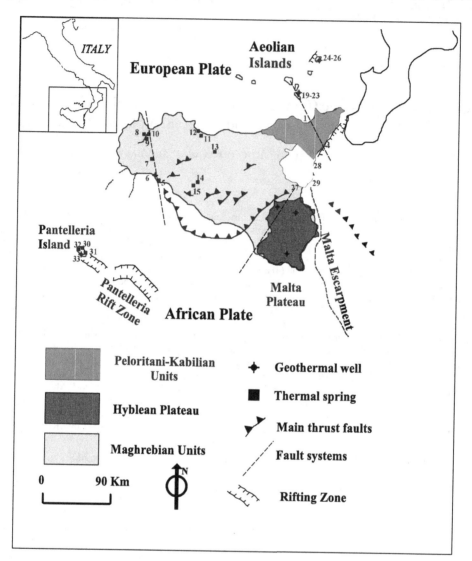

Figure 1
Sketch map of Sicily. Main geological outcropping are highlighted: Peloritani-Kabilian Units (dotted area) consisting of crystalline basement (1–4); Maghrebian Units (Grey area) made of Mesozoic to Cenozoic carbonate basins and platforms building the chain and Messinian evaporites, clays and marly limestones, from Miocene to Plio-Pleistocene; (5–10 Western Sicily thermal district; 11–15 Central Sicily thermal district); Hyblean plateau (16–18) an undeformed foreland, Mesozoic to Cenozoic in age, made mainly of a thick carbonate succession deposited with substantial intercalations of volcanic deposits (dashed area); Vulcano (19–23) and Stromboli (24–26) are the southernmost and the northermost islands of the Aeolian volcanic arc respectively. Mt. Etna (27–29) is a basaltic strato-volcano. Pantelleria Island (30–33) is a trachytes and peralkaline ryolites (pantellerites) strato-volcano. Main thrusts, fault systems and rifting zones are also showed together with the location of thermal springs and geothermal wells. Thermal discharges numbers as reported in table 1.

which are also active areas of recent earthquakes. The second group is linked to the active volcanic systems of Mt. Etna and the Islands of Pantelleria, Vulcano and Stromboli. In Figure 1 the location of our sampling sites is shown. Many authors have already studied these hydrothermal waters for various reasons (ALAIMO et al., 1978; DONGARRÀ and HAUSER 1982; CAPASSO et al., 1992; FAVARA et al., 1998; FAVARA et al., 1999; AIUPPA et al., 2000a; AIUPPA et al., 2000b; CAPASSO et al., 2001; FAVARA et al., 2001a; FAVARA et al., 2001b; CARACAUSI et al., 2004). The aim of this paper is to better define the characteristics of thermal springs from Sicily and its adjacent islands by means of the chemical and isotopic composition ($\delta^{13}C_{CO2}$) of the dissolved gaseous species and to evaluate the interaction processes occurring between gas and thermal reservoirs in different geodynamic environments as well as the origin of dissolved CO_2.

2. Geology and Hydrothermal Systems

Sicily represents the southernmost part of the Alpine orogenic belt at the convergent margin between the African and European plates (Fig. 1). This belt is formed by the overthrusting of nappes from different origins: (1) the Peloritani-Kabylian belt, representing the westernmost prolongation of the Calabrian arc and (2) the Maghrebian units, consisting mainly of carbonates platforms and basins (CATALANO et al., 1996). The emplacement of the nappes started during the Paleogene-Neogene and probably continued up the present.

The Peloritani-Kabylian units outcrop in the northeastern part of the island and consist of crystalline basement overlain by a thin Mesozoic cover. In this sector, two hydrothermal systems are known. The first hydrothermal system is located along the Tyrrhenian coast (T. Vigliatore Spring no.1) and it is probably connected to the Tindari-Giardini lithospheric fault system oriented NW-SE. The second thermal area, in proximity of the Ionian Sea, is characterized by three discharges: two wells (T. Marino 1 and 2, nos. 2 and 3, respectively) and one spring (G. Cassibile no. 4). This hydrothermal system is probably linked to the intersection between the above described NW-SE lithospheric fault and the Messina-Capo Vaticano regional fault, trending NE-SW.

The Maghrebian part of the chain extends from east to west in central and western Sicily and is made of Mesozoic to Cenozoic carbonate basins and platforms which were part of the African margin. In western Sicily a total of seven thermal discharges belonging to three hydrothermal systems (Sciacca, Montevago and Castellammare-Alcamo) have been sampled. In the central part of the island and proceeding from north to south, we sampled the thermal springs of T. Imerese 1 and 2 (nos. 11 and 12, respectively), Sclafani (no. 13) and A. Fitusa 1 and 2 (nos. 14 and 15, respectively).

The Hyblean plateau, undeformed foreland lies in the south-eastern corner of the island. It is an autochthonous sedimentary cover of a thick carbonate succession deposited mainly as a platform facies. These rocks are from Mesozoic to Cenozoic in age and contain substantial intercalations of volcanic deposits occurred during several episodes since Cretaceous time. Thermal discharges in this area include three wells (Bongiovanni no. 16, Pisana no. 17 and Naftia no. 18) with temperature below 30°C. Nonetheless Naftia sampling site lies within the Hyblean foreland, CARACAUSI *et al.* (2003), demonstrated that it belongs to the plumbing system of Mount Etna, because it showed synchronous geochemical variations with other gas manifestations around the volcanic edifice. This link seems to be due because this sampling site is located on the NE-SW Comiso-Etna regional fault system, that is thought to be the prolongation south-westward of the aforementioned Messina-Capo Vaticano lithospheric fault.

The Aeolian Archipelago is a subduction-related volcanic arc where the Adriatic-Ionian lithosphere is subducted below the Corsica-Sardinia block. Volcanic products form a complex series of rocks with a compositional trend from calc-alkaline to shoshonite (HORNIG-KJARSGAARD *et al.*, 1993).

Vulcano and Stromboli are two of the seven islands forming the Aeolian volcanic archipelago, located in the southern Tyrrhenian Sea, in front of Sicilian northern coast, where there are evidences of hydrothermal systems.

The last eruption in Vulcano Island took place during the period 1888–1890, while present activity is characterized by fumarolic activity located both in the "La Fossa" crater area ($100 < T°C < 500$) and on the Vulcano Porto beach (temperature of about 100°C). Several thermal wells are also present with temperatures up to 95°C. For this study, we have sampled five drilled wells (Bartolo no. 19, Currò no. 20, Discarica no. 21, Bambara no. 22 and C.Sicilia no. 23) belonging to the Volcano Porto thermal aquifer (CAPASSO *et al.*, 1992).

Persistent degassing and mild explosive activity throughout the years so-called "Strombolian activity" characterize volcanic activity at Stromboli volcano. Violent explosions and lava flow emissions have periodically occurred, thus interrupting its normal activity. In 1985–1986, a 141 days-eruption took place, the longest one during last century. Another intense eruptive period started at the end of December 2002 and finished at July 2003 (CARAPEZZA *et al.*, 2004). Present hydrothermal activity includes a fumarolic field in the crater area and some thermal wells ($T = 35–45°C$) widespread located along the base of the cone. Waters were collected from Zurro (no. 24), Fulco (no. 25) and Cusolito (no. 26) wells.

Along the northern margin of the Hyblean foreland is located Mount Etna, the largest active strato-volcano in Europe. Mount Etna volcano has been built-in a complex and geodynamically active zone in which extensional processes and collisional movements between the European plate and the African plate coexist (BARBERI *et al.*, 1974). Volcanism started between 0.2 and 0.5 Ma and is characterized by products ranging from tholeiites to alkaline basalts. In the last ten years, several eruptions occurred: in 1991–1993 (473 days) and more recently the last two eruptions took place during the summer of 2001 and during the period October 2002-January 2003. Around the Etnean volcanic edifice three low- temperature springs were sampled (A.Grassa no. 27, Ponteferro no. 28 and S.Venera no. 29).

Pantelleria Island, an active intraplate volcano, is connected to the a rift system within the Mediterranean Sea, caused by crustal thinning and mantle rising (MAHOOD and HILDRETH, 1986; CIVETTA et al., 1988). Located in the Sicily Channel, the most recent activity was a submarine eruption that occurred in 1891. At the present time, hydrothermal activity on the island includes fumaroles and thermal discharges with maximum temperatures close to 100°C. Waters from three thermal springs (Venus Lake no. 30, Gadir no. 31 and Polla 3 no. 32) and one deep well (Nika no. 33) were sampled. Although presently dormant, previous estimates on the CO_2 output from Pantelleria have revealed significant mantle degassing (FAVARA et al., 2001a).

3. Analytical Procedures

Thirty-three waters samples from thermal wells and springs throughout Sicily were collected. Water temperature, electrical conductivity, pH, and Eh have been measured directly in the field. Samples for laboratory analyses have been collected and stored in polyethylene containers. HCO_3 content was determined in the laboratory by volumetric titration with HCl 0.1 N. Chemical analyses of major constituents have been performed by ion-chromatography (Dionex 2000i) on filtered (0.45 μm) and acidified (100 μl HNO_3 Suprapure) samples (Na, K, Mg and Ca) and untreated samples (F, Cl, Br, NO_3, SO_4). Dissolved silica was determined by colorimetric techniques, using a spectrophotometer Shimadzu UV 1601. Isotope determinations (D/H and $^{18}O/^{16}O$) on water samples were performed by equilibration technique (EPSTEIN and MAYEDA, 1952 for oxygen) and water reduction (hydrogen production by using granular Zn, KENDALL and KOPLEN, 1982) respectively. Measurements were carried out using a Finnigan Delta Plus mass spectrometer (Hydrogen) and an automatic preparation system coupled with an AP 2003 IRMS (Oxygen). Analytical precision for each measurement is better than 0.2‰ for $\delta^{18}O$ and 1‰ for δD. In order to determine the chemical composition of the dissolved gases and the isotopic composition of total dissolved inorganic carbon (TDIC) in thermal waters we utilized the procedures proposed by CAPASSO and INGUAGGIATO (1998) and FAVARA et al., (2002) respectively. The first technique is based on the partitioning equilibrium of gaseous species between the liquid and the gas phase. The analytical determinations of dissolved gases were performed using a Perkin-Elmer 8500 gas chromatograph with argon as carrier and equipped with a double detector (TCD-FID). The method for determination of $\delta^{13}C_{TDIC}$ is based on the chemical and physical stripping of CO_2. The stripped gas is purified by means standard procedures. The isotopic values were measured using a Finnigan Delta Plus mass spectrometer and the results are reported in δ ‰ vs. V-PDB standard. The standard deviation of $^{13}C/^{12}C$ ratio is $\pm$ 0.2 ‰.

4. Chemical Composition of Thermal Waters

Table 1 reports the results of major and some minor chemical constituents dissolved in the collected samples together with physico-chemical parameters. Thermal waters range from low (19.0 °C) to high (83.7°C) temperature, TDS contents (Total Dissolved Solid) range between 783 and 34433 mg/l, pH values are in the range from slightly acid (5.4) to mildly alkaline (9.0), while redox conditions vary from slightly aerobic (75 mV) to extremely reducing (−351 mV). Such a variability in physical and chemical parameters corresponds to a wide compositional range in water chemistry, as can be observed from the Langelier-Ludwig diagram (Fig. 2) where major ions are plotted.

4.1 Na-Cl waters

Most of the collected samples are mainly of the Na-Cl-type thus falling within the chloride-sulphate alkaline field. For most of them, their peculiar chemical feature is to contain a noticeable contribution of seawater often due to location of hydrothermal systems very close to the coastal areas. The origin of solutes in this group of waters generally reflects a seawater dilution with low-salinity waters, such as meteoric waters even if, in some areas, other different mineralization processes should be invoked.

As regard Bartolo (no. 19) and C. Sicilia (no. 23) wells at Vulcano Island, they belong to the volcanic waters group defined by GIGGENBACH (1991) as already highlighted by CAPASSO et al., (1992) and AIUPPA et al. (2000b). They are fed by an aquifer that is likely the result of a mixing between meteoric waters and a deep hydrothermal end-member (T $\sim$ 200°C, BOLOGNESI and D'AMORE, 1993) derived from high-temperature magmatic vapors. On the contrary, CHIODINI et al. (1996) proposed a different hydrologic model explaining such chemical features as the result of a mixing between the meteoric end-member and fumarolic steam condensates flowing along the northern flank of the crater.

FAVARA et al., (2001b) indicated for thermal waters of Sciacca area a seawater contribution up to 50% (no. 6), while sample no. 5 keeping the same ion ratios, but lower salinity, is diluted by a factor 2 with meteoric waters. However, the same authors suggested that in these samples the modification of Ca/Cl, Mg/Cl, Br/Cl and SO_4/Cl molar ratios with respect to seawater composition is due to water-rock and biogenic processes occurring within the reservoir. On the basis of the sulfur isotopic

Table 1

No.	Samples	T	EC	pH	Eh	SiO₂	Na	K	Mg	Ca	F	Cl	Br	SO₄	Alk	TDS	$\delta^{18}O$	δD	$\delta^{13}C_{TDIC}$
1	T. Vigliatore[a]	32.9	4.4	6.72	-265	102	1100	30	201	132	4.0	123	0.2	56	4078	5825	-6.9	-27	-3.8
2	T.Marino 1[b]	31.2	11.3	6.15	-235	32	2546	104	316	666	6.4	4060	13.4	1651	1421	10816	-4.1	-23	-1.1
3	T.Marino 2[b]	23.8	1.7	6.14	-156	27	88	12	127	282	1.1	144	0.5	219	1147	2047	-6.6	-34	-3.6
4	G.Cassibile[a]	40.5	32.8	6.44	-250	29	7028	241	679	893	5.1	12394	43.5	1732	1415	24460	-1.0	-2	0.1
5	Molinelli[a]	32.1	16.7	6.43	-124	41	3238	171	220	603	5.9	5938	38.0	338	354	10946	-3.7	-22	-0.5
6	T. Selinuntine[b]	54.0	32.5	6.42	-268	43	6494	337	472	1249	7.2	12777	117.8	739	470	22705	-1.7	-14	-2.2
7	Montevago[a]	39.2	1.7	6.89	-303	22	179	15	84	189	3.2	312	1.0	480	275	1561	-6.7	-35	-4.0
8	T.Segestane[a]	44.2	2.6	6.81	-47	26	231	13	70	204	3.6	418	1.7	480	253	1699	-7.2	-38	-4.6
9	Gorga 1[a]	48.3	2.7	7.07	-145	28	220	12	56	177	4.4	351	1.2	500	238	1587	-7.5	-39	-4.4
10	Gorga 2[a]	49.6	3.0	7.15	-30	n.d.	395	16	83	236	1.5	663	2.1	528	296	2221	-7.5	-38	-4.9
11	T. Imerese 1[a]	41.0	22.4	7.04	-30	n.d.	3712	147	293	445	9.5	6469	31.2	813	287	12207	-3.9	-21	-3.4
12	T. Imerese 2[a]	41.3	22.5	7.02	-50	n.d.	3984	126	291	471	18.0	6736	27.2	745	287	12685	-3.9	-19	-3.7
13	Sclafani Bagni[a]	32.6	15.3	6.65	-351	14	3242	60	194	244	45.5	5398	43.3	214	281	9736	-5.0	-36	-1.8
14	A.Fitusa1[a]	25.2	4.6	7.22	-265	n.d.	805	18	14	25	6.5	738	4.7	254	799	2665	-4.5	-31	-1.5
15	A.Fitusa2[a]	20.7	3.3	7.49	-78	n.d.	596	14	19	60	3.6	491	5.7	433	567	2190	-3.9	-33	-2.0
16	Bongiovanni[b]	28.4	1.0	8.62	-24	43	217	9	4	4	1.1	103	0.6	85	305	776	-6.3	-35	-5.3
17	Pisana[b]	22.0	1.0	7.39	-277	41	146	5	34	43	1.4	157	0.1	45	390	861	-6.2	-31	-7.8
18	Naftia[b]	22.8	3.0	6.03	-37	102	657	15	105	134	0.2	374	2.9	237	1617	3244	-4.8	-25	-1.1
19	Bartolo[b]	84	n.m.	7.09	n.m.	n.d.	463	206	4	20	11.3	121	0.9	713	549	2087	-3.3	-26	1.4
20	Currò[b]	42.0	n.m.	6.43	n.m.	113	266	278	159	444	0.9	290	b.d.l.	1614	903	4068	-5.3	-27	0.6
21	Discarica[b]	42.6	n.m.	6.58	n.m.	n.d.	657	432	361	9	18.6	423	1.5	940	2147	4990	-6.4	-32	1.1
22	Bambara[b]	23.2	n.m.	5.41	n.m.	n.d.	68	27	20	128	3.2	67	b.d.l.	293	168	774	-5.5	n.m.	-2.9
23	C. Sicilia[b]	51.5	n.m.	7.43	n.m.	n.d.	1521	516	94	142	8.1	1787	b.d.l.	1669	403	6141	1.4	-8	0.0
24	Zurro[b]	34.8	42.5	7.78	75	55	10452	571	1221	609	14.6	18518	64.9	2435	336	34276	0.4	6	1.6
25	Fulco[b]	38.3	5.3	6.70	57	104	885	153	215	210	17.6	1377	8.7	469	1104	4544	-6.1	-31	2.6
26	Cusolito[b]	43.2	45.4	6.47	-206	n.d.	6617	407	791	459	42.3	12305	46.7	1580	525	22773	-1.7	-9	2.3
27	A Grassa[a]	18.0	1.7	6.13	n.m.	103	173	18	129	142	b.d.l.	77	b.d.l.	30	1427	2099	-8.6	-48	-0.7
28	Ponteferro[a]	19.6	1.8	7.40	52	63	274	46	100	78	1.7	188	0.2	241	848	1840	-7.9	-43	1.2
29	S. Venera[a]	22.5	10.6	7.11	2	24	2174	46	195	92	2.2	3531	17.1	221	689	6992	-6.0	-30	-10.0
30	Venus Lake[a]	19.0	31.6	9.02	73	32	6758	321	130	25	8.1	8317	34.0	718	2745	19442	2.2	9	2.8
31	Gadir[a]	53.1	8.6	6.36	-77	184	2667	134	122	75	11.2	3976	15.0	355	1147	8687	-4.6	-21	-2.1

Table 1

(Contd.)

No.	Samples	T	EC	pH	Eh	SiO$_2$	Na	K	Mg	Ca	F	Cl	Br	SO$_4$	Alk	TDS	δ^{18}O	δD	δ^{13}C$_{TDIC}$
32	Polla3[a]	55.5	9.6	6.29	n.m.	196	2963	254	116	75	12.1	4243	17.4	344	1525	9745	-2.7	-13	-2.7
33	Nika[b]	81.0	13.4	8.81	46	108	6406	143	3	10	121.1	8161	28.7	309	2184	17599	-4.3	-24	-6.0
	Seawater	n.m.	n.m.	n.m.	n.m.	n.m.	12252	469	1494	441	0.0	21802	79.1	3072	159	39767	1.0	10	-8.0

a = spring b = well. Water temperature is expressed in °C, EC (= electrical conductivity) in mS/cm, pH in pH units, Eh in mV, concentrations and TDS in mg/l. Alk = total Alkalinity reported as HCO3. Isotopic compositions are expressed in delta per mille vs. V-SMOW for d18O and dD and vs. V-PDB for d13C. n.m. = not measured; n.d. = not determined

1–4 Peloritani-Kabylian units; 5–10 Western Sicily thermal district; 11–15 Central Sicily thermal district; 16–18 Hyblean Plateau; 19–23 Vulcano Island; 24–26 Stromboli Island; 27–29 Etnean aquifer; 30–33 Pantelleria Island

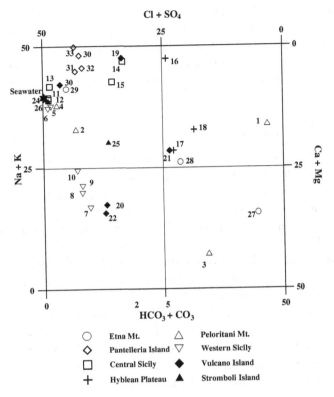

Figure 2
Langelier-Ludwig classification diagram.

ratios ($^{34}S/^{32}S_{SO4}$ = +31.7 ‰ and $^{34}S/^{32}S_{H2S}$ = +3.1 ‰) DONGARRÀ and HAUSER (1982) recognized that bacterial reduction of sulphate takes place, in waters of spring no.6 as confirmed also by an intense H_2S smell.

T. Imerese thermal waters (Nos. 11 and 12) show molar ratios close to those of seawater but Ca, Mg and alkalinity. On the basis of Br and Cl contents, behaving as conservative elements, a seawater contribution about 30% has been estimated. After scaling the concentrations related to the seawater composition, these springs are significantly depleted in Mg and, on the contrary, enriched in calcium. Dissolution of calcite and dolomitisation occurring within the carbonate reservoir feeding these springs seem to be the processes controlling the earth-alkaline element and bicarbonate chemistry, being Ca + Mg/Cl ratio close to that of seawater.

Samples nos. 30-33 collected at Pantelleria Island fall in a restricted field characterized by a relative Na + K and C-species enrichment with respect to seawater. Major cations distribution is due to dissolution/precipitation processes releasing Na from Na-rich volcanic rocks and removing Ca and Mg due to formation of secondary carbonate minerals (e.g. dolomite and aragonite) as supported by saturation indices (DUCHI et al., 1994; GIANNELLI and GRASSI, 2001). It is reasonable to think that carbon enrichment results from the dissolution of large amount of magmatic-derived CO_2-gas in the water (CO_{2aq}, HCO_3 and CO_3, depending on the activity of H^+).

4.2 Chloride-sulphate alkaline-earth waters

Waters from Castellammare-Alcamo and Montevago thermal districts are chloride-sulphate alkaline-earth, showing a narrow range in almost all chemical constituents except for Na and Cl. For their origin a mixing among three different end members have been invoked (FAVARA et al., 1998; 2001b): the deep thermal reservoir hosted within Tertiary platform carbonate rocks, is fed by a mixing between a Calcium-(Mg)-bicarbonate water type, originating from the dissolution of carbonate rocks and a sulphate-rich water deriving from the leaching of evaporite rocks, mainly made of gypsum. During the ascent towards surface, deep end member is slightly contaminated by seawater (1–3%).

Two wells (nos. 20 and 22) from Vulcano Island belong also to this group. Previous surveys on the origin of solutes in these groundwaters (CAPASSO et al., 1992; AIUPPA et al., 2000b) highlighted an intermediate composition between bicarbonate peripheral water-type and steam-heated water-type, recognized by GIGGENBACH (1991) in geothermal areas and in volcanic systems.

4.3 Bicarbonate-alkaline waters

Thermal waters belonging to this group (nos. 1, 17, 18, 21 and 28) are characterized by Na-HCO_3 chemistry, high dissolved carbon and silica concentrations. The main geochemical processes governing the composition of major constituents dissolved in these waters could be, at least an intense water-silicate minerals interaction coupled with an inflow of CO_2-rich gas in groundwater system. The latter process will enhance the weathering of silicate minerals as reported in the following hydrolysis reaction:

$$Silicates + CO_2 + H_2O <> Cations + HCO_3^- + H_4SiO_4 + Alteration\ Minerals$$

Furthermore, cation exchange processes between water and Na-feldspars and/or clay minerals (DREVER 1982) will lead to a relative enrichment in alkali metals with respect to alkaline-earth elements. Another possible explanation for the alkali metals enrichment could be also due to the removal of alkaline-earth elements from solution during formation of carbonate or silicate alteration minerals such as calcite, zeolite or others. This is the case of samples 1 and 18 showing very large amounts of carbon species as bicarbonate ions as well as dissolved carbon dioxide that have achieved saturation state with respect to both calcite and dolomite.

Following a Cl-SO$_4$-HCO$_3$ diagram (GIGGENBACH 1991) for geothermal waters classification, Discarica well waters (no. 21) have been ascribed as peripheral bicarbonate groundwaters (AIUPPA et al, 2000b). In this well, during weathering, almost all dissolved carbon resulting from the interaction of an external source of CO$_2$ with the aquifer is converted into bicarbonate. This is consistent with the location of the well in an area characterized by of volcanic-derived CO$_2$ anomalous degassing area. On the contrary, sample 17, hosted within the Hyblean plateau, is the result of a mixing between local bicarbonate-alkaline-earth water, typical for carbonate aquifers and a Na-Cl brine (GRASSA, 2002).

4.4 Bicarbonate-alkaline-earth waters

Only two samples, well no.3 in North-Eastern Sicily and spring no. 27 belonging to the Etnean aquifer fall within the bicarbonate-alkaline-earth field being the sum of HCO$_3$ and Ca + Mg > 78% in meq/l. These thermal discharges are characterized by water temperature less than 24°C, pH values slightly acidic (6.1) and TDS values < 2.0 g/l. In order to explain such low salinity and their isolated behavior with respect to the other sample we can invoke a weak water-rock interaction due to short residence time and/or low aquifer temperature. However, high dissolved carbon contents mainly as bicarbonate ion, reflect the noticeable interaction with CO$_2$-rich fluids.

5. Water Isotope Composition

The results of hydrogen and oxygen isotope analyses are reported in Table 1. Further, they are plotted in Figure 3 together with two reference lines: the first line is the so-called MWL (CRAIG, 1961) identifying the world meteoric water line while the second one is the Mediterranean meteoric water line (MMWL) defined by GAT and CARMI (1970). The latter differs because its deuterium excess value is equal to +22‰. As can be seen, almost all the samples fall within the two reference lines reported, thus indicating a marked meteoric origin. This group includes thermal manifestations from Castellammare-Alcamo and Montevago areas for which FAVARA et al., (2001b) found that they are fed by their respective local meteoric waters, thermal discharges from Hyblean Plateau (GRASSA, 2002) and from Etnean reservoirs (ANZA et al., 1989; ALLARD et al., 1997).

All the other samples, but one, lie to the right of the MWL reference lines, thus indicating a different origin of water and/or that other groundwaters have modified their pristine isotopic composition due to: (1) mixing between groundwaters and heavy isotope-rich waters such as Mediterranean seawater ($\delta D = +10$ e $\delta^{18}O = +1$), condensate fumarolic steam on Vulcano Island (CAPASSO et al., 1997); (2) secondary processes causing an isotopic fractionation as a consequence of

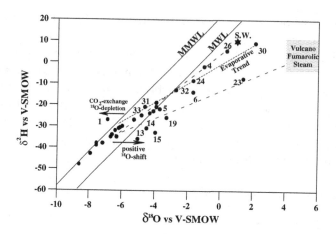

Figure 3

Oxygen-18 vs Deuterium plot showing Meteoric Water Line (MWL) of Craig (1961) and the Mediterranean Meteoric Water Line (MMWL) of GAT and CARMI (1970). Water stable isotopes (δ^2H and $\delta^{18}O$) suggest almost all thermal waters to be strictly meteoric in origin or resulting from a mixing between meteoric water and heavy-isotope end members such as seawater, evaporated waters or fumarolic condensates. In some samples, the effects of equilibrium processes between thermal waters and rocks (positive ^{18}O-shift) or thermal waters and CO_2 (negative ^{18}O-shift) on $\delta^{18}O$ values are also displayed.

evaporation process (Deuterium- and ^{18}O-enrichment); (3) ^{18}O-shift during water-gas rock interaction. The latter process seems to take place at A. Fitusa 1 and 2, and Sclafani springs (no.13, 14 and 15, respectively) showing only a moderate isotopic shift. In order to account for the heavy isotopes enrichment in the T. Selinuntine sample (No. 6) FAVARA *et al.* (2001b) proposed two realistic geochemical models. The first hypothesis predicts the presence of a deep thermal end-member, entirely marine in origin, which during its rise towards the surface is diluted by regional meteoric waters. On the contrary the second model suggests that the mixing between local meteoric recharge and seawater takes place within the deep reservoir. Due to long residence times within the thermal reservoir, these two proposed end-members (marine and mixed marine-meteoric) have modified their isotopic composition as a consequence of exchange processes of oxygen between water and minerals. Sample 5 belonging to Sciacca thermal district has been identified as a mixing term between T. Selinuntine thermal water-type and meteoric water.

As regards Venus Lake sample (no. 30), showing a δD value close to Mediterranean seawater but a slightly enriched 18-Oxygen, evaporation processes should be taken into account. In fact, the slopes of the least squares fitting line among four samples (nos. 30-33) collected at Pantelleria Island is close to 4, thus confirming the occurrence of non-equilibrium processes affecting their water isotopic composition.

CAPASSO *et al.*, (1992 and 1999) reported that chemical and isotopic features of waters from C. Sicilia well (no. 23) reflect the contribution of fumarolic steam

(CAPASSO *et al.*, 1997). Furthermore, the authors have shown that temporal variations in C. Sicilia well waters are clearly correlated with isotopic changes of the acid fumarolic condensate emitted from La Fossa crater on Vulcano Island. During 1996 well no. 23 changed its chemical and isotopic composition showing a trend towards the point representative of the fumarolic condensate as a consequence of an increase in the gas/steam ratio at the fumarolic field. The amount of fumarolic condensate has been estimated in the range between 80 and 90% (CAPASSO *et al.*, 1999). Bartolo well (no. 19) shows an intermediate composition between local meteoric waters and well no. 23 water-type.

As regard the only sample falling on the left with respect to the MMWL line (sample no. 1) this could be the result of an ^{18}O-depletion of waters probably due to an isotopic exchange between water and CO_2 at low-temperature. In fact, high CO_2 content dissolved (more than 1000 cc/liter STP) and the presence of CO_2-rich free gas would probably lead a depletion of heavy oxygen isotope in the water. DREVER (1982) and PAUWELS at al., (1997) found such an isotopic trend in CO_2-rich thermal waters from Tuscany (Italy) and Mont-Dore (France) respectively.

As previously pointed out, seawater input heavily influences water TDS, major ion content as well as the isotopic composition. Because δD is not usually involved in exchange processes between water and gas/rock, a seawater contribution can be easily identified from the δD-Cl diagram (Fig. 4). It clearly appears that samples from

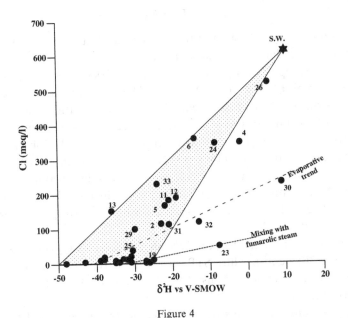

Figure 4

Chloride content vs. Deuterium plot: The dotted area indicates waters that are mixtures of meteoric water and sea water. Deuterium enrichment due to evaporative processes or to mixing with fumarolic steam are represented by dashed and dotted line respectively.

Sciacca thermal basin (nos. 5 and 6), T. Imerese (nos. 11 and 12) Stromboli Island (nos. 24–26), S. Venera spring (no. 29) belonging to the etnean aquifer, Nika well (no. 33) at Pantelleria Island as well as G.Cassibile (no. 4) and T. Marino 1 (no. 2) belonging to the metamorphic complex (North-Eastern Sicily) are terms of a simple mixing between meteoric water and seawater. In the same diagram, on one hand Venus Lake (no.30), Gadir spring (no. 31) and Polla 3 (no. 32) samples, and on the other hand C. Sicilia well (no. 23) follow two different isotopic trends thus confirming that their excess in 18-oxygen is due to processes other than seawater addition.

6. Dissolved Gas Geochemistry

The chemical composition of the dissolved gaseous species has been utilized in several geochemical studies as tracer of interaction processes occurring between deep fluids and thermal reservoirs (KING, 1986; THOMAS, 1988; D'ALESSANDRO *et al.*, 1997; CAPASSO *et al.*, 2001; FEDERICO *et al.*, 2002; INGUAGGIATO *et al.*, in press). In this study, we have analyzed the chemical composition of the dissolved gases collected from some thermal discharges both wells and springs in Sicily. In Table 2 are reported the concentrations of dissolved gaseous species expressed both as cc/litre STP and as partial pressures (Pi) of gases in equilibrium with water calculated by using Henry's law as follows:

$$Pi = K_{Hi}\chi_i$$

where K_{Hi} is the Henry's constant of the gas computed at sampling temperature and χ_I is its respective molar fraction in solution.

The same table reports also the concentration of dissolved gases in a water sample equilibrated with air (Air-Saturated Water = ASW). All the samples have dissolved CO_2 contents higher than those at equilibrium with the atmosphere (ASW), thus suggesting a direct input of CO_2-rich fluids. Figure 5 plots the content of dissolved O_2-N_2-CO_2 expressed as cc/litre STP together with the ASW. We observe that almost all the samples are characterized by an O_2/N_2 ratio lower than ASW. It is reasonable to think oxygen consumption due to reducing redox conditions is the process causing the relative N_2-enrichment. It is also possible to distinguish different degrees of interaction between thermal and CO_2-rich deep fluids, corresponding to two main different groups of samples:

Strong interaction leading to extremely high dissolved CO_2 contents in the range from 330 up to 3019 cm^3 per litre STP (nos. 1, 3, 4, 18, 20–22, 27 and 32);

Medium-low interaction with dissolved CO_2 contents less than 230 cm^3 per litre STP (all the other samples).

Such a wide range in the dissolved CO_2 contents can be related to geodynamic and structural context affecting the CO_2 discharges as well as the hydrogeological, physical and chemical characteristics of the aquifer feeding the springs.

Table 2

No.	Sample	Hecc/litre STP	H2cc/litre STP	O2cc/litre STP	N2cc/litre STP	COcc/litre STP	CH4cc/litre STP	CO2cc/litre STP	pHe	pH2	pO2	pN2	pCO	pCH4	pCO2
1	T.Vigliatore	b.d.l.	b.d.l.	0.36	2.70	b.d.l.	2.66E-03	1007	—	—	1.26E-02	1.80E-01	—	8.47E-05	1.42E+00
3	T.Marino2	b.d.l.	b.d.l.	0.78	3.68	b.d.l.	7.92E-03	1080	—	—	2.38E-02	2.17E-01	—	2.15E-04	1.22E+00
4	G.Cassibile	b.d.l.	b.d.l.	0.18	2.25	b.d.l.	b.d.l.	330	—	—	6.87E-03	1.61E-01	—	—	5.44E-01
5	Molinelli	4.80E-03	6.56E-03	0.95	10.83	2.57E-05	1.42E-02	68	4.89E-04	3.39E-04	3.24E-02	7.07E-01	1.15E-06	4.40E-05	9.20E-02
6	T.Segestane	3.97E-03	b.d.l.	0.10	10.59	9.76E-05	3.69E-02	23	3.98E-04	—	3.82E-03	7.84E-01	5.01E-06	1.38E-03	4.00E-02
7	Montevago	4.32E-03	4.26E-04	0.82	17.88	9.69E-05	1.39E-02	21	4.37E-04	2.26E-05	3.09E-02	1.27E+00	5.20E-06	4.86E-04	3.40E-02
8	Gorga 1	3.02E-03	b.d.l.	0.09	15.90	1.27E-04	4.61E-02	18	3.00E-04	—	3.60E-04	1.22E+00	6.81E-06	1.83E-03	3.40E-02
9	Gorga 2	3.93E-03	5.31E-04	0.30	11.04	1.46E-04	5.71E-02	30	3.90E-04	2.86E-05	1.25E-02	8.47E-01	8.10E-06	2.27E-03	5.80E-02
10	T. Selinuntine	4.85E-04	3.88E-01	0.20	12.59	1.52E-04	4.73E-02	74	4.76E-05	2.09E-02	8.89E-03	9.89E-01	7.68E-06	3.94E-04	1.56E-01
11	T. Imerese 1	2.89E-03	b.d.l.	0.66	12.20	2.11E-04	1.08E-02	38	2.91E-04	—	2.57E-02	8.91E-01	1.06E-05	1.45E-04	6.45E-02
12	T. Imerese 2	2.40E-03	b.d.l.	1.34	13.18	b.d.l.	3.97E-03	33	2.42E-04	—	5.21E-02	9.58E-01	—	1.45E-04	5.56E-02
13	Sclafani Bagni	b.d.l.	3.96E-04	b.d.l.	7.79	b.d.l.	1.07E-02	20	—	2.06E-05	—	5.16E-01	—	3.41E-04	2.80E-02
14	A.Fitusa1	b.d.l.	1.49E-03	0.31	8.71	b.d.l.	8.34E-02	15	—	7.46E-05	9.60E-03	5.24E-01	—	2.32E-03	1.80E-02
15	A.Fitusa2	3.22E-03	b.d.l.	0.12	4.36	2.52E-04	1.85E-01	34	3.25E-04	—	3.51E-03	2.45E-01	1.06E-05	4.71E-03	3.60E-02
16	Bongiovanni	b.d.l.	b.d.l.	3.94	16.07	b.d.l.	6.17E+01	0.7	—	—	1.26E-01	9.94E-01	—	4.89E-05	8.76E-04
17	Pisana	3.72E-03	b.d.l.	b.d.l.	17.77	b.d.l.	4.25E-01	18	3.77E-04	—	—	1.04E+00	—	1.65E+00	1.99E-02
18	Naftia 1	2.01E-02	b.d.l.	0.35	23.09	b.d.l.	8.38E-03	3019	2.03E-03	—	1.07E-02	1.35E+00	—	1.14E-02	3.37E+00
19	Bartolo	b.d.l.	2.87E-03	1.22	14.90	b.d.l.	3.63E-04	24	—	1.44E-04	6.59E-02	1.18E+00	—	4.09E-04	7.59E-02
20	Currò	1.00E-04	b.d.l.	0.16	9.56	b.d.l.	2.54E-04	367	7.90E-06	—	6.00E-03	6.90E-01	—	1.32E-05	6.10E-01
21	Discarica	6.10E-05	b.d.l.	0.22	4.38	b.d.l.	3.22E-03	506	6.00E-06	—	9.00E-03	3.30E-01	—	9.30E-06	9.80E-01
22	Bambara	7.20E-05	b.d.l.	0.16	9.57	b.d.l.	3.49E-03	692	7.30E-06	—	5.00E-03	6.10E-01	—	8.76E-05	9.00E-01
23	C. Sicilia	1.60E-04	b.d.l.	0.42	10.86	b.d.l.	4.26E-03	17	1.60E-05	—	1.80E-02	8.50E-01	—	1.66E-04	4.00E-02
24	Zurro	b.d.l.	b.d.l.	b.d.l.	11.01	b.d.l.	b.d.l.	139	—	—	—	8.08E-01	—	—	2.41E-01
25	Fulco	b.d.l.	b.d.l.	0.73	13.78	b.d.l.		230	—	—	2.74E-02	9.69E-01	—	1.57E-04	3.62E-01
26	Cusolito	2.37E-04	6.92E-03	0.08	11.11	b.d.l.	9.39E-04	19	2.46E-05	3.62E-04	2.92E-03	7.55E-01	—	3.09E-05	2.80E-02
27	A Grassa	b.d.l.	b.d.l.	0.54	2.96	3.69E-05	1.84E-03	1103	—	—	2.45E-03	1.80E-01	1.39E-06	5.02E-05	1.21E+00
28	Ponteferro	b.d.l.	b.d.l.	4.38	14.71	b.d.l.	3.89E-04	23	—	—	1.24E-01	8.13E-01	—	9.66E-06	2.30E-01
29	S.Venera	b.d.l.	b.d.l.	b.d.l.	4.14	b.d.l.	2.73E-01	31	—	—	—	2.50E-01	—	7.15E-01	3.35E-02
30	Venus Lake	b.d.l.	b.d.l.	1.60	8.53	b.d.l.	4.86E-03	3	—	—	5.11E-02	5.25E-01	—	1.39E-04	4.23E-03
31	Gadir	6.84E-04	b.d.l.	2.13	6.17	b.d.l.	8.91E-04	194	7.63E-05	—	1.05E-01	5.47E-01	—	4.16E-05	4.55E-01

Table 2
(Contd.)

No.	Sample	Hecc/ litre STP	H2cc/ litre STP	O2cc/ litre STP	N2cc/ litre STP	COcc/ litre STP	CH4cc/ litre STP	CO2cc/ litre STP	pHe	pH2	pO2	pN2	pCO	pCH4	pCO2
32	Polla3	b.d.l.	b.d.l.	1.75	6.48	b.d.l.	2.39E-03	580	-	-	7.79E-02	5.12E-01	-	1.01E-04	1.25E+00
33	Nika'	b.d.l.	1.84E-01	0.06	10.07	1.61E-03	1.87E-02	6	-	9.37E-03	3.16E-03	8.08E-01	1.00E-04	9.07E-04	1.89E-02
	ASW	4.55E-05	-	6.60	12.30	-	-	0.31	5.50E-06	-	2.10E-01	7.80E-01	-	-	3.90E-03

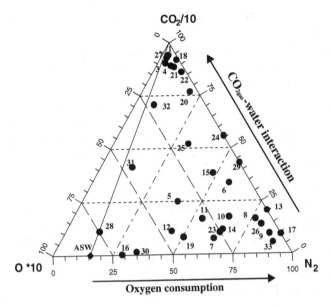

Figure 5

CO_2-N_2-O_2 triangular diagram of the dissolved gas in the thermal waters from Sicily. Data have been plotted as cc of dissolved gas per litre of water at STP conditions (T = 25°C and P = 1 atm). ASW = air-saturated water. N_2/O_2 ratios higher than ASW suggest dissolved oxygen consumption. High CO_2 contents (up to 3000cc/litre STP) dissolved in the thermal waters indicate the presence of an external source of carbon dioxide-rich gas.

Almost all the waters belonging to first group are connected to active volcanic areas and their high dissolved CO_2 contents are the consequence of intense interaction processes between groundwaters and deep volcanic fluids emitted from anomalous soil CO_2-degassing zones (ANZÀ et al., 1989; ALLARD et al., 1997; BADALAMENTI et al., 1988; D'ALESSANDRO and PARELLO, 1997; FAVARA et al., 2001a). The highest dissolved CO_2 contents, were observed at Naftia well, located within a zone characterized by high soil degassing extending along the preferential NE-SW direction and following the Comiso-Etna deep fault (De GREGORIO et al., 2002).

Among thermal discharges connected to active volcanic areas, only few samples (nos. 19, 23, 24, 28–30 and 33) display low pCO_2 values. As regards Zurro well (no. 24), Ponteferro (no. 28) and S.Venera (no. 29) springs, this is probably due to their location in volcanic areas having low CO_2 fluxes. Moreover, sample no. 28 is fed by a huge aquifer, thus causing a dilution effect on the CO_2 dissolution process. On the contrary, Bartolo (no. 19) and C. Sicilia (no. 23) wells at Vulcano Island seem to reflect the significant contribution of fumarolic condensate containing only low amount of dissolved gases.

Although the carbon dioxide is an extremely reactive gas having high solubility in water (up to 760 cc per litre of pure water at STP conditions), physico-chemical conditions of groundwater can affect also the dissolution process of CO_2 into the thermal aquifers being the relative abundance of the dissolved inorganic carbon species (CO_{2aq}, HCO_3^- and CO_3^{2-}) strongly depending on pH values of the waters. For these reasons, in waters with alkaline pH values, the CO_2 dissolution processes can be masked due to the conversion of carbon dioxide into dissolved bicarbonate (HCO_3^-) and carbonate (CO_3^{2-}) ions. This is the case of two thermal discharges (Venus Lake, no.30 pH = 9.0 and Nikà, no.33 pH = 8.8) located within the volcanic island of Pantelleria. Although they show the lowest pCO_2 values, their high dissolved inorganic carbon (DIC) contents (51.0 and 38.1 mmol/l for Venus Lake and Nikà well respectively) suggest the dissolution process of CO_2 in these two samples is also marked.

It is noteworthy that CO_2 is the dominant dissolved gas phase (up to 1080 cc/litre STP) also in three thermal discharges (samples nos 1, 3 and 4) located in northeast Sicily, a non-volcanic area, where the occurrence of thermal basins is connected to the existence of a lithospheric fault system. Deep tectonic discontinuities seem to play an important role in controlling the rising both of thermal waters and deep gases towards the surface. In this area, the presence of moderate aquifers probably coupled with high gas fluxes determines high dissolved CO_2 contents.

On the contrary, thermal waters in Western Sicily (Sciacca, Montevago and Castellammare-Alcamo thermal basins), hosted within carbonate aquifers with thickness of about 3000 m show lower amounts of dissolved CO_2 (up to 74 cc/litre STP) probably as a consequence of a dilution effect, rather than due to low gas fluxes, as better described below.

On the CH_4-N_2-CO_2 ternary diagram (Figure 6), water samples are arranged in two main groups: the first group including all the hydrothermal systems hosted within carbonate reservoirs as well as S.Venera spring (no. 29) highlights the presence of light hydrocarbons-rich gases such as methane produced under favorable low-potential redox conditions. The second group consists of CO_2-rich thermal waters both belonging to volcanic areas and hosted within metamorphic reservoirs.

Dissolved He contents greater than ASW (Tab. 2) recognized in many samples could be related to the adding of non-atmospheric helium radiogenically-produced within the crust and/or He degassed from the mantle.

Dissolved helium isotope ratios in the range from 4.06 to 4.23 Ra in the thermal waters from Stromboli Island suggests a clear mantle-derived origin (INGUAGGIATO and RIZZO, 2004). ^3He and ^4He contents for Etnean samples (nos. 27 and 28) indicate a simple mixing trend between ASW and a magmatic endmember with a negligible crustal contamination, having helium isotope ratio of 6.7 ± 0.4 Ra (ALLARD *et al.*, 1997).

Measured $^3He/^4He$ ratios (from 0.7 to 2.8 Ra) and the estimated mantle-derived He flux have revealed for the hydrothermal systems in Western Sicily (Castellammare-Alcamo, Montevago and Sciacca thermal basins) the occurrence of a mantle-derived degassing from magmas probably intruded in the crust (CARACAUSI *et al.*, 2004).

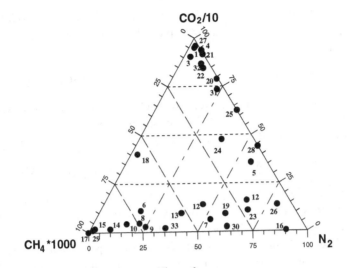

Figure 6

CH$_4$-N$_2$-CO$_2$ ternary diagram: thermal waters are arranged in two main alignments: hydrothermal systems hosted within carbonate reservoirs and S. Venera spring (no. 29) highlights the presence of significative amount of dissolved methane contents, produced under favorable low-potential redox conditions. CO$_2$-rich samples having low dissolved hydrocarbons contents include thermal waters hosted both within volcanic areas and within metamorphic reservoirs.

7. Carbon Isotope Geochemistry

The δ^{13}C of Total Dissolved Inorganic Carbon (TDIC) is a useful geochemical parameter representing the result of interaction processes among water, carbonate rocks and CO$_2$ coming from different sources. The primary natural CO$_2$-sources such as magmatic/mantle, metamorphism, carbonates and biogenic have different isotopic markers even if they are partly overlapping. For these reasons many geochemical investigations on volcanic and geothermal systems have been focused on the δ^{13}C$_{TDIC}$ to understand the physico-chemical processes occurring between thermal reservoirs and deep CO$_2$-rich fluids and to better define the origin of the gas.

The isotopic composition of δ^{13}C$_{TDIC}$ in the studied thermal waters is reported in Table 1. The ^{13}C/^{12}C ratios, expressed as delta values (δ^{13}C), show extremely different values ranging between $-10.0‰$ (no. 29) and $+2.8‰$ (no. 30) with respect to PDB international standard. Total amount of dissolved inorganic carbon (DIC) also shows a wide range from 4.5 to more than 145 mmol/Kg. These insights indicate that probably more than one C-source interact with the investigated thermal aquifers and that water-gas interaction process occurs under different environmental conditions.

The isotopic composition of TDIC represents the average of the isotopic composition of the carbon species, weighted on the respective contents of the inorganic carbon compounds. Main dissolved carbon species in solution are in the form of dissolved carbon dioxide (CO$_{2aq}$), bicarbonate (HCO$_3^-$) and carbonate ions

(CO_3^{2-}), because their relative concentrations are strongly pH-dependent. Therefore, the $\delta^{13}C_{TDIC}$ can be expressed as an isotopic balance of dissolved carbon species and it can be written as follows:

$$\delta^{13}C_{TDIC} = (\delta^{13}C_{CO2aq} * \chi_{CO2aq} + \delta^{13}C_{HCO3} * \chi_{HCO3} + \delta_{13}C_{CO3} * \chi_{CO3})/M_{Tot} \quad (1)$$

where χ expresses the molar fractions of each inorganic carbon species in water $(CO_{2aq}, HCO_3^-$ and $CO_3^{2-})$. Because of all samples, but two (nos. 30 and 33) have negligible CO_3 contents, the equation 1 begins:

$$\delta^{13}C_{TDIC} = (\delta^{13}C_{CO2aq} * \chi_{CO2aq} + \delta^{13}C_{HCO3} * \chi_{HCO3})/M_{Tot} \quad (2)$$

Therefore, $\delta^{13}C_{CO2aq}$ and $\delta^{13}C_{HCO3}$ can be computed considering the isotope enrichment factors between CO_{2gas} and HCO_3 (ε_a) and CO_{2gas} and CO_{2aq} (ε_b) defined by DEINES *et al.*, (1974) and MOOK *et al.* (1974).

In this case, making explicit equation 2 with respect to $\delta^{13}C_{CO2}$, the carbon isotope composition of corresponding CO_2 gas phase in equilibrium with collected thermal waters at respective sampling temperatures, have been calculated as follows:

$$\delta^{13}C_{CO2} = \delta^{13}C_{TDIC} - [\varepsilon_a * X_{HCO3} - \varepsilon_\beta * X_{CO2}] \quad (3)$$

In order to better discriminate the origin of carbon dioxide in the studied thermal manifestations, the samples have been arranged in two groups: the first group includes hydrothermal systems hosted within non-carbonate aquifers (i.e. volcanic and metamorphic reservoirs) while the second one includes only thermal discharges circulating within carbonate aquifers.

7.1 Volcanic and Metamorphic Thermal Reservoirs

Thermal waters collected within non-carbonate aquifers generally have a TDIC content higher than those of the geothermal waters fed by carbonate reservoirs. $\delta^{13}C_{CO2}$ values computed following the equation 3 range from -17.3 and $-0.9‰$. Being negligible C derived both from the atmosphere and from the dissolution of carbonate minerals, the origin of carbon dioxide dissolved in this group of thermal discharges could be assessed by means of a mixing process between an isotopically heavy C-source (e.g., inorganic) and a ^{12}C-enriched carbon dioxide (organic origin). Many samples are in good agreement with theoretical curves (Fig. 7) simulating the hypothesized mixing process. The organic end member is assumed to have a $\delta^{13}C_{CO2}$ $= -24‰$ (DEINES, 1974) and to contribute to TDIC for 0.001 and 0.005 mol/Kg. The $\delta^{13}C$ values assumed for the inorganic deep-CO_2 component are in the range $+2‰$ (solid lines) to $-2‰$ (dashed lines). These values include as the upper limit a CO_2 deriving from thermal decomposition of marine carbonates while as the lower limit the CO_2 degassed from mantle. For this end member, we used a value relative to the Mediterranean area as suggested by several authors (GIAMMANCO *et al.*, 1998;

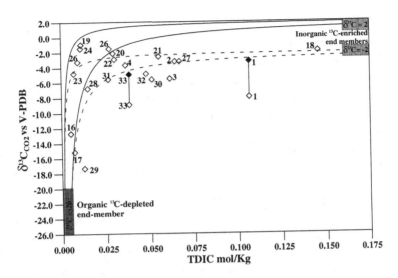

Figure 7

TDIC (Total Dissolved Inorganic Carbon) contents vs. $\delta^{13}C_{CO2gas}$ (calculated) for thermal waters hosted within non-carbonate aquifers. The carbon isotope composition of CO_2 gas interacting with thermal waters have been calculated assuming the achievement of full equilibrium conditions between aquifer and gas phase at the sampling temperatures (open diamonds) and only for two samples (nos 1 and 33) also at the estimated temperatures (solid diamonds). For detailed calculations see the text. Solid curves represent the theoretical mixtures resulting from the progressive addition of deep inorganic CO_2 with $\delta^{13}C_{CO2} = 2\%$ vs. V-PDB deriving from thermal decomposition of marine carbonates ($\delta^{13}C = 0\%$) to an organic CO_2 having a $\delta^{13}C_{CO2} = -24\%$ vs. V-PDB. Dashed lines represent the theoretical mixtures obtained from the increasing input of Mediterranean magmatic CO_2 ($\delta^{13}C_{CO2} = -2\%$ vs. V-PDB, GIAMMANCO et al., 1998; CAPASSO et al., 1997, INGUAGGIATO et al., 2000) to the same organic component ($\delta^{13}C_{CO2} = -24\%$ vs. V-PDB). In both cases, the organic end-member contributes 0.001 to 0.005 mol/Kg. Almost all the samples are in good agreement with theoretical curves simulating the suggested mixing process with the exception of two samples (nos 1 and 33) for which CO_2 and aquifers are probably equilibrated at temperatures higher than sampling one (solid diamonds).

CAPASSO et al., 1997, INGUAGGIATO et al., 2000). Volcanic gases released from active volcanoes belonging to this area showed $\delta^{13}C_{CO2}$ values in the range from -3% to 0% slightly heavier than those typical for MORB source ($\delta^{13}C_{CO2}$ from -5% to -8%) proposed by SANO and MARTY (1995). The enrichment in heavy carbon isotopes is thought to be due to a contaminated mantle at depth rather to reflect shallow contamination processes within the crust.

All the analytical data fit very well the obtained theoretical mixing lines but only two samples (no. 1 and 33) showing $\delta^{13}C$ values significantly lower with respect to theoretical ones. This could be due to the fact that CO_2 and aquifers are equilibrated at temperatures higher with respect to those recorded during sampling, rather than to invoke a larger organic contribution. Such an explanation is confirmed if we plot, in the same diagram, the $\delta^{13}C_{CO2}$ newly calculated at 110°C and 130°C as assessed using both silica (FOURNIER, 1981) and cations geothermometers (FOUILLAC and MICHARD, 1981; FOUILLAC, 1983, FOURNIER and TRUESDELL, 1973).

7.2 Carbonate hydrothermal systems

In thermal carbonate aquifers, the contribution of carbon deriving from the dissolution of host rocks, has to be included. For this reason, water/gas/rock interaction processes in these aquifers have been evaluated following the geochemical modeling proposed by CHIODINI *et al.* (2000). The authors simplified such processes excluding, in a first time, a deep origin of gas, and thus taking into account only three main C-sources: atmosphere, organic and carbon deriving from the dissolution of carbonate host rocks, each of them having an isotope marker. Another assumption implies that no carbon sinks such as CO_2-degassing and calcite precipitation take place.

In Figure 8, are reported the theoretical evolution curves of TDIC of a hypothetical local meteoric water initially in equilibrium with the atmosphere ($\delta^{13}C_{CO2atm} = -7\%$) that during infiltration interacts with carbon dioxide organic in origin. For this source, a range of isotopic values ($\delta^{13}C_{CO2org} = -18 \div -26\%$) has been chosen. During this step, groundwaters were kept always the equilibrium with calcite ($\delta^{13}C_{CO2rock} = 0\%$). The distribution of species in solution and the equilibration with mineralogical phases have been calculated by means of the PHREEQC computer program (PARKHURST, 1995).

For comparison, five cold groundwaters flowing within the main carbonate aquifers in western Sicily have been also plotted (CAPASSO *et al.*, in press). It is easy to observe that cold springs follow the obtained curves, thus showing any deep fluids contribution. In contrast with this behavior, all the thermal samples lie above the theoretical curves arranging towards more positive $\delta^{13}C_{TDIC}$ values. Therefore, we have supposed the contribution of a deeply-derived CO_2 having $\delta^{13}C_{CO2}$ values in the range from -2% to $+2\%$ corresponding to a mantle-derived CO_2 in Mediterrenean areas (lower value) and a carbon dioxide produced from thermometamorphic decomposition of carbonate rocks (higher value) respectively. The obtained theoretical curves fit very well with the measured values, thus implying that the addition of a deeply-derived carbon dioxide, enriched in heavy isotopes have to be inferred. This finding agrees with the model proposed by CARACAUSI *et al.*, (2004) for Castellammare-Alcamo, Montevago and Sciacca thermal reservoirs which suggested the occurrence of a significant contribution of mantle-derived helium degassed from magmas probably intruded in the crust and that deep fluids are therefore transported through faults (i.e. advective transport). In this scenario, lithospheric discontinuities could act as preferential pathways also for other mantle volatile components such as CO_2.

8. Conclusions

Thermal waters from Sicily and its adjacent island belong to four main water types: (1) a Na-Cl type; (2) a Ca-Mg > Na-SO_4-Cl type; (3) a Ca-Mg-HCO_3 type and

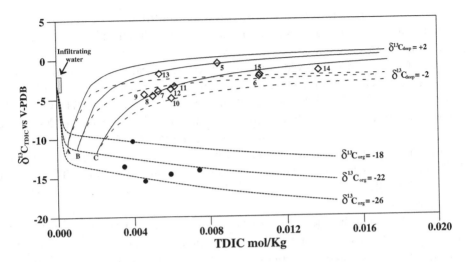

Figure 8

TDIC (Total Dissolved Inorganic Carbon) contents vs. $\delta^{13}C_{TDIC}$ for thermal waters hosted within carbonate aquifers (open diamonds). The same plot shows the theoretical curves of TDIC evolution obtained applying the geochemical modeling proposed by Chiodini *et al.* (2000). Infiltrating water (dotted area) has been assumed as local meteoric water initially in equilibrium with the atmosphere ($\delta^{13}C_{CO2atm} = -7‰$). During infiltration (dotted lines) it interacts with carbon dioxide, organic in origin, for which a range of the isotopic composition ($\delta^{13}C_{CO2org} = -18 \div -26‰$) has been chosen. During this step, groundwaters were also kept always the equilibrium with calcite ($\delta^{13}C_{CO2rock} = 0‰$). Only five cold groundwaters (solid circles) flowing within the main carbonate aquifers in western Sicily and plotted for comparison follow the theoretical curves. In contrast, all the thermal samples lie above the dotted curves arranging towards more positive $\delta^{13}C_{TDIC}$ values, thus inferring the presence of a deeply-derived carbon dioxide enriched in heavy isotopes. Two groups of theoretical curves have been obtained by modelling of the addition of a deep component having the Mediterranean magmatic CO_2 signatures ($\delta^{13}C_{CO2} = -2‰$ vs. V-PDB, GIAMMANCO *et al.*, 1998; CAPASSO *et al.*, 1997, INGUAGGIATO *et al.*, 2000; dashed lines) or deriving from thermometamorphic decomposition of carbonate rocks ($\delta^{13}C_{CO2} = 2‰$ vs. V-PDB; solid lines). Three points (A, B, and C) have been chosen as starting values having different initial TDIC contents (0.5, 1 and 2 mmol/kg, respectively).

(4) a Na-HCO$_3$ type water. Water isotope composition indicates that most waters are meteoric in origin or resulting from mixing between meteoric water and heavy-isotope end members. In some samples $\delta^{18}O$ values reflect the effects of equilibrium processes between thermal waters and rocks (positive ^{18}O-shift) or thermal waters and CO_2 (negative ^{18}O-shift). Concentrations of dissolved gas from thirty-two thermal discharges from Sicily show that all the samples have O_2/N_2 ratios lower than a water equilibrated with the atmosphere (ASW), and dissolved CO_2 contents higher than ASW. These are the result of two main processes: (1) prevailing reducing conditions (Eh < 0) resulting from oxygen consumption processes, thus causing the relative N$_2$-enrichment and (2) a marked interaction between thermal waters and CO_2-rich deep fluids. Most of the collected samples are also enriched in Helium and

methane with respect to ASW. This suggests the contribution of non-atmospheric helium and that methanogenesis occurs under low-potential redox condition thus producing significative amount of dissolved CH_4.

High CO_2 contents dissolved in thermal discharges located within active volcanic areas reflect intense interaction processes between waters and deep volcanic fluids degassed from magmatic body. At Pantelleria Island, in high-pH waters the conversion of dissolved carbon dioxide mainly into bicarbonate (HCO_3^-) and carbonate (CO_3^{2-}) ions is favored thus masking the effective CO_2 dissolution.

In non-volcanic areas high amounts of dissolved CO_2 have been found in some hydrothermal systems connected to regional tectonic structures. This is due because lithospheric faults are preferential structures for advective transporting of deep gas.

$\delta^{13}C_{TDIC}$ values has been used to assess the origin of carbon dioxide dissolved in the studied thermal waters. As regard non-carbonate reservoirs, we have calculated the isotopic composition of the carbon dioxide assuming for each thermal discharge that this gas has reached equilibrium conditions with thermal waters at their respective sampling temperatures. The obtained $\delta^{13}C_{CO2}$ values indicate a mixing process between a ^{12}C-enriched carbon dioxide (organic origin) and a prevalent isotopically heavy C-source (e.g. inorganic), inferred to originate from thermal decomposition of marine carbonates ($\delta^{13}C = +2\%_0$), or mantle-derived CO_2 with $\delta^{13}C_{CO2} = -2\%_0$ as magmatic CO_2 in the Mediterranean area (CAPASSO *et al.*, 1997; GIAMMANCO *et al.*, 1998; INGUAGGIATO *et al.*, 2000).

Thermal waters in carbonate reservoirs show TDIC CO_2 contents lower than thermal discharges fed by non-carbonate reservoirs, probably as result of a dilution effect due to the large amount of water stored in these aquifers and/or of low CO_2 degassing rates. $\delta^{13}C_{TDIC}$ values for this group of thermal manifestations has highlighted the presence of a deep CO_2-rich contribution, enriched in heavy isotopes, consistent with an end member having the same compositional range previously considered ($\delta^{13}C_{CO2deep} = -2 \div +2\%_0$). Recent investigations on helium fluxes from mantle in Western Sicily, higher than those typical for a continental crust, seem to indicate the presence of mantle-derived degassing from a magmatic body probably intruded in the crust (CARACAUSI *et al.*, 2004). In this area, a lithospheric fault system allowing the advective transport of other mantle volatile components such as CO_2 has been hypothesized.

Acknowledgements

The authors are indebted with Dr. F. Ganci who kindly collaborated collecting the samples. Thanks are also due to the colleagues at INGV laboratories for their assistance during chemical and isotopic analyses. The authors also thank two anonymous reviewers for their critical review of the manuscript.

REFERENCES

AIUPPA, A., ALLARD, P., D'ALESSANDRO, W., MICHEL, A., PARELLO, F., TREUIL, M., and VALENZA, M. (2000a), *Mobility and fluxes of major, minor and trace metals during basalt weathering and groundwater transport at Mt. Etna volcano (Sicily)*, Geochim. Cosmochim. Acta 64, 1827–1841.

AIUPPA, A., DONGARRÀ, G., CAPASSO, G., and ALLARD, P. (2000b), *Trace elements in the thermal groundwaters of Vulcano Island (Sicily)*, J. Volcanol. Geotherm. Res. 98, 189–207.

ALAIMO R., CARAPEZZA M., DONGARRÀ G., and HAUSER S. (1978), *Geochimica delle sorgenti termali siciliane*, Rend. Soc. Mineral. Ital. 34, 577–590.

ALLARD, P., JEAN-BAPTISTE, P., D'ALESSANDRO, W., PARELLO, F., PARISI, B., and FLEHOC, C. (1997), *Mantle- derived helium and carbon in groundwaters and gases of Mount Etna, Italy*, Earth Plan. Sci. Lett. 148, 501–516.

ANZÀ, S., DONGARRÀ, G., GIAMMANCO, S., GOTTINI, V., HAUSER, S., and VALENZA, M. (1989), *Geochimica dei fluidi dell'Etna: le acque sotterranee*, Miner. Petrogr. Acta 32, 231–251.

BADALAMENTI, B., GURRIERI, S., HAUSER, S., PARELLO, F., and VALENZA, M. (1988), *Soil CO_2 output in the Island of Vulcano during the period 1984–1988: Surveillance of gas hazard and volcanic activity*, Rend. Soc. It. Min. Petrol. 43, 893–899.

BARBERI, F., CIVETTA, L., GASPARINI, P., INNOCENTI, F., and SCANDONE, R. (1974), *Evolution of a section of the African–Europe plateboundary: Paleomagnetic and volcanological evidence from Sicily*, Earth Planet. Sci. Lett. 22, 123–132.

BOLOGNESI, L. and D'AMORE, F. (1993), *Isotopic variation of the hydrothermal system on Vulcano Island Italy*, Geochim. Cosmochim. Acta 57, 2069–2082.

CAPASSO, G., D'ALESSANDRO, W., FAVARA, R., INGUAGGIATO, S., and PARELLO, F. (2001), *Interaction between the deep fluids and the shallow groundwaters on Vulcano Island (Italy)*, J. Volcanol. Geotherm. Res. 108, 187–198.

CAPASSO, G., DONGARRÀ, G., HAUSER, S., FAVARA, R., and VALENZA, M. (1992), *Isotope composition of rain water, well water and fumarole steam on the island of Vulcano, and their implication for volcanic surveillance*, J. Volcanol. Geotherm. Res. 49, 147–155.

CAPASSO G., FAVARA R., and INGUAGGIATO S. (1997), *Chemical features and isotopic composition of gaseous manifestations on Vulcano Island, Aeolian Islands, Italy: An interpretative model of fluid circulation*. Geochim. Cosmochim. Acta 61, 3425–3440.

CAPASSO G., FAVARA R., FRANCOFONTE, S., and INGUAGGIATO S. (1999), *Chemical and isotopic variations in fumarolic discharge and thermal waters at Vulcano Island (Aeolian Islands, Italy) during 1996: evidence of resumed volcanic activity*, J. Volcanol. Geotherm. Res. 88, 167–175.

CAPASSO, G. and INGUAGGIATO, S. (1998), *A simple method for the determination of dissolved gases in natural waters, An application to thermal waters from Vulcano Island*. Appl. Geochem. 13, 631–642.

CARACAUSI, A., FAVARA, R., GIAMMANCO, S., ITALIANO, F., NUCCIO, P.M., PAONITA, A., PECORAINO, G., and RIZZO, A. (2003), *Mount Etna: Geochemical signals of magma ascent and unusually extensive plumbing system*, Geophys. Res. Lett, 30(2), 1057, doi10.1029/2002GL015463.

CARACAUSI, A., FAVARA, R., ITALIANO, F., NUCCIO, P.M., PAONITA, A., and RIZZO, A. (2004), *Evidence of mantle derived fluid contributions to the thermal basins of Western Sicily: geotectonic and geodynamic implications*, Porecceding of the WRI-11 International Symposium. Saratoga Springs, Wanty & Seal II eds. 91–94.

CARAPEZZA, M.L., INGUAGGIATO, S., Brusca, L., and LONGO, M., (2004), *Geochemical precursors of the activity of an open-conduit volcano: The Stromboli 2002–2003 eruptive events*, Geophys. Res. Lett., 31, L07620, doi: 10.1029/2004GL019614.

CATALANO, R., Di STEFANO, P., SULLI, A., and VITALE, F.P. (1996), *Paleogeography and structure of the central Mediterranean: Sicily and its offshore area*, Tectonophys. 260, 291–323

CAPASSO, G., FAVARA, R., GRASSA, F., INGUAGGIATO, S., and LONGO, M. *On-line technique for preparation and measuring stable carbon isotope of total dissolved inorganic carbon in water samples ($\delta^{13}C_{TDIC}$)*, Annals of Geophysics, (in press)

CHIODINI, G., FRONDINI, F., CARDELLINI, C., PARELLO, F., and PERUZZI, L. (2000), *Rate of diffuse carbon dioxide degassing estimated from carbon balance of regional aquifers: The case of central Apennine*, Italy. J. of Geophys. Res. *105*, 8423–8434.

CHIODINI, G., FRONDINI, F., and RACO, B. (1996), *Diffuse emissions of CO2 from the Fossa crater Vulcano Island (Italy)*, Bull. Volcan. *58*, 41–50.

CIVETTA, L., CORNETTE, Y., GILLOT, P.Y., and ORSI, G., (1988), *The eruptive history of Pantelleria Sicily Channel in the last 50 ka*, Bull. Volcanol. *50*, 47–57.

Craig, H. (1961), *Isotopic variations in meteoric waters*, Science *133*, 1702–1703.

D'ALESSANDRO, W., DONGARRÀ, G., GURRIERI, S., PARELLO, F., and VALENZA, M. (1994), *Geochemical characterization of naturallly occurring fluids on the island of Pantelleria (Italy)*, Mineral. Petrogr. Acta *37*, 91–102.

D'ALESSANDRO, W., DEGREGORIO, S., DONGARRÀ, G., GURRIERI, S., PARELLO, F., and PARISI, B. (1997), *Chemical and isotopic characterization of the gas of Mt. Etna (Italy)*, J. Volcan. Geotherm. Res. *78*, 65–76.

D'ALESSANDRO, W., and PARELLO, F. (1997), *Soil gas prospection of He, ^{222}Rn and CO_2: Vulcano Porto area, Aeolian Islands, Italy*, Appl. Geochem. *12*, 213–224.

DEINES, P., LANGMUIR, D., and RUSSELL, S. (1974), *Stable carbon isotope ratios and the existence of a gas phase in the evolution of carbonate groundwaters*, Geochim. Cosmochim. Acta *38*, 1147–1164

DONGARRÀ, G., and HAUSER, S. (1982), *Isotopic composition of dissolved sulphate and hydrogen sulphide from some thermal springs of Sicily*, Geothermics *11*, 193–200.

DREVER, J.J. *The Geochemistry of Natural Waters* (Prentice-Hall, N.Y. 1982).

DUCHI, V., CAMPANA, M.E., MINISSALE, A., and THOMPSON, M. (1994) *Geochemistry of thermal fluids on the volcanic Isle of Pantelleria*, Southern Italy. Appl. Geochem. *9*,147–160.

EPSTEIN, S., and MAYEDA, T. (1953) *Variation of ^{18}O content of water from natural sources*, Geochim. Cosmochim. Acta *4*, 213–224.

FAVARA, R., GIAMMANCO, S., INGUAGGIATO, S., and PECORAINO, G. (2001a), *Preliminary estimate of CO2 output from Pantelleria Island volcano (Sicily, Italy): evidence of active mantle degassing*, Appl. Geochem. *16*, 883–894.

FAVARA, R., GRASSA, F., and INGUAGGIATO, S. (1999), *Chemical and isotopic features of dissolved gases from thermal springs of Sicily, Italy*, GES-5 1999, Reykjavik (Iceland), 495–498.

FAVARA, R., GRASSA, F., INGUAGGIATO, S., and D'Amore, F. (1998), *Geochemical and hydrogeological characterization of thermal springs in Western Sicily, Italy*, J. Volcan. Geotherm. Res. *84*, 125–141.

FAVARA, R., GRASSA, F., INGUAGGIATO, S., Pecoraino, G., and CAPASSO, G. (2002), *A simple method to determine the $\delta^{13}C$ of Total Dissolved Inorganic Carbon*, Geofisica International *41*, 313–320

FAVARA, R., GRASSA, F., INGUAGGIATO, S., and VALENZA, M. (2001b), *Hydrogeochemistry and stable isotopes of thermal springs: earthquake-related chemical changes along belice fault (Western Sicily)*, Appl. Geochem. *16*, 1–17.

FEDERICO, C., AIUPPA, A, ALLARD, P., Bellomo, S., JEAN-BAPTISTE, P., PARELLO, F., and VALENZA M. (2002), *Magma-derived gas influx and water-rock interactions in the volcanic aquifer of Mt. Vesuvius, Italy*, Geochim. Cosmochim. Acta *66*, 963–981.

FOUILLAC, C. (1983), *Chemical geothermometer in CO_2-rich thermal waters. Example of the Frenc Massif Central*, Geothermics *12*, 149–160.

FOUILLAC, C. and MICHARD, G, (1981), *Sodium/lithium ratio in water applied to geothermometry of geothermal reservoir*, Geothermics *10*, 55–70

FOURNIER, R.O. (1981), *Application of water geochemistry to geothermal exploration and reservoir engineering*, In: Geothermal systems: Principles and case Histories. (eds. Rybach and Muffler) (John Wiley, New York 1981) pp. 109–143.

FOURNIER, R.O. and TRUESDELL, A.H. (1973), *An empirical Na-K-Ca geothermometers for natural waters*, Geochim. Cosmochim. Acta *37*, 1255–1279.

GAT, J.R. and CARMI, I. (1970), *Evolution in the isotopic composition of atmospheric waters in the Mediterranean Sea area*, J. Geophys. Res. *75*, 3039–3048.

GIAMMANCO, S., INGUAGGIATO, S., and VALENZA, M. (1998), *Soil and fumarole gases of Mount Etna: geochemistry and relations with volcanic activity*, J. Volcan. Geotherm. Res. *81*, 297–310.

GIANNELLI, G., and GRASSI, S., (2001), *Water-rock interaction in the active geothermal system of Pantelleria, Italy*, Chem. Geol. *181*, 113–130.

GIGGENBACH, W.F. (1991), *Chemical techniques in geothermal exploration*. In *Applications of Geochemistry in Geothermal Reservoir Development*, (ed. D'Amore, F.) (UNITAR/UNDP Centre on Small Energy Resources, Rome 1991) pp. 1–95.

GRASSA, F. (2002), *Geochemical processes governing the chemistry of groundwater hosted within the Hyblean aquifers (Southeastern Sicily, Italy)*, Ph.D. Thesis, Univ. of Palermo, 113 pp.

HORNIG-KJARSGAARD, I., KELLER, J., KOBERSKI, U., STADLBAUER, E., FRANCALANCI, L., LENHART, R., (1993), *Geology, stratigraphy and volcanological evolution of the island of Stromboli, Aeolian arc, Italy*, Acta Vulcanol. *3*, 21–68.

INGUAGGIATO, S., MARTIN-DEL POZZO, A.L., AGUAYO, A., CAPASSO, G., and FAVARA, R. *Isotopic, chemical and dissolved gas constraints on spring water from Popocatepetl Volcano (Mexico)*, Evidence of gas-water interaction between magmatic component and shallow fluids. J. Volcan. Geotherm. Res, in press.

INGUAGGIATO, S., PECORAINO, G., and D'AMORE F., (2000), *Chemical and isotopical characterisation of fluid manifestations of Ischia Island (Italy)*, J. Volcan. Geotherm. Res. *99*, 151–178.

INGUAGGIATO, S. and RIZZO, A., (2004), *Dissolved Helium isotope ratios in ground-water: a new technique baed on gas-water re-equilibration and its application to volcanic areas*, Appl. Geoch. *19*, 665–673.

KENDALL, C. and COPLEN T.B., (1985), *Multisample conversion of water to hydrogen by zinc for stable isotope determination*, Anal. Chem. *57*, 1437–1440.

KING, C.Y. (1986), *Gas geochemistry application to earthquake prediction: an overview*, J. Geophys. Res., *91*, 12,269–12,289.

MAHOOD, G.A. and HILDRETH, W., (1986), *Geology of the peralkaline volcano at Pantelleria, Strait of Sicily*, Bull. Volcanol. *48*, 143–172.

MOOK, W.G., BOMMERSON, J.C., and STAVERMAN, W.H. (1974), *Carbon isotope fractionation between bicarbonate and gaseous carbon dioxide*, Earth Plan. Sci. Lett. *22*, 169–176.

PARKHURST, D.L. (1995), *A computer program for speciation, reaction path, advective transport, and inverse geochemical calculations*, USGS Water-Resources Investigations Report *95*, 4227.

PAUWELS, H., FOUILLAC, C., GOFF, F., and VUATAZ, F.D. (1997), *The isotopic and chemical composition of CO_2-rich thermal waters in the Mont-Dorè region (Massif-Central, France)*, Appl. Geochem. *12*, 411–427.

SANO, Y. and MARTY, B. (1995), *Origin of carbon in fumarolic gas from island arc*, Chem. Geol., *119*, 265–74.

THOMAS, D. (1988), *Geochemical precursor to seismic activity*, PAGEOPH., *126*, 241–266.

(Received: June 19, 2003, revised: September 20, 2004, accepted: October 23, 2004)

To access this journal online:
http://www.birkhauser.ch

Pure appl. geophys. 163 (2006) 809–823
0033–4553/06/040809–15
DOI 10.1007/s00024-006-0042-1

© Birkhäuser Verlag, Basel, 2006

❘Pure and Applied Geophysics

Geochemical Changes in Spring Water Associated with the 2000 Eruption of the Miyakejima Volcano, Japan

Tsutomu Sato,[1] Isao Machida,[2] Makoto Takahashi,[1] and Taro Nakamura[1]

Abstract—Water was sampled from eight springs and a lake in volcanic Miyakejima Island of Japan after the 2000 eruption. Major chemical and isotopic compositions of the water were analyzed. Significant increases of sulfate ion are observed in several springs where the thickness of ejecta exceeds 32 mm. A good relationship of Cl/S mole ratios between spring water and leachate of the ejecta is observed. Sulfur isotopic compositions of the spring water become close to that of leachate of the ejecta as time elapses after the eruption. Consequently the sources of the added sulfate ion in the spring water after the eruption are interpreted to be anhydrite and adhered sulfur of the ejecta.

Key words: Spring water, Miyakejima volcano, volcanic eruption, geochemical composition, isotopic composition, erupted products.

1. Introduction

Significant geochemical changes in groundwater may occur during or after volcanic eruption (e.g., Wakita *et al.*, 1988). The main cause of the changes lies in mixing of groundwater and chemically active fluid related to the volcanic activity. To investigate the nature of fluid emitted during the volcanic activity, it is important to monitor groundwater. For example, a significant rise of groundwater temperature, more than 50°C, and concentration of bicarbonate ion were observed in Izu Oshima Island, Japan, in the 1986 volcanic eruption (Takahashi *et al.*, 1991). The cause was thought to be the mixing of bicarbonate-ion-rich thermal water influenced by CO_2-rich volcanic gas with groundwater. In other words, the groundwater study revealed that the fluid emitted during the volcanic activity was high-temperature CO_2-rich volcanic gas. The groundwater was an important water resource in this island, and such volcanic contamination became a serious problem.

[1] Geological Survey of Japan, 1-1-1, Higashi, Tsukuba, Ibaraki, 305-8567, Japan. E-mail: sugar@ni.aist.go.jp
[2] Chiba University, 1-33, Yayoi-cho, Inage-ku, Chiba-shi, Chiba, 263-8522, Japan

The eruption of the Miyakejima volcano in 2000 began in June 2000. To investigate the influence of the volcanic activity on groundwater, we had collected spring and lake water once or twice every two months since September 2000, and analyzed the major chemical and stable isotopic compositions.

2. *The Miyakejima Volcano*

Miyakejima Island is an active volcanic island situated 180 km south of Tokyo, circular in plan with a diameter of 8 km (Fig. 1). Volcanic deposits cover the entire island. The hydraulic conductivity of the surface soil was estimated to be 10^{-3} to 10^{-2} cm/s by ARAI (1978).

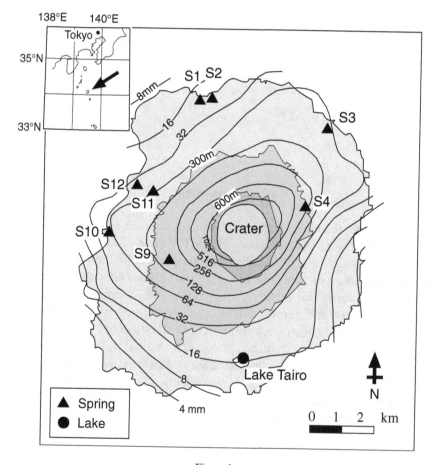

Figure 1

Map of Miyakejima Island. Filled triangles and a circle show the sampling points of spring water and lake water, respectively. Isopach shows the thickness of ejecta by GESHI *et al.* (2002).

Those values are so high that there is no year-round surface stream in the island in spite of the annual precipitation being about 3000 mm, twice as much as that of Tokyo. Because of strong dependence on groundwater for the domestic water supply, many hydrological studies have been carried out. Recently, MACHIDA (1999) investigated major chemical compositions of twelve springs. MACHIDA (2000b) also analyzed hydrogen and oxygen isotopes of the spring water, and discussed effects of altitude on isotopic compositions.

The eruption of the Miyakejima volcano in 2000 became distinctively more active from June 26. Following the summit eruption on July 8, the collapse of the summit area occurred forming a new big crater (Fig. 1). The diameter and volume of the crater were estimated to be about 1.5 km and 0.6 km^2, respectively (NAKADA et al., 2001). Volcanic gas has been emitted continuously from the bottom of the crater, the altitude of which is about 250 m above sea level. Three major eruptions with a total amount of ejecta, consisting of ash, scoria and volcanic bomb, of over $10^6 m^3$ occurred on July 14–15, August 18, and August 29. Figure 1 shows an isopach map of thickness distribution of the ejecta from July 8 to August 30. During this period the island was entirely covered with the ejecta with a minimum thickness of 4 mm. The total volume of the ejecta is calculated to be $11 \times 10^6 m^3$ by NAKADA et al. (2001).

After the main eruptions in July–August 2000, gas emission became predominant. Average SO_2 flux for five months from September 2000 is estimated to be 48 kt/d (KAZAHAYA et al., 2001).

3. Experimental

Water samples have been collected from eight springs (S1–S4 and S9–S12 in Fig. 1) once or twice every two months since September 2000. The all springs are small as the flow rate is 0.1–10 L/min, and the aquifer lies on the local geologic volcanic ash layer. Those spring numbers are the same as in MACHIDA (1999, 2000b). We collected water from five springs, S1–S2 and S10–S12, in 2000, and additional three springs, S3, S4 and S9, in May 2001, July 2001 and January 2002, respectively. Other springs described by MACHIDA (1999), S5–S8, could not be located probably because they were buried under volcanic mudflow.

We have also collected lake water from Lake Tairo in the southern part of the island. Because there is no river flowing into the lake, the lake water is believed to be recharged by ground- and rainwater. Thus geochemical changes in the lake water may reflect those in the ground- and rainwater.

Temperature, pH and electric conductivity (EC) were measured in the field using portable instruments (Table 1). All other determinations have been done in the laboratory. First, the water samples were filtered by 0.45 μm filter. Concentrations of bicarbonate ion were calculated using alkalinity measured by sulfuric acid titration to a final value of pH of 4.8. Major anions of F^-, Cl^-, Br^-, NO_3^-, and SO_4^{2-} and major cations of Na^+, K^+, Ca^{2+}, and Mg^{2+} were analyzed by ion chromatography (Type DX-100, Dionex Co.) and atomic absorption spectrophotometry (Type AA-670, Shimadzu Co.), respectively. SiO_2 was measured by molybdenum yellow method.

Selected samples were analyzed for their stable isotopic ratios of δD, $\delta^{18}O$, $\delta^{13}C$, and $\delta^{34}S$ (Table 1). δD and $\delta^{18}O$ were determined using the zinc reduction method (COLEMAN *et al.*, 1982) and the CO_2 equilibration method (YOSHIDA and MIZUTANI, 1986), respectively. The dissolved inorganic carbon (DIC) was extracted as CO_2 by acidification, and $\delta^{13}C$ of CO_2 was analyzed with a mass-spectrometer using the method of GLEASON *et al.* (1969). The dissolved sulfate ion was precipitated as barium sulfate, and $\delta^{34}S$ of SO_2 that was thermally decomposed was analyzed with mass-spectrometer using the method of YANAGISAWA and SAKAI (1983). All stable isotopic data are expressed in the usual δ per mille ($\delta‰$) notation, where $\delta = (R/R_{std} - 1) \times 1000$ (‰). R represents either D/H, $^{18}O/^{16}O$, $^{13}C/^{12}C$, or $^{34}S/^{32}S$ ratio of the sample and R_{std} is the isotopic ratio of the standard mean ocean water (SMOW) for δD and $\delta^{18}O$, whereas the Peedee belemnite (PDB) is used as standard for $\delta^{13}C$, and the Canyon Diablo troilite (CDT) standard for $\delta^{34}S$. Analytical errors for $\delta^{18}O$, $\delta^{18}O$, and $\delta^{34}S$ are $\pm 0.2 ‰$ and $\pm 2 ‰$ for δD.

4. Results and Discussions

4.1 Major Chemical Compositions

The analytical results are shown in Table 1. The results for major compositions are visually expressed by stiff diagrams in Figure 2. Figures 2a, 2b and 2c show the results for July 2001, January and August 2002, respectively. In all figures, data by MACHIDA (1999) in 1988 are also shown as hatched diagrams.

Before the eruption, major anion and cation constituents of most spring water were Cl^- and Na^+. The concentrations of Cl^- and Na^+ were 25–173 and 13–82 mg/L, respectively. They tended to be higher near the coast because of contamination from seawater spray. On the other hand, the major anion and cation constituents of the lake water were HCO_3^- at 208 mg/L and Mg^{2+} at 41 mg/L, respectively. The concentrations of SO_4^{2-} of spring and lake water were relatively low at 3.8–25 mg/L in spring water and 81 mg/L in lake water. The all springs were classified into Cl^- or HCO_3^- type in terms of the anionic composition. The concentration ratio of SO_4^{2-} to total anionic composition is less than 20% in the all springs before the eruption (large marks in Fig. 4).

In July 2001, a year after the eruption, marked changes occurred in the shape of the stiff diagrams at S3 and S4 as shown in Figure 2a. In both springs, concentrations of SO_4^{2-} increased more than ten times, changing SO_4^{2-} to be the major anion constituent. In addition, concentrations of all cationic constituents increased.

Table 1

Chemical and isotopic compositions of spring and lake waters from Miyakejima Island, Japan

Location	Date	T °C	pH	E.C. mS/m	HCO$_3^-$ mg/L	F$^-$ mg/L	Cl$^-$ mg/L	Br mg/L	NO$_3^-$ mg/L	SO$_4^{2-}$ mg/L	Na$^+$ mg/L	K$^+$ mg/L	Ca^{2+} mg/L	Mg^{2+} mg/L	SiO$_2$ mg/L	δD ‰	δ^{18}O ‰	δ^{13}C ‰	δ^{34}S ‰
Spring water																			
S1	19 Feb.98 *	15.6	8.1	26.0	44.4	–	56.8	–	12.5	14.2	32.4	2.3	18.2	8.7	–	-34.0	-5.9	–	–
	28 Oct.00	17.7	7.8	41.5	46.4	<0.1	78.6	0.3	18.8	22.3	38.5	2.6	22.0	10.1	52.1	-30.8	-6.3	-17.2	+11.4
	30 Jan.01	14.2	7.6	42.8	47.6	<0.1	74.7	0.4	18.5	24.1	39.2	2.6	21.1	10.2	40.5	-31.9	-6.2	–	–
	24 Jul.01	20.9	7.8	39.7	53.6	<0.1	69.7	0.3	19.2	20.4	37.8	2.3	20.0	9.4	40.6	-32.4	-6.4	-16.3	+8.4
	16 Jan.02	16.5	7.7	40.1	52.5	<0.1	67.7	0.2	19.1	27.8	33.8	2.3	19.9	9.6	51.1	-32.8	-6.3	-16.6	+7.9
	6 Aug.02	20.1	7.7	42.5	55.4	<0.1	65.6	0.2	18.7	41.0	41.3	2.5	21.8	10.6	49.5	-36.7	-7.0	-15.8	+6.1
S2	19 Feb.98 *	11.9	7.6	24.0	23.4	–	62.8	–	10.8	15.2	31.5	1.8	18.7	9.5	–	-27.1	-5.3	–	–
	24 Jul.01	26.2	7.9	37.3	39.7	<0.1	56.0	0.2	9.2	50.0	38.1	1.8	19.2	8.6	38.4	-33.3	-6.5	-9.5	+5.4
	18 Jan.02	9.1	6.7	39.7	34.6	<0.1	57.9	0.2	9.1	59.7	34.6	1.6	19.4	9.3	45.1	-33.9	-6.5	-9.1	+5.2
	7 Aug.02	22.3	7.5	48.9	37.3	<0.1	59.5	0.2	11.8	99.5	43.5	2.1	26.7	13.2	46.9	-40.4	-7.6	-14.7	+4.4
S3	20 Feb.98 *	14.7	7.4	44.2	21.5	–	140	–	4.2	25.0	67.7	3.4	15.6	16.7	–	-29.1	-5.5	–	–
	23 Jul.01	24.4	7.3	106	30.8	0.1	95.2	0.5	3.3	361	100	5.3	40.6	46.0	26.5	-38.8	-7.1	-16.8	+5.7
	15 Jan.02	13.9	7.4	157	20.3	0.1	108	0.5	3.7	637	121	6.2	72.8	81.9	32.6	-34.6	-6.7	-15.0	+4.0
	6 Aug.02	21	7.3	174	22.2	0.2	116	0.4	2.3	758	141	7.6	82.8	103	34.0	-36.6	-7.3	-14.0	+5.1
S4	18 Feb.98 *	–	7.8	10.0	24.4	–	25.0	–	0.6	3.8	13.2	1.3	8.3	3.6	–	-29.2	-5.7	–	–
	23 Jul.01	25.2	4.3	75.4	<0.1	0.2	76.5	0.3	9.9	236	32.8	3.7	70.8	21.8	43.2	-38.1	-6.9	–	+3.5
	19 Jan.02	5.6	4.6	111	2.4	0.2	95.9	0.2	20.9	395	35.7	2.7	107	39.1	43.3	-41.2	-7.4	–	+3.7
	7 Aug.02	25.2	4.6	213	4.4	<0.1	146	0.4	47.3	999	66.4	7.72	257	100	56.8	-37.0	-6.9	–	+2.6
S9	19 Feb.98 *	12.3	7.5	19.0	49.8	–	43.5	–	<0.1	5.4	21.1	1.1	18.2	8.3	–	-34.6	-6.3	–	–
	18 Jan.02	8.1	7.7	82.4	54.7	0.1	73.4	0.3	<0.1	268	40.2	1.1	70.0	31.0	60.1	-41.1	-7.5	-18.3	+5.2
	7 Aug.02	25.5	7.9	77.0	41.5	0.1	78.5	0.2	<0.1	231	46.8	1.3	59.6	28.1	61.1	-40.2	-7.8	-15.5	+5.1
S10	19 Feb.98 *	13.7	7.8	58.0	54.7	–	173	–	11.4	21.9	81.9	3.7	30.7	21.6	–	-34.2	-6.0	–	–
	15 Jan.02	15.9	8.0	88.9	63.7	0.1	170	0.6	8.3	117	89.0	3.5	31.8	22.7	57.6	-32.7	-6.7	-6.0	+7.6
	7 Aug.02	23.9	8.4	110	61.0	0.1	186	0.7	7.9	207	109	4.5	42.7	33.9	54.4	-38.2	-7.4	-8.5	+5.8
S11	19 Feb.98 *	13.7	7.6	17.0	49.8	–	30.7	–	8.7	4.6	18.3	1.4	15.4	7.9	–	-33.4	-6.1	–	–
	15 Sep.00	19.3	7.2	18.9	48.3	<0.1	20.4	0.1	12.8	8.1	16.5	1.1	10.5	5.3	45.3	-33.7	-7.0	-17.9	+10.9
	28 Oct.00	17.7	7.7	21.1	50.0	<0.1	23.1	0.1	12.6	13.7	17.6	1.3	12.7	6.1	45.3	-33.3	-6.8	–	–
	30 Jan.01	14.4	7.6	23.3	52.5	<0.1	25.3	<0.1	15.0	18.7	17.8	1.5	14.7	7.2	35.5	-35.7	-6.8	–	–

Table 1
(Contd.)

Location	Date	T °C	pH	E.C. mS/m	HCO₃⁻ mg/L	F⁻ mg/L	Cl⁻ mg/L	Br mg/L	NO₃⁻ mg/L	SO₄² mg/L	Na⁺ mg/L	K⁺ mg/L	Ca²⁺ mg/L	Mg²⁺ mg/L	SiO₂ mg/L	δD ‰	δ¹⁸O ‰	δ¹³C ‰	δ³⁴S ‰
Spring water																			
	30 May.01	17.1	7.4	25.7	47.6	<0.1	29.9	0.1	14.4	22.4	18.5	1.7	15.2	7.4	35.9	−37.3	−7.0	−13.7	+8.8
	24 Jul.01	18.9	7.3	23.4	54.4	<0.1	30.3	0.1	11.5	14.8	18.5	1.4	13.9	6.8	35.5	−36.6	−7.1	−11.0	+7.8
	17 Jan.02	14.0	7.7	33.4	52.0	<0.1	41.5	0.2	12.9	38.8	22.0	1.6	21.2	10.5	45.3	−40.0	−7.4	−18.4	+7.8
	7 Aug.02	18.3	7.1	45.8	41.2	<0.1	46.5	0.2	12.8	103	30.7	1.8	32.9	16.2	43.0	−41.3	−7.9	−14.8	+6.5
S12	19 Feb.98 *	14.9	7.3	37.0	25.4	–	114	–	7.0	12.6	48.4	2.1	22.7	13.5	–	−34.7	−6.1	–	–
	30 Jan.01	–	7.8	37.3	40.3	<0.1	65.3	0.2	2.2	42.8	46.4	1.7	13.2	8.1	32.8	–	–	–	–
	30 May.01	16.8	7.6	42.2	32.2	<0.1	64.7	0.3	2.0	62.5	47.0	1.4	15.6	9.6	32.6	−41.9	−7.3	–	–
	24 Jul.01	19.0	7.6	42.9	44.0	<0.1	67.7	0.3	1.4	57.9	47.4	1.8	17.4	10.1	32.9	−38.8	−7.2	−13.7	+8.8
	17 Jan.02	13.8	7.7	58.4	32.7	<0.1	76.8	0.3	3.0	128	50.9	2.0	27.1	15.2	41.7	−38.8	−7.2	−11.0	+7.8
	7 Aug.02	19.7	7.5	88.6	31.7	<0.1	84.0	0.4	2.0	293	68.1	2.7	54.0	31.2	39.3	−43.8	−8.0	−12.6	+6.5
Lake Water																			
Tairo	10 Apr.98 *	19.2	8.8	66.0	208	–	88	–	<0.1	80.9	56.4	4.1	35.5	40.7	–	−15.0	−2.6	–	–
	26 Oct.00	22.1	7.5	106	297	0.1	110	0.3	0.1	150	76.8	5.9	70.7	49.2	24.8	−25.9	−4.5	+1.5	+11.4
	31 Jan.01	8.9	8.0	113	346	0.2	129	0.2	<0.1	172	85.1	6.0	79.3	58.9	28.1	−27.6	−5.0	–	–
	23 Jul.01	30.1	8.3	117	356	0.2	123	0.4	<0.1	165	84.0	5.7	65.7	58.7	33.1	−24.9	−4.8	+5.3	+7.7
	15 Jan.02	11.7	8.2	125	381	0.2	143	0.5	<0.1	172	86.0	6.0	75.9	64.5	34.9	−25.8	−4.8	+4.2	+8.5
	5 Aug.02	29.5	8.2	123	341	0.1	141	0.4	<0.1	175	81.5	5.9	66.7	65.3	29.7	−28.6	−5.3	+5.0	+7.6

* The data of chemical and isotopic composition are after MACHIDA (1999) and MACHIDA (2000b), respectively

Increases of SO_4^{2-} were also observed at S2 and S12. On the other hand, a different type of change occurred at Lake Tairo; that is, increases in concentrations of all anionic and cationic constituents.

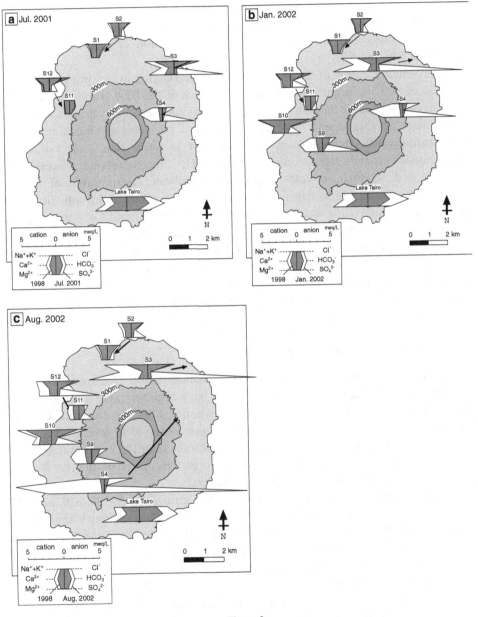

Figure 2

Stiff diagrams showing the major chemical composition of the spring and lake waters on a) July 2001, b) January 2002 and c) August 2002. Hatched diagrams show the data in 1998 by MACHIDA (1999).

In January 2002, in addition to S3 and S4, a marked change was observed at S9, the change being very similar to that of S4 (Fig. 2b). The changes at S3 and S4 have been continuing in the same direction. Increases of SO_4^{2-} were also seen at S10 and S11 in addition to S2 and S12. However, the lake water seemed to resist changes in the major ionic compositions in comparison with the data in July 2001.

In August 2002, further changes in all springs were observed as compared to the data for January 2002 (Fig. 2c). Increases of all cation concentrations became marked at S11 and S12.

The common feature of the observed geochemical change in the spring water is an increase of SO_4^{2-} concentration. In the early stage of the 2000 eruption, the island was covered by a large amount of ejecta that contained as much sulfur as anhydrite and adhered gas components. Furthermore, volcanic gas including abundant SO_2 has been continuously emitted from the crater. It is likely that such sulfur has contaminated groundwater as ionic constituent, SO_4^{2-}, through several processes; for example, as direct mixing of volcanic gas with groundwater, as infiltration of leachate of ejecta, or as acid rain. Especially the mixing of leachate from anhydrite with groundwater would cause simultaneous enrichment of SO_4^{2-} and Ca^{2+} in the spring water. On the other hand, acidification of groundwater would make the minerals in aquifer rocks soluble to groundwater. It is conceivable that the varieties of increased cationic constituents are related to the enrichment process of SO_4^{2-} or rocks containing groundwater aquifer. Simultaneous increases of SO_4^{2-} and Ca^{2+} observed at S4 and S9 seem to indicate a possibility of the mixing of leachate from anhydrite with groundwater.

4.2 Increased Concentrations of Excess SO_4^{2-}

Figure 3 shows the temporal variations of concentration of excess SO_4^{2-}. The concentration of excess SO_4^{2-} was calculated by subtracting SO_4^{2-} concentration before the eruption (Table 1) from the concentration after the eruption. Gradual increases of the concentration of excess SO_4^{2-} were observed at several springs. At S3, S4 and S9, the concentration already reached more than 100 mg/L at the time of the first sampling. We classified the springs into two groups using the threshold of 100 mg/L of concentration of excess SO_4^{2-} in 2002.

The first group, over 100 mg/L, consists of five springs, S3, S4, S9, S10, and S12. The spatial distribution of the first group in Figure 1 correlates closely with thickness of the ejecta; all springs located in the area where the thickness of the ejecta is over 32 mm belong to the first group except for S11. Furthermore, the concentration of excess SO_4^{2-} seems to be related to the thickness of the ejecta; the thicker the ejecta, the higher the concentration. The highest value of concentration of excess SO_4^{2-}, about 1000 mg/L, was observed at S4, where the ejecta is thickest at over 256 mm. This suggests that the excess SO_4^{2-} in the spring water originated as leachate of the ejecta.

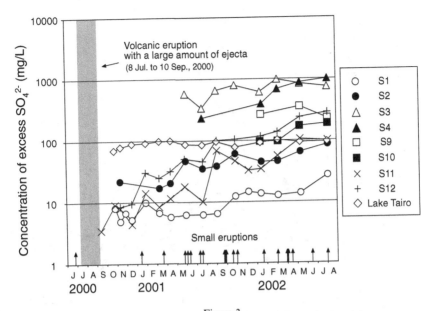

Figure 3

Temporal variations of concentration of excess sulfate ion. The excess concentration is calculated as subtracting data by MACHIDA (1999) in 1998 from data in this study. Arrows show the time of small eruptions. A hatched period shows the period of active eruption with a large amount of ejecta.

On the assumption that the origin of the excess SO_4^{2-} is leachate of the ejecta, mean residence time of the spring water may be estimated. If the concentration of excess SO_4^{2-} significantly increased at the spring, the time taken from the eruption to the time of increase may be taken as residence time. For example, at S11, the concentration of excess SO_4^{2-} increased suddenly in September 2001, about a year after the eruption. Thus the mean residence time of the spring water at S11 is thought to be one year.

Furthermore, if the specific flux, q, of the groundwater flow of S11 is given, we can estimate the mean length of the groundwater flow path. The q is roughly estimated using Darcy's law, $q = -K\, dH/dL$, where K is the hydraulic conductivity and dH/dL is the hydraulic gradient. Using the hydraulic gradient of 0.15, which is assumed to be equal to the gradient of the ground surface, the q is calculated to be from 0.13 to 1.3 m/day for the K of 10^{-3} to 10^{-2} cm/sec by ARAI (1978). Thus the mean length of the groundwater flow path of S11 is estimated to be from 50 to 500 m.

4.3 Ratios of Major Anion Compositions

Figure 4 shows the weight percent ratios of major anion constituents, Cl^-, HCO_3^-, and SO_4^{2-}. Before the eruption, SO_4^{2-} was less than 20 wt% as shown by larger marks in Figure 4. The ratio of SO_4^{2-} however, increased at all springs

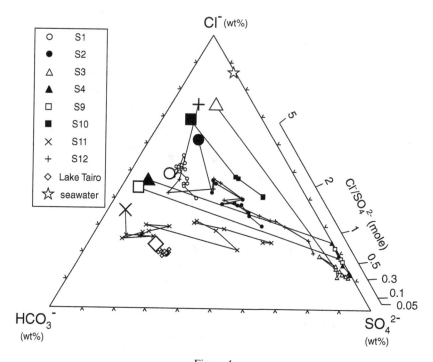

Figure 4
Weight ratios of the major anion constituents. Larger marks show the data by Machida (1999) in 1998.

becoming over 20 wt%, in particular over 70 wt% at S3, S4, S9 and S11. These changes were accompanied with the increases of concentrations of excess SO_4^{2-} shown in Figure 3. On the other hand, there was no significant change in the ratio at Lake Tairo.

It is very interesting that each plot seems to move toward a common end member (Fig. 4). The weight % ratio of the end member is located within a range from 75 wt % SO_4^{2-} / 25 wt % Cl^- to 90 wt % SO_4^{2-} /10 wt% Cl^-, which corresponds to 0.3–1 of Cl/S mole ratios.

Cl/S mole ratios have been determined in various samples such as volcanic gas, leachate of the ejecta, and acid rain in relation to the 2000 eruption. Mori *et al.* (2001) measured HCl/SO_2 mole ratio of volcanic gases with a remote measurement method using an FT-IR spectral radiometer and found that there were 0.06. Shinohara *et al.* (2002) also estimated HCl/SO_2 mole ratios of volcanic gases at 0.08 by the alkaline-trap method. The results were consistent with those by Mori *et al.* (2001). Satoh *et al.* (2001) measured Cl/S mole ratios of leachate of the ejecta, and obtained ranges of 0.01–0.05 for August 18, 0.1–0.14 for August 29, and 0.5–1.5 for September 2000. Sato *et al.* (2003) measured Cl^-/SO_4^{2-} mole ratios of acid rain collected at Lake Tairo in 2002 and obtained results in a range of 2.2–6.6. In comparison to these results on Cl/S mole ratios, the results obtained for leachate of the ejecta were consistent with the end member in Figure 4. Thus the hypothesis that leachate, after infiltrating the ejecta, mixed with groundwater explains the trend of changes seen in Figure 4 well.

4.4 Hydrogen and Oxygen Isotopic Compositions

The Relationship between δD and $\delta^{18}O$ of the spring and lake water is shown in Figure 5. Local meteoric water lines (LMWL) with a fixed gradient of eight are calculated for the spring water. The calculations are performed separately for before and after the eruption.

In a comparison of the plots in Figure 5 showing trends for before and after the eruption, both δD and $\delta^{18}O$ values decreased after the eruption. The value of the intercept of LMWL also changed from 14.8 to 18.6. The cause of these changes is not clear. It is possible that such changes occur due to long-term fluctuation of δD and $\delta^{18}O$ of rainwater. However, no data are available for isotopic compositions of rainwater after the eruption. MACHIDA (2000a) analyzed $\delta^{18}O$ of rainwater from

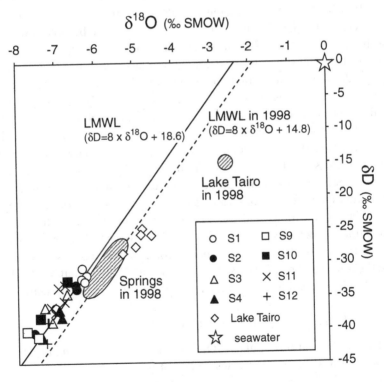

Figure 5
Relationship between hydrogen and oxygen stable isotopic composition. Hatched area shows the distribution of data by MACHIDA (2000b) in 1998. Dashed and solid lines shows the local meteoric water line (LMWL) for data in 1998 and data in this study, respectively.

1994 to 1995. In order to validate the long-term fluctuation hypothesis, it is necessary to collect rainwater for a long period and compare the data with those by MACHIDA (2000a).

Changes in Figure 5, however, are hardly caused by the volcanic activity such as the mixing of volcanic fluid with groundwater. If such volcanic fluid originates in the magma, the δD and $\delta^{18}O$ values of the groundwater should shift to those values of fluid in the magma. Considering andesite water by GIGGENBACH (1992) as the origin of the volcanic fluid, the δD and $\delta^{18}O$ values of the volcanic fluid are expected to be $-30‰$ of δD and $+7‰$ of $\delta^{18}O$. In a comparison of the δD and $\delta^{18}O$ values with those of the spring water, the $\delta^{18}O$ value of the volcanic fluid is too high although there is little contrast in the δD value. Thus, we believe that mixing of the volcanic fluid with groundwater causes a significant increase in the $\delta^{18}O$ value of the spring water. The $\delta^{18}O$ value of the spring water, however, decreased after the eruption, showing little possibility that the volcanic activity changed the δD and $\delta^{18}O$ values of the spring water.

The springs, S1 and S2, occur closely and the altitude is almost the same (Fig. 1). The δD and $\delta^{18}O$ values of S2, however, are lower than those in S1 as shown in Figure 5. It is likely that the difference is the reflecting isotopic altitude effect of rainwater. Generally the δD and $\delta^{18}O$ values of rainwater become lower at higher altitude due to the isotopic altitude effect. Therefore it is necessary for the δD and $\delta^{18}O$ values of groundwater also to become lower as the groundwater is recharged at higher altitude. It is thought that the mean altitude of the recharge area of S2 is higher than that of S1, resulting in relatively low values of δD and $\delta^{18}O$ for S2.

The difference in concentrations of excess SO_4^{2-} in the S1 and S2 spring water is noteworthy; the concentration of S2 is higher than that of S1 (Fig. 3). The difference may also be related to the difference in mean altitude of the recharge area. The thickness of the ejecta increases at higher altitude in the island (Fig. 1), suggesting a possibility that the recharge area of the high altitude spring contains higher excess SO_4^{2-}. If so, it is reasonable that the difference of mean altitude of the recharge area is responsible for those isotopic compositions and excess SO_4^{2-} in spring water at the same time. A similar relationship between S1 and S2 is also seen between S11 and S12. These springs are also located nearby. However, the isotopic composition of S12 is lower than that of S11, and the concentration of excess SO_4^{2-} is higher at S12 than that of S11. This is also interpreted to be in the difference in the mean altitude of the recharge area.

4.5 Temporal Variations of Isotopic Compositions

The temporal variations of δD, $\delta^{18}O$, $\delta^{13}C$, and $\delta^{34}S$ are shown in Figure 6. These temporal variations are small as a whole except for $\delta^{34}S$.

As for δD, $\delta^{18}O$, and $\delta^{13}C$, no significant changes in the temporal variations occur. Gradual decrease of δD with time can be seen. However, the cause is not clear. As discussed in Section 4.4, it is possible that a long-term decrease of δD in rainwater

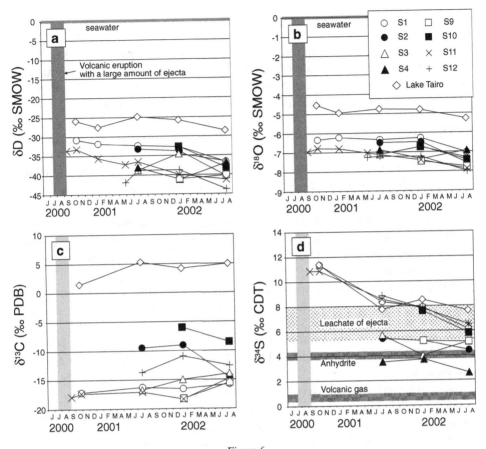

Figure 6

Temporal variations of a) hydrogen, b) oxygen, c) carbon, and d) sulfur stable isotopic compositions. Hatched period shows the period of active eruption with a large amount of ejecta. Hatched values in d show those of leachate of the ejecta (SATOH *et al.*, 2001), anhydrite and volcanic gas (SHINOHARA *et al.*, 2002).

may be the reason. The $\delta^{18}O$ values have slightly decreased as well with time, which might also be due to the long-term fluctuation of rainwater.

Significant decreases of $\delta^{34}S$ values with time are observed at S1, S10, S11, S12, and Lake Tairo. The $\delta^{34}S$ values which were about 11‰ in October 2000, changed to 6–8‰ in August 2002 simultaneously at S1, S11, and Lake Tairo. The value arrived at for $\delta^{34}S$ corresponds well to that of the leachate of the ejecta emitted from July to September 2000, which is 5.5–8‰ by SATOH *et al.* (2001). On the other hand, relatively small changes in $\delta^{34}S$ are observed at S3, S4 and S9. The values of $\delta^{34}S$ fluctuate around that of anhydrite, which is 4‰ by SHINOHARA *et al.* (2002). At these springs, significantly increased concentrations of excess SO_4^{2-} were seen in Figure 3. Especially at S4 and S9, significant increases of the Ca^{2+} concentrations were also observed. Therefore, it is conceivable that the origin of the excess SO_4^{2-} is anhydrite included in the ejecta.

5. Summary

In relation to the eruption of the Miyakejima volcano in 2000, several geochemical changes were observed at eight springs and a lake. In particular, remarkable increases of concentrations of SO_4^{2-} were seen in several springs; the concentration of which reached 1000 mg/L. These results are summarized as follows:

1) The concentration of the excess SO_4^{2-} of the spring water tend to increase in the area where thick ejecta cover occurs.
2) Changes in the weight balance of anion concentrations show that the end member has the Cl/S mole ratio of 0.3–1, the values of which are consistent with those of leachate of ejecta.
3) Springs with low values of δD and $\delta^{18}O$ tend to show high concentrations of excess SO_4^{2-}. According to isotopic altitude effect, the recharge area of the springs with low values of δD and $\delta^{18}O$ is thought to be a high altitude area, where thick ejecta are distributed.
4) The $\delta^{34}S$ values of several spring water changed gradually from 11 to 6–8‰ with time, the latter values of which are consistent with that of leachate of ejecta.

Consequently it is interpreted that the remarkable increases of SO_4^{2-} concentrations of the spring water are derived from the mixing of leachate of the ejecta and groundwater.

Accompanied with the increase of the SO_4^{2-} concentration, hardness of spring water, closely related to the Ca^{2+} and Mg^{2+} concentrations, also increased significantly. This is a serious problem for the inhabitants, because some of the springs are used for domestic purposes, including drinking. However, our results predict that these geochemical changes will decrease with time as elimination of the ejecta is taking place by mudflow or engineering works.

Acknowledgements

We would like to thank the Japan Meteorological Agency, the Miyake Village Office, the Tokyo Metropolitan Government, the Tokyo Fire Department, the Metropolitan Police Department and the Japan Self-Defense Force for their support for the survey. We thank Dr. N. Matsumoto, Dr. H. Shinohara, Dr. R. Ohtani, Dr. M. Yasuhara and Dr. N. Koizumi for supporting water sampling and analysis.

REFERENCES

ARAI, T. (1978), *Water balance of lake Tairo in Miyake Island*, Geograph. Rev. Japan *51*, 704–720 (in Japanese with English abstract).

COLEMAN, M.L., SHEPHERD, T.L., DURHAM, J.J., ROUSE, J.E., and MOORE, G.R. (1982), *Reduction of water with zinc for hydrogen isotope analysis*, Anal. Chem. *54*, 993–995.

GESHI, N., SHIMANO, T., NAGAI, M., and NAKADA, S. (2002), *Magma plumbing system of the 2000 eruption on Miyakejima volcano*, Japan, Bull. Volcanol. Soc. Japan *47*, 419–434 (in Japanese with English abstract).

GIGGENBACH, W.F. (1992), *Isotopic shifts in waters from geothermal and volcanic systems along convergent plate boundaries and their origin*, Earth Planet. Sci. Lett. *113*, 495–510.

GLEASON, J.D., FRIEDMAN, I., and HANSHAW, B.B. (1969), *Extraction of dissolved carbonate species from natural water for carbon-isotope analysis*, U. S. Geol. Survey Prof. Paper 650-D, D248–D250.

KAZAHAYA, K., HIRABAYASHI, J., MORI, H., ODAI, M., NAKAHORI, Y. NOGAMI, K., NAKADA, S., SHINOHARA, H., and UTO, K. (2001), *Volcanic gas study of the 2000 Miyakejima volcanic activity: Degassing environment deduced from adhered gas component on ash and S02 emission rate*, J. Geography, *110*(2), 271–279 (in Japanese with English abstract).

MACHIDA, I. (1999), *Study on the hydrological circulations of Miyakejima Island and Hachijojima Island, Izu Volcanic Islands, Tokyo, Japan*, Geochemical and Isotopical Investigation, Graduate School of Science and Technology, Chiba Univ., p.38.

MACHIDA, I. (2000a), *Spatial and temporal changes in oxygen and hydrogen isotope ratio of precipitation on Miyakejima Island, Tokyo, Japan*, Japan Soc. Hydrol. Water Resour. *13*(2), 103–113 (in Japanese with English abstract).

MACHIDA, I. (2000b), *Study on groundwater in Miyakejima Island by using water quality, oxygen and hydrogen stable isotopes*, EOS Trans. *81*(22), WP49.

MORI, T., SUMINO, H., and NOTSU, K. (2001), *Remote measurements of volcanic gas chemistry in the plume of Miyakejima volcano*, Abstr. 2001 Japan Earth Planet. Sci. Joint Meeting, V0-029.

NAKADA, S., NAGAI, M., YASUDA, A., SHIMANO, T., GESHI, N., OHNO, M., AKIMASA, T., KANEKO, T., and FUJII, T. (2001), *Chronology of the Miyakejima 2000 eruption: Characteristics of summit collapsed crater and eruption products*, J. Geography *110* (2), 168–180 (in Japanese with English abstract).

SATO, T., NAKAMURA, T., TAKAHASHI, M., ITO, J., and GESHI, N. (2003), *Acid rain along the coastal loop road in Miyakejima Island, Japan –in relation to the 2000 eruption*, Abstr., 2003 Japan Earth Planet. Sci. Joint Meeting, V055-P019.

SATOH, H., SHIMIZU, T., and SHINOHARA, H. (2001), *S and O isotopic systematics of the 2001 eruptions of Miyakejima volcano*, Abstr., 2002, Japan Earth Planet. Sci. Joint Meeting, V0-P018.

SHINOHARA, H., KAZAHAYA, K., SAITO, G., MATSUSHIMA, N., NISHI, Y., SATO, H., FUKUI, K., and ODAI, M. (2002), *Observation of volcanic plume from Miyakejima Volcano (September 2000 – May 2001)*, Rep. Coordinating Commit. Predict. Volc. Erup., No. 78, 109–112 (in Japanese).

TAKAHASHI, M., ABE, K., NODA, T., KAZAHAYA, K., ANDO, N., ENDO, H., and SOYA, T. (1991), *Remarkable temperature rising of groundwater observed in Izu Oshima Island*, Bull. Volcanl. Soc. Japan *36*, 403–417 (in Japanese with English abstract).

YANAGISAWA, F. and SAKAI, H. (1983), *Thermal decomposition of barium sulfate-vanadium pentaoxide-silica glass mixtures for preparation of sulfur isotope ratio measurements*, Anal. Chem. *55*, 985–987.

YOSHIDA, N. and MIZUTANI, Y. (1986), *Preparation of carbon dioxide for oxygen-18 determination of water by use of a plastic syringe*, Anal. Chem. *58*, 1273–1275.

WAKITA, H., NOTSU, K., NAKAMURA, Y., and SANO, Y. (1988), *Temporal variation in geochemical parameters of gas from stream well and hot spring water associated with the 1986 eruption of Izu –Oshima volcano*, Bull. Volcanol. Soc. Japan *33*, S285-S289 (in Japanese with English abstract).

(Received: July 8, 2003; revised: November 21, 2005; accepted: November 24, 2005)

To access this journal online:
http://www.birkhauser.ch

Pure appl. geophys. 163 (2006) 825–835
0033–4553/06/040825–11
DOI 10.1007/s00024-006-0051-0

© Birkhäuser Verlag, Basel, 2006

❙Pure and Applied Geophysics

Monitoring Quiescent Volcanoes by Diffuse CO_2 Degassing: Case Study of Mt. Fuji, Japan

KENJI NOTSU,[1] TOSHIYA MORI,[1] SANDIE CHANCHAH DO VALE,[1,2] HIROYUKI KAGI,[1] and TAKAMORI ITO[1,3]

Absrtract—Since the 8th century, more than seventeen eruptions have been recorded for the Mt. Fuji volcano, with the most recent eruption occurring in 1707 (Hoei eruption). For the past 300 years the volcano has been in a quiescent stage and, since the early 1960s, has exhibited neither fumarolic nor thermal activity. However, the number of low-frequency earthquakes with a hypocentral depth of 10–20 km increased significantly beneath the northeastern flank of Mt. Fuji in 2000–2001, suggesting a possible resumption of magmatic activity. In this study, diffuse CO_2 efflux and thermal surveys were carried out in four areas of the volcano in 2001–2002 in order to detect possible signs of the upward movement of deep magma. At all survey points, the CO_2 efflux was below the detection limit with the exception of a few points with biological CO_2 emission, and ground temperatures at a depth of 20–30 cm were below ambient, indicating no surface manifestations of gas or heat emission. Should magma rise into the subsurface, the diffuse CO_2 efflux would be expected to increase, particularly along the tectonically weakened lineation on the Mt. Fuji volcano, allowing for the early detection of pre-eruptive degassing.

Key words: Quiescent volcanoes, CO_2 degassing, Mt. Fuji.

1. Introduction

Recent developments in the monitoring of active volcanoes have revealed that CO_2 and He degassing can signal the upward movement of magma. At the time of the 1986 eruption of the Izu-Oshima volcano in Japan, the $^3He/^4He$ ratio of steam emitted from a well 3 km away from a vent extruding new magma increased from 1.7 to 5.5 Ra (Ra: unit of atmospheric $^3He/^4He$ = 1.40×10^{-6}) over six months, reflecting the upward magma movement, and then decreased gradually corresponding to drain-back of the magma (SANO *et al.*, 1995). Similar changes were also observed at Unzen volcano, Japan (NOTSU *et al.*, 2001) and at Mammoth Mountain,

[1]Laboratory for Earthquake Chemistry, Graduate School of Science, The University of Tokyo Hongo, Bunkyo-ku, Tokyo, 113-0033, Japan
[2]Laboratoire de Geologie, Ecole Normale Superieure, rue Lhomond, 75231, Paris Cedex 05, France.
[3]Applied Earthquake Measurement Co., Saitama, 336-0015, Japan

USA (SOREY *et al.*, 1998). In the case of Mammoth Mountain, soil CO_2 efflux was also found to be a good indicator of magma behavior (GERLACH *et al.*, 2001). Prior to the 2000 eruption of Usu volcano in Japan, integrated soil CO_2 efflux for the entire area of the summit caldera increased significantly, and then decreased following the eruption (HERNÁNDEZ *et al.*, 2001a). Anomalous levels of soil CO_2 degassing have also been found to correlate with faults and eruptive fissures at Mt. Etna, Italy (GIAMMANCO *et al.*, 1998).

These examples show that the degassing of He and CO_2 can be useful for monitoring subsurface magma movement. In the cases of Izu-Ozhima and Unzen volcanoes, magma is extruded to the surface as lava flows or lava domes, whereas at Mammoth Mountain, a dike is thought to have been emplaced at a depth of about 2 km (SOREY *et al.*, 1998). From the viewpoint of eruption prediction, the development of a method for monitoring magma behavior at depths greater than 10 km could be useful, because such deep events are difficult to be detected by geodetic, gravitational, electromagnetic or thermal observations. At present, the observation of volcanic earthquakes and tremors is the most reliable method for detecting deep magma behavior.

Recent volcanic activity at Mt. Fuji volcano in Japan provides a good example of seismic-based monitoring. The number of low-frequency earthquakes taking place at a depth of 10–20 km beneath the northeastern flank of this volcano increased in 2000–2001, suggesting renewed deep magma activity (UKAWA, 2003). However, no other manifestations suggestive of magma movement, such as thermal anomalies, geodetic change or magnetic anomalies, have been observed. Considering that large amounts of magmatic CO_2 can be released from deep magma reservoirs via the volcanic edifice without thermal manifestations, as observed at Hakkoda volcano, Japan (HERNÁNDEZ *et al.*, 2003) and Mammoth Mountain (GERLACH *et al.*, 1998), CO_2 efflux surveys can be expected to provide information on the current conditions of deep magma below Mt. Fuji volcano. Thus far, studies of this type on the features of diffuse degassing from quiescent volcanoes have been conducted only infrequently. In the present study, the potential utility of CO_2 efflux surveys to evaluate the volcanic activity at quiescent volcanoes is discussed.

2. Fuji Volcano and its Eruption History

Mt. Fuji is the highest mountain (3776 m asl) in Japan and is located in the northernmost reaches of the Izu-Ogasawara arc, which is formed by subduction of the Pacific Plate. More than seventeen eruptions are recorded in historical documents since the 8th century AD (METEOROLOGICAL AGENCY, 1995). Among them, large, disastrous eruptions took place in 800, 864–865 and 1707 AD. In the 864–865 eruptions, more than 1 km^3 of lava flowed from the northwestern flank of the volcano, bisecting a large lake. In the latest 1707 eruption, approximately 0.7 km^3 of scoria and ash was emitted from new vents (Hoei craters) on the southeastern flank 3 km from the summit (MIYAJI, 1988).

According to historical documents, it is known that intense fuming from the summit crater was a continuous feature during that period (TSUJI, 1992). However, significant summit fuming has not been seen for 300 yrs, although small-scale steaming within the summit crater was observed from the crater rim up until the early 19th century. After the 1854 Ansei Tokai Earthquake (M 8.4) off the southern coast of the Tokai area, fumarolic and geothermal activity migrated to the eastern edge of the summit crater (around "Aramaki", see Fig. 2), where the temperature decreased from 82°C in 1897, to 80°C in 1928, 54°C in 1954, and only slightly warm in 1963. Geothermal manifestations have also been observed around Hoei craters and on the eastern flank until about 1950 (SUWA, 1982). However, neither geothermal nor fumarolic manifestations have been reported for any region of Mt. Fuji for the last 30 years, indicating that the volcano is currently in a quiescent stage.

From October 2000 to May 2001, the number of low-frequency earthquakes at a depth of 10–20 km below the northeastern flank increased significantly, suggesting a possible resumption of magmatic activity (UKAWA, 2003). During the period 1980–1999, the total number of these events was 274, while the numbers in 2000 and 2001were 180 and 172, respectively. AIZAWA (2004) suggested the existence of an active hydrothermal system deep beneath the summit crater based on the positive self-potential (SP) measured in 2001 and 2002.

3. Observation

Diffuse CO_2 efflux and thermal surveys were conducted in four areas (summit area, around Hoei craters on the southeastern flank, along paths on the northeastern flank, and along a path on the eastern flank), including previous geothermal or fumarolic sites, in order to detect possible signs of magmatic activity (Fig. 1). Diffuse CO_2 efflux was measured by an accumulation chamber method (CHIODINI et al., 1998). The system consists of a cylindrical chamber opened at the bottom with a fan to improve gas mixing and an NDIR (non-disperse infrared) spectrophotometer with an accuracy of approximately 5%. The reproducibility for the range of 100 to 10,000 $gCO_2/m^2/day$ is 10%.

The summit area of Mt. Fuji is characterized by a summit crater 600 m in diameter and 200 m in depth. Diffuse CO_2 efflux and soil temperature measurements were carried out at 18 sites along the circular path around the rim of the summit crater on July 10, 2001. Figure 2 shows the location of sampling sites. Several measurements were conducted in a small area near Aramaki on the eastern edge of the summit crater, where fumarolic and geothermal activity was observed until the 1960s. Soil gas samples were also collected at depths of 25–30 cm using a commercial soil gas sampling probe, and the CO_2, N_2 and O_2 concentrations were determined by standard

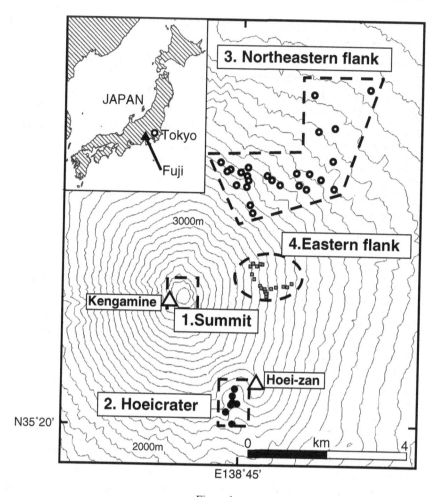

Figure 1

A map of Fuji volcano showing 4 surveyed areas. Observation sites are shown as solid circles (around Hoei craters), open circles (along the Takizawa and Yoshidaguchi paths on the northeastern flank) and solid squares (along the Subashiriguchi path on the eastern flank). For the summit area, observation sites are shown in Figure 2.

gas chromatography. A column filled with Porapak-Q and He as a carrier gas were used for CO_2 determination, whereas a molecular sieve column and Ar carrier gas were used for other species.

Soil CO_2 efflux and temperature measurements were carried out at 7 sites around the Hoei I crater, which was formed by a large phreatic eruption in 1707, on the southeastern flank on September 23, 2001. Along the Takizawa and Yoshidaguchi paths on the northeastern flank, soil CO_2 efflux measurements were carried out at 25 sites on July 23, 2002. This area corresponds to the epicentral region of a low-frequency

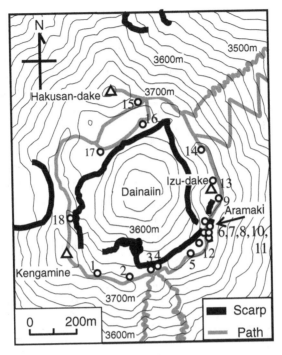

Figure 2
Observation sites in the summit area of Mt. Fuji. Numbers correspond to those in Table 1.

earthquake swarm that occurred at a depth of 10–20 km in 2000–2001. Soil CO_2 efflux measurements were carried out at 16 sites along the Subashiriguchi path on the eastern flank on September 19, 2002. This region was later reported to exhibit a negative SP anomaly (AIZAWA, 2004).

4. Results

Tables 1, 2 show the data obtained in the summit area and around Hoei craters on the southeastern flank. For all of these survey points, CO_2 effluxes were below the detection limit of 5 $gCO_2/m^2/day$, and the ground temperature at 20–30 cm was below the ambient temperature, indicating no gas or heat emission. As the summit area and Hoei crater region are free of vegetation, it is reasonable to assume negligible CO_2 efflux due to biological activity occurs. Thus, chemical composition of the soil gas is assumed to be atmospheric, with 500–1000 ppm CO_2, implying that any CO_2 efflux due to magma degassing, if it occurs, should be detected immediately.

Although steam flow at the Aramaki site ceased around 1963 and the present ground temperature at 25–30 cm is 5–10°C indicating no surface thermal

Table 1

CO_2 efflux, soil temperature (25–30 cm deep) and chemical composition of soil gases in the summit region of Mt. Fuji

Location	CO_2 efflux (g m^{-2} d^{-1})	Soil temperature (°C)	Chemical composition of soil gas			
			CO_2(ppm)	N_2 (%)	O_2(%)	N_2/O_2
fuji01	n.d.	9.0	680	77.9	21.1	3.69
fuji02	n.d.	9.1	–	77.1	20.9	3.68
fuji03	n.d.	3.4	–	–	–	–
fuji04	n.d.	6.3	–	77.3	21.0	3.68
fuji05	n.d.	10.2	540	77.9	21.1	3.69
fuji06	n.d.	8.8	880	77.8	21.2	3.68
fuji06-1	n.d.	8.0	–	–	–	–
fuji06-2	n.d.	7.5	–	–	–	–
fuji07	n.d.	9.0	–	–	–	–
fuji08	n.d.	8.0	–	–	–	–
fuji09	n.d.	11.7	1000	77.8	21.2	3.68
fuji10	n.d.	12.0	650	77.8	21.2	3.68
fuji11	n.d.	–	–	–	–	–
fuji12	n.d.	7.6	690	77.8	21.2	3.68
fuji13	n.d.	1.1	710	77.8	21.2	3.68
fuji14	n.d.	9.0	680	77.9	21.1	3.68
fuji15	n.d.	12.5	950	77.8	21.1	3.68
fuji16	n.d.	7.5	740	77.9	21.1	3.68
fuji17	n.d.	8.7	590	77.9	21.1	3.68
fuji18	n.d.	12.0	740	77.8	21.1	3.68

Location numbers correspond to those in Figure 2. n.d.: not detected (below 5 g m^{-2}d^{-1}) –: not determined

Table 2

CO_2 efflux and soil temperature around Hoei craters

Location	CO_2 efflux (g m^{-2} d^{-1})	Soil temperature (depth) (°C (cm))
hoei01	n.d.	18.2 (20)
hoei02	n.d.	12.9 (50)
hoei03	n.d.	–
hoei04	n.d.	10.3 (50)
hoei05	n.d.	11.5 (–)
hoei06	n.d.	13.9 (25)
hoei07	n.d.	12.8 (25)

n.d.: not detected (below 5 g m^{-2}d^{-1}) –: not determined

manifestations related to magmatic activity, AIZAWA (2004) suggested the existence of a deep hydrothermal system based on a positive SP anomaly in the summit region. Above 3000 m, Mt. Fuji hosts a permafrost layer (KAIZUKA, 1986), and the steaming that persisted up until the 1960s is considered to have escaped via a defrost path through this permafrost layer. However, the path is now closed, presumably due to a decrease in the steam pressure and/or temperature, and no further thermal manifestations or diffuse CO_2 efflux occur at the surface in this region. The deep SP anomaly source associated with the suggested hydrothermal system might be sealed by the overlying permafrost layer, effectively preventing any effect on surface conditions. In contrast, the large amount of porous scoria around Hoei craters readily allows mixing with ambient air, resulting in a dilution of the gas and decreasing the CO_2 efflux.

Along the paths on the northeastern flank corresponding to the epicenter region of the 2000–2001 low-frequency earthquake swarm (UKAWA, 2003), all CO_2 effluxes were below the detection limit of 5 $gCO_2/m^2/day$ except a few points located in woodlands where CO_2 emissions reached 87 $gCO_2/m^2/day$ due to biological contribution. Judging from the depth of the swarm, magma might have moved to a depth of 10–20 km, too deep to trigger significant diffuse soil degassing without deep fissures and allow direct transport of gas to the surface. Along the path on the eastern flank, a negative SP anomaly has been reported and interpreted as a manifestation of the downward percolation of groundwater in weak and permeable ground (AIZAWA, 2004). All 16 measurement points on this path exhibited CO_2 effluxes below the detection limit of 5 $gCO_2/m^2/day$. Although characteristic CO_2 efflux related to the negative SP anomaly was expected since the negative SP region coincides with the locations of fumaroles of 40 years ago (AIZAWA, 2004), no efflux was observed in this region.

5. Discussion

In the present survey, relatively low CO_2 effluxes were measured, suggesting the absence of magmatic gas in the diffuse emanations along the surface environment of Mt. Fuji. However, based on the hypocentral depth of the 2000–2001 low-frequency earthquake swarm, magma appears to be pooling at a depth of 10–20 km below the volcano. Furthermore, the $^3He/^4He$ ratios of bubbling gases released in the hot spring discharges on the northeastern and southeastern slopes of Mt. Fuji indicate a significant contribution of magmatic helium. The maximum $^3He/^4He$ ratio recorded in these areas is 6.43 Ra, obtained at the Suyama hot spring about 15 km southeast of the summit (OHNO et al., personal communication). The positive SP anomaly suggests the existence of a deep hydrothermal system below Mt. Fuji (AIZAWA, 2004), implying that magmatic fluids and gases are currently being supplied at deep levels, although the exact depth remains unclear. These gases are stored in a deep chamber and only helium in them might seep up to the surface.

Previous studies on diffuse CO_2 efflux have revealed that high efflux is typically observed on volcanoes when active plume degassing is low, corresponding to a low-activity stage, as exemplified by the Hakkoda volcano (HERNÁNDEZ et al., 2003) and

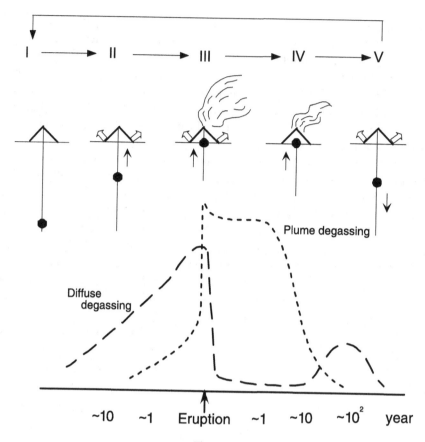

Figure 3

Schematic illustration showing an evolutionary model of gas release from volcanoes. Solid circle is magma rising and descending along the vent. Emission patterns of both diffuse and plume degassing are roughly illustrated as two broken lines. Time scale of horizontal axis is tentatively indicated in case the recurrence time of eruptions is $10^2 \sim 10^3$ years. The time scale generally differs considerable volcano by volcano, because the recurrence time of eruptions varies from several years to more than 10^3 years and the duration of each stage in a cycle varies case by case.

Mammoth Mountain (GERLACH *et al.*, 2001). On the other hand, it is not unusual to observe a very week efflux on volcanoes exhibiting intense plume activity from a central crater. In the case of White Island, New Zealand, plume CO_2 efflux was measured to be 2570–2650 ton/day, whereas diffuse CO_2 efflux was only 8.7 ton/day (WARDELL *et al.*, 2001). Similar contrasts have also been reported for the Nisyros volcano, Greece (BROMBACH *et al.*, 2001). The Popocatepetl volcano in Mexico provides an extreme example of negligible diffuse CO_2 emissions of magmatic origin throughout the extensive volcanic edifice (VARLEY and ARMIENTA, 2001) despite huge emissions of up to 390,000 ton/day of CO_2 from the summit crater (GOFF *et al.*, 2001).

Considering these observations on the two types of volcanic gas release, it is possible to propose a five-stage evolutionary model for the release of volcanic gas, as follows. Figure 3 shows the sketch of this model, demonstrating the evolutionary change in the correspondence between magma behavior and degassing pattern.

Stage I: When magma pools at a significant depth (e.g., > 10 km), neither plume degassing nor diffuse degassing of volcanic gas is typically observed.

Stage II: As magma rises into the subsurface, diffuse degassing through the permeable ground begins, and the efflux at the surface increases as the magma migrates upward. Such increases in efflux have been observed at Mt. Usu, Japan, prior to the 2000 eruption (HERNÁNDEZ *et al.*, 2001a). Another example of Stage II is observed at Mammoth Mountain, where very significant diffuse soil CO_2 degassing appeared after shallow intrusion of magma in 1989 (SOREY *et al.*, 1998; GERLACH *et al.*, 2001), although the magma intrusion has not lead to eruption in the following fifteen years or more. If magma comes into contact with the subsurface groundwater layer, new fumaroles will appear.

Stage III: When magma reaches the surface, volcanic eruptions occur associated with lava extrusion or violent explosions accompanied by an instantaneous release of a large amount of volcanic gas and ash as a plume. This causes a sudden drop in gas pressure within the volcanic body and results in a rapid decrease in diffuse degassing throughout the volcanic edifice. This was observed just after the 2000 Usu eruption (HERNÁNDEZ *et al.*, 2001a).

Stage IV: While new magma is being supplied to the surface, intense pluming continues and diffuse degassing remains negligible. Examples of this stage have been observed at Mt. Popocatepetl (VARLEY and ARMIENTA, 2001) and at White Island (WARDELL *et al.*, 2001).

Stage V: As magma begins to drain away from the surface, the efflux of plume degassing decreases. After degassing from the volcanic vent stops, if gases are still released from the descending magma at greater depth, diffuse degassing increases considerably due to an increase in the gas pressure of the volcanic body. Diffuse degassing from quiescent volcanoes such as that at Mt. Hakkoda (HERNÁNDEZ *et al.*, 2003) may represent post-eruptive degassing. This diffuse degassing will gradually recede as the volcanic system returns to Stage I.

According to this evolutionary model, Mt. Fuji can be considered to be in Stage I at the present time. Parasitic cones of Mt. Fuji are distributed in NW-SE direction following the direction of tectonic stress originating from the movement of the Philippine Sea plate (NAKAMURA *et al.*, 1984). The major eruptions in 864–865 and 1707 took place on this parasitic cone belt, and the highest $^3He/^4He$ ratios have been observed at the SE extent of this belt (OHNO *et al.*, personal communication). Based on CO_2 efflux measurements at Mt. Etna, GIAMMANCO *et al.* (1998) suggested that only zones of strain are capable of channeling deep gases to the surface. Prior to the 2000 Miyakejima eruption, soil CO_2 efflux was observed only in the summit area (HERNÁNDEZ *et al.*, 2001b), the site of a major collapse during the eruption (GESHI *et al.*, 2002). Since gas components seem to migrate, finding an easy passageway to the surface via faults, fractures, or other zones of weakness, soil CO_2 efflux can be expected to appear in the summit area or in the NW-SE lineation of Mt. Fuji when magma does begin to rise into the subsurface. However, it is a difficult proposition to predict exactly where the gas will reach the surface. At present, no indications of anomalous soil CO_2 efflux have been observed in either the summit region or the southwest region. However, continuous measurement of soil CO_2 efflux is considered a necessary aspect of any monitoring program to detect pre-eruptive activity at the Mt. Fuji volcano.

Acknowledgements

This work was partially supported by the Grant-in-Aid for Scientific Research (Grant No. 13874064) from the Ministry of Education, Culture, Sports, Science and Technology, Japan. Helpful reviews of the manuscript were provided by Dr. K. McGee and Dr. P.A. Hernández. We thank Dr. S. Tezuka, Head of Fujisan Meteorological Station and Dr. Y. Igarashi of the Meteorological Research Institute, Japan Meteorological Agency, for providing us the facility of the observation station on the summit of Mt. Fuji. We also thank Dr. Y. Dokiya of Edogawa University for encouragement of this work. Mr. S . Matsumoto and Mr. F. Barahona were acknowledged for their assistance during the field work.

REFERENCES

AIZAWA, K. (2004), *A large self-potential anomaly and its changes on the quiet Mt. Fuji, Japan*, Geophys. Res. Lett. *31*, L05612, doi:10.1029/2004GL019462.

BROMBACH, T., HUNZIKER, J.C., CHIODINI, G., CARDELLINI, C., and MARINI, L. (2001), *Soil diffuse degassing and thermal energy fluxes from the southern Lakki plain, Nisyros (Greece)*, Geophys. Res. Lett. *28*, 69–72.

CHIODINI, G., CIONI, R., GUIDI, M., RACO, B., and MARINI, L. (1998), *Soil CO_2 flux measurements in volcanic and geothermal areas*, Appl. Geochem. *13*, 543–552.

GERLACH, T.M., DOUKAS, M.P., McGEE, K.A., and KESSLER, R. (2001), *Soil efflux and total emission rates of magmatic CO_2 at the Horseshoe Lake tree kill, Mammoth Mountain, California, 1995–1999*, Chem. Geol. *177*, 101–116.

GESHI, N., SHIMANO, T., CHIBA, T., and NAKADA, S. (2002), *Caldera collapse during the 2000 eruption of Miyakejima volcano, Japan*, Bull. Volcanol. *64*, 55–68.

GIAMMANCO, S., GURRIERI, S., and VALEMZA, M. (1998), *Anomalous soil CO_2 degassing in relation to faults and eruptive fissures on Mount Etna (Sicily, Italy)*, Bull. Volcanol. *60*, 252–259.

GOFF, F., LOVE, S.P., WARREN, R.G., COUNCE, D., OBENHOLZNER, J., SIEBE, C., and SCHMIDT, S.C. (2001), *Passive infrared remote sensing evidence for large, intermittent CO_2 emissions at at Popocatepetl volcano, Mexico*, Chem. Geol., *177*, 133–156.

HERNÁNDEZ, P.A., NOTSU, K., SALAZAR, J.M., MORI, T., NATALE, G., OKADA, H., VIRGILI, G., SHIMOIKE, Y., SATO, M., and PÉREZ, N.M. (2001a), *Carbon dioxide degassing by advective flow from Usu volcano, Japan*, Science *292*, 83–86.

HERNÁNDEZ, P.A., SALAZAR, J.M., SHIMOIKE, Y., MORI, T., NOTSU, K., and PÉREZ, N.M. (2001b), *Diffuse emission of CO_2 from Miyakejima volcano, Japan*, Chem. Geol. *177*, 175–185.

HERNÁNDEZ, P.A., NOTSU, K., TSURUMI, M., MORI, T., OHNO, M., SHIMOIKE, Y., SALAZAR, J., and PÉREZ, N. (2003), *Carbon dioxide emissions from soils at Hakkoda, north Japan*, J. Geophys. Res. *108*(B4): 2210, doi:10.1029/2002JB001847.

KAIZUKA, S. (1986), *Mountainous country, Japan: an introduction*. In *Mountain in Japan*, (Iwanami-shoten, Tokyo), pp. 1–28 (in Japanese).

METEOROLOGICAL AGENCY (1995), *National Catalogue of the Active Volcanoes in Japan*, Second Edition, 500 pp. (in Japanese).

MIYAJI, H. (1988), *History of younger volcano*, J. Geol. Soc. Japan *94*, 433–452 (in Japanese).

NAKAMURA, K., SHIMAZAKI, K., and YONEKURA, N. (1984), *Subduction, bending and eduction. Present and Quaternary tectonics of the northern border of the Philippine Sea plate*, Bull. Soc. geol. France *26*, 221–243.

NOTSU, K., NAKAI, S., IGARASHI, J., ISHIBASHI, J., MORI, T., SUZUKI, M., and WAKITA, H. (2001), *Spatial distribution and temporal variation of $^3He/^4He$ in hot spring gas released from Unzen volcanic area, Japan*, J. Volcanol. Geotherm. Res. *111*, 89–98.

SANO, Y., GAMO, T., NOTSU, K., and WAKITA, H. (1995), *Secular variations of carbon and helium isotopes at Izu-Oshima volcano, Japan*, Earth Planet. Sci. Lett. *64*, 83–94.

SOREY, M.L., EVANCE, W.C., KENNEDY, B.M., FARRAR C.D., HAINSWORTH, L.J., and HAUSBACK, B. (1998), *Carbon dioxide and helium emissions from a reservoir of magmatic gas beneath Mammoth Mountain, California*, J. Geophys. Res. *103*, 15303–15323.

SUWA, A. (1982), *Volcanic Activity in Japan* (Kyoritu-shuppan, Tokyo), 112 pp. (in Japanese).

TSUJI, Y. (1992), *Eruptions of Mt. Fuji* (Tsukiji-Shokan, Tokyo), 260 pp. (in Japanese).

UKAWA, M. (2003), *Activity of Fuji volcano*. In *Reports on Volcanic Activities and Volcanological Studies in Japan for the Period from 1999 to 2002*, Volcanol. Soc. Japan and National Committee for VCE, pp. 18–22.

VARLEY, N.R. and ARMIENTA, M.A. (2001), *The absence of diffuse degassing at Popocatepetl volcano, Mexico*, Chem. Geol. *177*, 157–173.

WARDELL, L.J., KYLE, P.R., DUNBAR, N., and CHRISTENSON, B. (2001), *White Island volcano, New Zealand: Carbon dioxide and sulfur dioxide emission rates and melt inclusion studies*, Chem. Geol. *177*, 187–200.

(Received: August 30, 2004; revised: November 28, 2005; accepted: November 30, 2005)

To access this journal online:
http://www.birkhauser.ch

Pure appl. geophys. 163 (2006) 837–851
0033–4553/06/040837–15
DOI 10.1007/s00024-006-0036-z

© Birkhäuser Verlag, Basel, 2006

❙ Pure and Applied Geophysics

Puhimau Thermal Area: A Window into the Upper East Rift Zone of Kīlauea Volcano, Hawaii?

K.A. McGee,[1] A.J. Sutton,[2] T. Elias,[2] M.P. Doukas,[1] and T.M. Gerlach[1]

Abstract—We report the results of two soil CO_2 efflux surveys by the closed chamber circulation method at the Puhimau thermal area in the upper East Rift Zone (ERZ) of Kīlauea volcano, Hawaii. The surveys were undertaken in 1996 and 1998 to constrain how much CO_2 might be reaching the ERZ after degassing beneath the summit caldera and whether the Puhimau thermal area might be a significant contributor to the overall CO_2 budget of Kīlauea. The area was revisited in 2001 to determine the effects of surface disturbance on efflux values by the collar emplacement technique utilized in the earlier surveys. Utilizing a cutoff value of 50 g m^{-2} d^{-1} for the surrounding forest background efflux, the CO_2 emission rates for the anomaly at Puhimau thermal area were 27 t d^{-1} in 1996 and 17 t d^{-1} in 1998. Water vapor was removed before analysis in all cases in order to obtain CO_2 values on a dry air basis and mitigate the effect of water vapor dilution on the measurements. It is clear that Puhimau thermal area is not a significant contributor to Kīlauea's CO_2 output and that most of Kīlauea's CO_2 (8500 t d^{-1}) is degassed at the summit, leaving only magma with its remaining stored volatiles, such as SO_2, for injection down the ERZ. Because of the low CO_2 emission rate and the presence of a shallow water table in the upper ERZ that effectively scrubs SO_2 and other acid gases, Puhimau thermal area currently does not appear to be generally well suited for observing temporal changes in degassing at Kīlauea.

Key words: Carbon dioxide, volcanic gases, soil, efflux, puhimau, kīlauea.

Introduction

Puhimau thermal area, known also as the Puhimau Hot Spot, is a small irregularly shaped area of hot ground approximately 16 ha in size located about 400 m northwest of Koʻokoʻolau Crater and about 200 m south of Puhimau Crater in the upper ERZ of Kīlauea volcano, Hawaii (Figs. 1 and 2). T. A. Jaggar (1938) described the area as a new hot area in the forest where the ground temperature was as high as 85°C and speculated that the previous vegetation had died between 1936 and 1938. Thus, although not well constrained, it appears that the Puhimau thermal area began to form about 1936 as a result of heat and/or gases released from an intrusion beneath

[1] U.S. Geological Survey, Cascades Volcano Observatory, Vancouver, Washington, USA
[2] U.S. Geological Survey, Hawaiian Volcano Observatory, Hawaii National Park, Hawaii

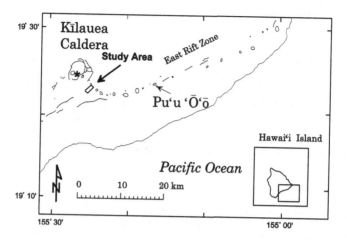

Figure 1
Location map for Kīlauea volcano and the East Rift Zone showing the area of study. The small circle
within the boundaries of Kīlauea caldera is Halema'uma'u; the adjacent asterisk marks the approximate
location where the majority of the CO_2 from Kīlauea is vented to the atmosphere.

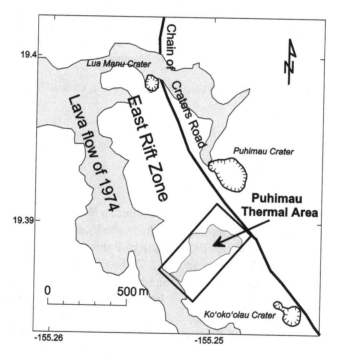

Figure 2
Location map for Puhimau thermal area in Kīlauea's upper East Rift Zone. The study area is bounded on
the northeast by the Chain of Craters road and on the southwest by the Lava Flow of 1974.

that area killing much of the nearby vegetation in the surrounding ohia forest. A self-potential (SP) survey carried out in 1975 revealed a large anomaly that suggested a stocklike intrusion under the Puhimau thermal area (ZABLOCKI, 1978). On the surface, the area is characterized by an inner region of hot barren ground consisting mostly of clay formed by alteration of a surface layer of lapilli and ash from the 1790 A.D. Keanakāko'i eruption (CASADEVALL and HAZLETT et al., 1983; McPHIE, 1990). Temperatures today are still high, ranging up to 87 °C at a depth of only 5 cm. Diffuse steam emanates from many places but is particularly noticeable in the shallow holes that frequently still contain ohia tree stumps from the earlier forest. Moving outward from the hot inner area, lichens and moss begin to appear followed by short grasses. Further out, tall grasses dominate until the edge of the ohia forest is reached where ferns and trees dominate. Significant ground cracks are noticeable at the northwestern boundary of the Puhimau thermal area. The southwestern boundary is marked by a thick aa lava flow emplaced in July 1974. Although originally restricted to the west side of Chain of Craters Road, a recent photograph of the area (Fig. 3) shows its probable extension to the east side of the road.

Besides steam, the gases released from the hot inner area of Puhimau thermal area contain magmatic CO_2 (GERLACH and TAYLOR, 1990). Earlier investigators also report finding about 100 ppb of total sulfur (HINKLE, 1978), although the exact chemical form was not identified. A study of condensate samples from Puhimau thermal area between 1987–1989 utilizing neutron activation/gamma-ray spectroscopy techniques was inconclusive for sulfur with the results falling at or

Figure 3
The extent of altered ground and lack of vegetation is clearly visible in this oblique aerial photograph of Puhimau thermal area taken in 2002. View is to the southwest. The anomaly now appears to extend across Chain of Craters road (foreground) and is truncated by the Lava Flow of 1974 at the far end. Photo courtesy of D. A. Swanson, U. S. Geological Survey.

below the detection limit of the analytical method (10 ppm) (BARNARD *et al.*, 1990). The odor of SO_2 or H_2S was not evident during any of our visits to the Puhimau thermal area.

Puhimau thermal area has also been identified as a significant point source of mercury. HINKLE (1978) found mercury in excess of 10 ppm in the central steaming area, more than 100 times the concentrations observed in peripheral areas, except for high concentrations near the fractures on the northwest edge of the area. Using mercury found in soil lichen and in the leaves of ohia trees, CONNOR (1979) estimated that vegetation as far away as 3 km downwind was being impacted by mercury outgassing from Puhimau thermal area. The concentration of mercury in ohia leaves ranges from about 500 ppb at the edge of the anomaly to about 200 ppb three km away (CONNOR, 1979). In a more recent study utilizing lichens as a biomonitor, DAVIES and NOTCUTT (1996) also found anomalously high values of mercury, up to twice the island background value, at the Puhimau thermal area. They attribute these localized high values to hot circulating ground waters dissolving mercury from rocks with the resulting mercury vapor carried to the surface by steam. Helium has also been found in concentrations ranging up to slightly more than 1.5 ppm above ambient, with the higher values mostly around the periphery of the hot ground and, like mercury, near the fractures on the northwest edge of the Puhimau thermal area (HINKLE, 1978).

Puhimau thermal area is a scant 2.5 km down the ERZ from the summit crater of Kīlauea. Besides being one of the few areas in the upper ERZ with well developed soils capable of supporting soil efflux measurements, Puhimau thermal area is an area of obvious degassing and heat flow, estimated by DUNN and HARDEE (1985) to be about 250 W m^{-2}. It is also an area where evidence of the passage of magma down the ERZ has, at times, been observed. In 1980, there were five intrusive events recorded in the upper ERZ of Kīlauea and, fortuitously, during at least two of those events, transient SO_2 was detected during routine sampling at Puhimau thermal area (DECKER, 1987; L. P. GREENLAND, unpublished analyses). In addition, in late December 1982, McGEE (1987) detected a widespread but transient gas event at several continuous monitoring stations at the summit of Kīlauea and along the ERZ, including Puhimau thermal area, using a chemical sensor that detects H_2, SO_2, and H_2S. This proved to be precursory to the east rift eruption that began a few days later on January 3, 1983 and continues today.

Thus we began this study in order to determine if the Puhimau thermal area could potentially provide a window into the upper ERZ through which we could gain information that might allow a better understanding of how Kīlauea volcano degasses. Here, we present the results of our work that include the first soil gas efflux surveys at Puhimau thermal area and an estimate of the total CO_2 emission rate for this important site on Kīlauea's upper ERZ.

Procedures and Methods

Two soil CO_2 efflux surveys were carried out at the Puhimau thermal area on June 11–13, 1996 and June 2–6, 1998. The instrumentation and methodology was identical in each survey except that 112 points were measured in 1996 and 167 points were measured in 1998. An existing grid, established earlier for other studies, and marked on the ground by rebar posts, was adapted for use in both surveys. The spacing between survey measurement points was regular (30 m) and based upon a square grid pattern. Additional points extending beyond the existing grid were located by measuring tape. The area covered by the surveys includes the thermal anomaly and its margins and is bounded on the northeast by the Chain of Craters road and on the southwest by the July 1974 lava flow.

Both surveys employed a LI-COR system based upon the closed chamber method for measuring soil CO_2 efflux (NORMAN et al., 1992, 1997). The measurement system includes a nondispersive infrared CO_2 analyzer (LI-6252), a flow control unit (LI-670), a closed circulation soil respiration chamber (LI-6000-09), and a laptop computer for data acquisition. The infrared analyzer was calibrated with CO_2 standards mixed in air. The soil chamber was placed on a PVC collar inserted into the ground and the air in it circulated to the infrared analyzer in a closed loop at a rate of 2 L min^{-1} and returned. Air flowing back to the soil chamber from the infrared analyzer passes through a perforated manifold designed to mix chamber air without causing pressure gradients or excessive ventilation of the soil (WELLES et al., 2001). A magnesium perchlorate scrubber tube removed any water vapor in the air flowing from the soil chamber to the infrared analyzer. CO_2 concentrations in the air flowing from the soil chamber through the sample cell of the infrared analyzer were averaged over 1-s intervals. As the partial pressure of CO_2 is sensitive to total pressure, CO_2 concentrations were automatically adjusted for ambient pressure by a pressure transducer mounted within the infrared analyzer. The CO_2 concentration measurement accuracy of the LI-COR analyzer is ± 1 ppm with a precision (signal noise) of ± 0.15 ppm (max). The internal temperature of the soil chamber was measured by a digital thermometer during every survey measurement. Soil temperatures were also measured with a Type E soil temperature probe hooked to a digital readout and inserted to a depth of 5 cm in the soil at each efflux measurement point.

The CO_2 soil efflux at each measurement point was determined by the standard LI-COR protocol for such measurements (NORMAN et al., 1992) where the rate of increase in CO_2 concentration in the soil chamber is proportional to the CO_2 efflux from the ground. Determination of this rate of increase was optimized by following a CO_2 scrubbing protocol (NORMAN et al., 1992, 1997; WELLES et al., 2001) whereby the rate of CO_2 build-up in the chamber is determined as the concentration of CO_2 in the chamber rises through the ambient ground level CO_2 concentration. The soil CO_2

efflux (F_{CO2}) is calculated from the following equation given in GERLACH *et al.*(2001):

$$F_{CO_2} = \frac{PV}{RTA}\left(\frac{dX_{CO_2}}{dt}\right), \tag{1}$$

where dX_{CO_2}/dt is the rate of change of the mole fraction CO_2 concentration in a chamber with footprint area A, R is the universal gas constant, and P, V, and T are the pressure, temperature, and volume of the chamber. It is assumed that the chamber pressure and temperature remain nearly constant during the short measurement interval and that pressure is equal to ambient barometric pressure.

To check the reproducibility of the soil efflux measurements, a series of measurements were made over a short time interval (< 1 hour) at a control point in order to compute a coefficient of variation (CV), the ratio of the standard deviation to the mean. The average CV for these measurements expressed in percent is 10%. Thus our assumed one-standard-deviation precision for the soil efflux measurements at Puhimau thermal area, reflecting uncontrolled variables and possibly small natural variations in efflux, is 10%.

An investigation into the effects of surface disturbance on efflux values determined during earlier soil surveys at the Horseshoe Lake tree kill in California showed that soil surface disruption due to collar emplacement can produce misleadingly high efflux values (GERLACH *et al.*, 2001). In order to determine whether such a phenomenon might have influenced the measurements in the 1996 and 1998 soil surveys reported here, a subset of representative points at the Puhimau thermal area was measured again on June 13, 2001. Two measurements were made at each survey point. First, the soil chamber was placed on the ground with no surface disturbance or preparation. Then, a second measurement was made at exactly the same spot using a collar emplacement strategy identical to the earlier surveys. The efflux values for the surface placement measurements (F_{sp}) where the soil surface was undisturbed were consistently lower than those by collar emplacement (F_{ce}) indicating that soil disturbance by collar emplacement did influence the results of the earlier surveys. The average ratio (F_{sp}/F_{ce}) in this study is 0.65, which is essentially the same as that obtained by GERLACH *et al.* (2001) for the survey at the Horseshoe Lake tree kill. Accordingly, all of the 1996 and 1998 Puhimau efflux values reported here have been corrected for the effects of collar emplacement (GERLACH *et al.*, 2001).

Results

The CO_2 efflux contour maps for the 1996 and 1998 surveys at Puhimau thermal area are shown in Figure 4. Ten soil efflux measurements were made at random

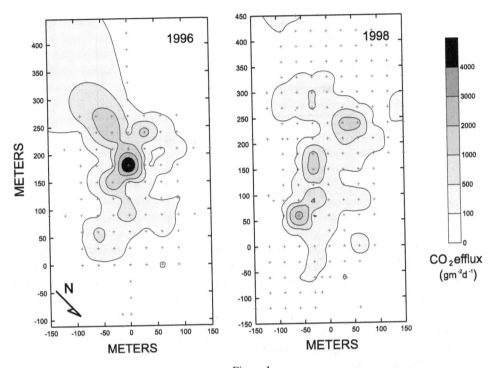

Figure 4

Contour maps of soil CO_2 efflux for the 1996 and 1998 surveys at Puhimau thermal area. Crosses locate measurement points. The total area represented in each plot is 17.25 ha. After subtracting the background efflux of the surrounding forest, the CO_2 emission rates for the study area were 27 ± 3 t d^{-1} in 1996 and 17 ± 2 t d^{-1} in 1998.

locations adjacent to the Puhimau thermal area to determine typical values for nearby forest soils so that the measured efflux in the area of study could be corrected for the local background efflux. Efflux values of CO_2 in the nearby forest soils, presumed to be biogenic in origin, ranged from 2 to 42 g m^{-2} d^{-1} so we assumed a conservative 50-g m^{-2} d^{-1} value for the forest background efflux. We applied gridding and contouring software (Surfer v. 7, Golden Software Inc.) to the 1996 and 1998 survey data to constrain the CO_2 emission rates for the Puhimau anomaly (GERLACH et al., 2001). Utilizing volume and area integration algorithms in Surfer to a grid produced by a linear kriging model yielded CO_2 emission rates of 27 ± 3 t d^{-1} (1996) and 17 ± 2 t d^{-1} (1998) using the 50-g m^{-2} d^{-1} cutoff value to separate the efflux of the Puhimau thermal area from the background forest soil efflux. Raising the background cutoff to a more conservative 100 g m^{-2} d^{-1} lowers emission rates by only about 3 t d^{-1} demonstrating that the Puhimau emission rates are not overly sensitive to the cutoff value used for the surrounding forest background efflux. The distribution of data is non-normal (Fig. 5), however using Surfer's model for

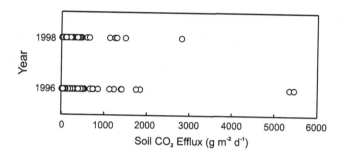

Figure 5
Point plot distributions of CO_2 efflux measurements for the 1996 and 1998 Puhimau thermal area surveys showing the non-normal distribution of the data.

triangulation with linear interpolation for contouring produces results that do not differ noticeably from the kriging model. Refer to GERLACH *et al.* (2001, p. 107) for more discussion on gridding and contouring non-normal soil efflux data and the non model-dependent nature of emission-rate estimates of this type. The uncertainties for the emission rates (~10%) are estimates of two standard deviations based on the sensitivity of the rates to the precision of individual efflux measurements.

The area covered by the Puhimau CO_2 anomaly (≥ 50 g m^{-2} d^{-1}) is about 8 ha for both the 1996 and 1998 surveys. Although the areas are about the same, the CO_2 data suggest that not only did the peak efflux values in the central core area decline but the anomaly as a whole may have migrated slightly to the NNE toward the Chain of Craters road between 1996 and 1998 (Fig. 4). In addition, it appears that the central core area of high CO_2 efflux mapped in 1996 has, by 1998, broken apart into three separate areas with the 1998 high about 100 m NE of the 1996 high. The coverage of measurement points in the central part of the anomaly is nearly identical for both surveys and thus the detail in the central portion of the anomaly could reflect real changes in degassing between 1996 and 1998; the additional points in the 1998 survey are mostly around the edges of the anomaly. Shallow faulting or a change in groundwater flow might plausibly cause such a change in the surface expression of the anomaly. Alternatively, the variation between the two surveys could simply be normal temporal and spatial variations due to the effects of meteorological forcing on individual grid points; such variations in soil CO_2 have been observed elsewhere (GERLACH *et al.*, 2001; McGEE and GERLACH, 1998; ROGIE *et al.*, 2001).

Soil temperatures measured at all grid stations during the 1996 survey ranged from about 19 °C on the periphery of the anomaly to 87 °C in the center while in 1998 they ranged from 17 °C to a high of 84 °C. All of the temperature measurements were taken at a soil depth of 5 cm to minimize wind and other surface effects. Figure 6 displays the temperature data after gridding and contouring

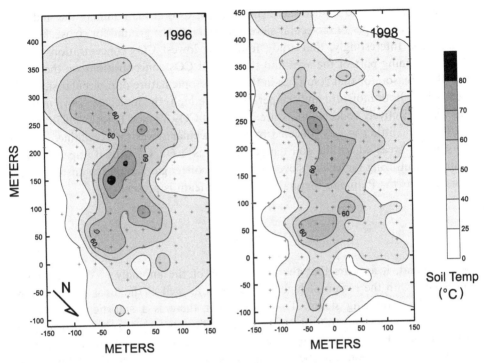

Figure 6

Contour maps of soil temperature for the 1996 and 1998 surveys at Puhimau thermal area. Crosses locate measurement points. The area and grid pattern are the same as for the CO_2 efflux measurements shown in Figure 4.

in the same manner as the CO_2 efflux data. For both surveys, the size of the anomaly where the soil temperature is $\geq 25°C$ is around 16 ha, about double the size of the CO_2 anomaly, and clearly extends outside of the area of study. As with the CO_2 anomaly, the temperature anomaly in 1998 appears to be more distributed than in 1996 with distinct fingers extending to the NNE, W, and S. The new areas to the W and S may simply be the result of better resolution in 1998 in those portions of the study area. However, the migration of hotter temperatures to the NNE is unambiguous because there are several measurement points in common in that portion of the study area between both years. Thus the migration of the temperature anomaly NNE toward the Chain of Craters Road between 1996 and 1998 appears to be unmistakable and Figure 3 suggests that the process may be continuing. Because the temperature measurements were taken in such a way as to minimize surface effects, they are less likely to be affected by meteorological forcing than the CO_2 measurements. This suggests that for the temperature measurements at least, the variation between the 1996 and 1998 surveys is more likely due to other shallow processes such as shallow faulting or a change in the local groundwater system.

HINKLE (1978) also studied the distribution of CO_2 in the Puhimau thermal area by inserting envelopes of molecular sieve material into the ground for six weeks prior to analysis. Interestingly, that study found the lowest CO_2 concentrations in the central steaming part of the area and the highest CO_2 concentrations on the edges. The likely reason for that result lies in the hygroscopic nature of molecular sieves. In areas where both moisture and temperature were high, the molecular sieve material may have selectively collected moisture instead of CO_2. In fact, HINKLE (1978) points out that very little gas was collected by the molecular sieves that contained > 10% moisture in areas where the soil temperature was ≥ 60 °C. Thus it appears that a strategy of utilizing molecular sieve material to collect gas samples is unsuitable for areas such as the Puhimau thermal area where steam is present.

Discussion

In general, it appears that the hotter areas of the anomaly (> 60 °C) tend to correspond with the region of higher CO_2 efflux as well as the areas that are more likely to be producing steam. To test whether there is a statistical relationship between efflux and temperature, correlation coefficients (r) were computed for each of the surveys. For the 1996 survey, $r = 0.55$, while for the 1998 survey, $r = 0.46$. This positive correspondence begs the question as to whether the CO_2 efflux results might be affected by either the hotter temperatures or the presence of water vapor. For each soil efflux measurement, the temperature of the gas inside the soil chamber was recorded during the measurement from a thermocouple suspended in the chamber. That temperature reading was then utilized in equation (1) to account for any temperature variations during the survey. However, when moisture is present in soil gases, as in the case of the steaming ground at Puhimau thermal area, water vapor will dilute the CO_2 concentration in the chamber. This, in turn, will result in a slower rate of increase of CO_2 in the chamber (dX_{CO_2}/dt) and produce efflux rates that are too low. The problem of water vapor dilution is discussed in detail by WELLES *et al.* (2001) who point out that the resulting error is proportionally larger when efflux rates are low. It is possible to correct for the effects of water vapor dilution if the mole fraction of water vapor is measured concurrently with CO_2. As we did not measure water vapor, we are unable to make that specific correction for the Puhimau thermal area surveys. Instead, we utilized a magnesium perchlorate scrubber to remove water in the air stream from the soil chamber to the CO_2 analyzer and obtain CO_2 values on a dry air basis, thus mitigating the problem of errors due to water vapor dilution in the soil chamber.

Although no isotopes were measured in this study, it is well established that the Puhimau thermal area gases are magmatic in origin (FRIEDMAN *et al.*, 1987; GERLACH and TAYLOR, 1990). The carbon isotope composition of the Puhimau gases, like all the summit fumarole gases at Kīlauea, reflects degassing of CO_2 from the summit magma chamber (GERLACH and TAYLOR, 1990). Clearly, the small amount of CO_2 measured in this study indicates that the Puhimau thermal area is not a significant contributor to the overall CO_2 budget of Kīlauea volcano. It is well established that most of the CO_2 present in Kīlauea's primary magma is vented to the atmosphere during quiescent residence in the shallow summit reservoir after having arrived there as exsolved vapor in a CO_2-rich fluid. This leaves only CO_2-depleted magma for injection into the rift system (GERLACH and GRAEBER, 1985; GREENLAND et al., 1985). Recent measurements made by GERLACH et al. (2002) over a period of 4 years reveal a CO_2 emission rate of 8500 t d^{-1} for the summit of Kīlauea that is considerably larger than earlier estimates, while the CO_2 emission rate at Pu'u 'Ō'ō is only 240–300 t d^{-1} (GERLACH et al., 1998). However, after loss of 94–97% of its CO_2 in summit degassing, magma injected into the rift system typically retains most of its remaining volatiles (H_2O, SO_2, HCl, HF). In 1995, for example, MCGEE and GERLACH (1998) measured an SO_2 emission rate of 2,160 metric tons per day (t d^{-1}) at the erupting Pu'u 'Ō'ō vent on the East Rift while SO_2 emissions at Kīlauea's summit averaged only about one-tenth that amount, 210 t d^{-1}, during the same period (SUTTON et al., 2001) indicating that a significant amount of SO_2 remains dissolved in magma traveling down the ERZ.

Thus it appears that SO_2-rich magma is passing through the upper ERZ at a shallow depth with no evidence of its passage at the surface. Given the high solubility of SO_2 in tholeiitic basalt and its tendency to exsolve only at very low pressures, ERZ magma may still be too deep to release much SO_2. Alternatively, any SO_2 that may be released would be susceptible to scrubbing by water. There is abundant evidence for a shallow ground water system (0.5–2 km below the surface) in the summit and upper ERZ (INGEBRITSEN and SCHOLL, 1993; KAUAHIKAUA, 1993; TILLING and JONES, 1991). In fact, an earlier self-potential (SP) study suggested an anomalously shallow water table in the upper ERZ likely resulting from impoundment by dikes within the rift zone (JACKSON and KAUAHIKAUA, 1987). Their SP profile crossing the upper ERZ just south of Puhimau thermal area at Ko'oko'olau Crater showed the apparent water table in that location to be as shallow as 230 m below ground surface. Thus the evidence suggests that the main magma conduit in the upper ERZ lies beneath the local water table. Given the propensity of SO_2 to undergo hydrolysis reactions upon contact with meteoric water and/or hydrothermal fluids (SYMONDS et al., 2001), there is little wonder that SO_2 is seldom detected along the ERZ except in eruption gases. Unfortunately, there are no data on dissolved chemical species for the groundwater at Puhimau thermal area.

KLEIN et al. (1987) describe the upper ERZ as a multi tiered and complex honeycomb of conduits with the main magma conduit located at about 3 km depth. The dikes and fractures that make up the ERZ describe a zone that is perhaps 2–3 km wide with a somewhat larger vertical dimension. The preponderance of geophysical evidence suggests that a narrow pipe-like conduit supplies magma from the mantle to a shallow summit reservoir complex residing at a depth of 2–7 km near the south edge of Kîlauea caldera that, in turn, intermittently injects magma into the rift system (KLEIN *et al.*, 1987; YANG *et al.*, 1992; TILLING and DVORAK, 1993). A recent three-dimensional velocity structure study of Kîlauea using *P*- and *S*-wave earthquake arrival times revealed two high V_p/V_s zones, thought to be magma reservoirs, at 1–4 km depth under the southern caldera rim and under Puhimau thermal area; the two zones appear to be connected by an inverted v-shaped passageway (DAWSON *et al.*, 1999). Yet, an earlier detailed controlled-source audiofrequency magnetotelluric survey at the Puhimau thermal suggests the anomaly there is actually a dike-like feature that may contain either hot, mineralized water or magma (BARTEL and JACOBSON, 1987). Regardless of the exact nature of the anomaly beneath Puhimau thermal area, it is evident from the work of GERLACH *et al.* (2002) that Kîlauea's summit region receives magma rich in exsolved CO_2 that is vented to the atmosphere at an area just east-southeast of Halema'uma'u; CO_2-poor and SO_2-rich magma is then fed into the upper ERZ in the vicinity of Puhimau thermal area.

Despite the fact that Puhimau thermal area is a persistently hot surface feature in the upper ERZ, regular monitoring of CO_2, SO_2 and other gases there seems of little use during the ongoing ERZ eruption that began in January 1983. The magma pathway from the summit chamber to the erupting Pu'u 'Ō'ō vent is well established and any escaping SO_2 is effectively removed by the high water table. However, once the current ERZ eruption eventually ceases and the activity at Kîlauea returns again to the summit region with periodic reservoir inflation and subsequent intrusive injection of fresh hot magma down one rift zone or another, the possibility of detecting transient fugitive gases, such as SO_2 or other acid gases, at upper ERZ locations such as Puhimau thermal area increases. The likelihood of detecting much CO_2 at Puhimau thermal area remains remote, however, barring a fundamental plumbing change that shifts the supply of CO_2-rich mantle magma from the summit magma chamber into the Puhimau chamber.

Conclusions

Two soil CO_2 surveys at the Puhimau thermal area in 1996 and 1998 using the accumulation chamber method show conclusively that Puhimau thermal area is not a significant contributor to the total CO_2 budget of Kîlauea. The size of the CO_2 anomaly is essentially the same for both years of the study, although the anomaly appears to have shifted slightly to the NNE. The soil temperature anomaly, covers about twice the area of the CO_2 anomaly, suggesting that the vegetation kill is probably due to temperature, not gas, and there is evidence that the temperature anomaly has also migrated to the NNE as well as to the west and south. The temporal and spatial differences for both soil CO_2 efflux and temperature between the two surveys may represent real persistent change due, perhaps, to some fundamental change in local conditions, or, alternatively may simply reflect the normal fluctuation of a dynamic system due to meteorological forcing, local shallow faulting, or hydrologic variability.

Although likely rich in other volatiles, magma reaching this region of the upper ERZ from the summit area contains little CO_2. SO_2 and other acid gases are only infrequently detected at Puhimau thermal area because of removal by hydrolysis reactions due to the high water table in the area. Without a significant change in subsurface plumbing either at the summit or in the upper ERZ, the likelihood of detecting much gas of any kind, except water vapor, remains low. Thus it does not appear that Puhimau thermal area currently provides a clear window into the upper ERZ through which to study Kīlauea's degassing behavior.

Acknowledgments

This work was supported by the U.S. Geological Survey through the Volcano Hazards Program. We thank Jennifer McGee and Rich Kessler for field assistance during the surveys. We are grateful to Dan Dzurisin, John Rogie, and Pedro Hernández Pérez for helpful reviews and comments.

REFERENCES

BARNARD, W.M., HALBIG, J.B., and FOUNTAIN, J.C. (1990), *Geochemical study of fumarolic condensates from Kilauea Volcano, Hawaii,* Pacific Science *44*, 197–206.

BARTEL, L.C. and JACOBSON, R.D. (1987), *Results of a controlled-source audiofrequency magnetotelluric survey at the Puhimau thermal area, Kilauea Volcano, Hawaii,* Geophysics *52*, 665–677.

CASADEVALL, T.J. and HAZLETT, R.W. (1983), *Thermal areas on Kilauea and Mauna Loa volcanoes, Hawaii,* J. Volcanol. Geotherm. Res. *16*, 173–188.

CONNOR, J.J. (1979), *Geochemistry of ohia and soil lichen, Puhimau thermal area, Hawaii,* The Science of the Total Environment *12*, 241–250.

DAVIES, F. and NOTCUTT, G. (1996), *Biomonitoring of atmospheric mercury in the vicinity of Kilauea, Hawaii,* Water, Air and Soil Pollution *86*, 275–281.

DAWSON, P.B., CHOUET, B.A., OKUBO, P.G., VILLASENOR, A., and BENZ, H.M. (1999), *Three-dimensional velocity structure of the Kilauea caldera, Hawaii,* Geophys. Res. Lett. *26*, 2805–2808.

DECKER, R.W. (1987), *Dynamics of Hawaiian volcanoes: An overview.* In *Volcanism in Hawaii* (Decker, R., Wright, T., and Stauffer, P. eds.), U.S. Geological Survey Professional Paper 1350, pp. 997–1018.

DUNN, J.C. and HARDEE, H.C. (1985), *Surface heat flow measurements at the Puhimau hot spot,* Geophysics *50*, 1108–1112.

FRIEDMAN, I., GLEASON, J., and JACKSON, T. (1987), *Variation of $\delta^{13}C$ in fumarolic gases from Kilauea volcano.* In *Volcanism in Hawaii* (Decker, R., Wright, T., and Stauffer, P. eds.), U.S. Geological Survey Professional Paper 1350, pp. 805–807.

GERLACH, T.M. and GRAEBER, E.J. (1985), *Volatile budget of Kilauea Volcano,* Nature *313*, 273–277.

GERLACH, T.M. and TAYLOR, B.E. (1990), *Carbon isotope constraints on degassing of carbon dioxide from Kilauea Volcano,* Geochim. Cosmochim. Acta *54*, 2051–2058.

GERLACH, T.M., MCGEE, K.A., SUTTON, A.J., and ELIAS, T. (1998), *Rates of volcanic CO_2 degassing from airborne determinations of SO_2 emission rates and plume CO_2/SO_2: Test study at Pu'u 'O'o cone, Kilauea volcano, Hawaii,* Geophys. Res. Lett. *25,* 2675–2678.

GERLACH, T.M., DOUKAS, M.P., MCGEE, K.A., and KESSLER, R. (2001), *Soil efflux and total emission rates of magmatic CO_2 at the Horseshoe Lake tree kill, Mammoth Mountain, California, 1995–1999,* Chem. Geol. *177,* 101–116.

GERLACH, T.M., MCGEE, K.A., ELIAS, T, SUTTON, A.J., and DOUKAS, M.P. (2002), *Carbon dioxide emission rate of Kīlauea volcano: Implications for primary magma and the summit reservoir,* J. Geophys Res. *107*(B9), 2189, doi:10.1029/2001JB000407.

GREENLAND, L.P., ROSE, W.I., and STOKES, J.B. (1985), *An estimate of gas emissions and magmatic gas content from Kilauea volcano,* Geochim. Cosmochim. Acta *49,* 125–129.

HINKLE, M.E. (1978), *Helium, mercury, sulfur compounds, and carbon dioxide in soil gases of the Puhimau Thermal Area, Hawaii Volcanoes National Park, Hawaii,* U. S. Geological Survey Open-File Report No. 78–246, 15 pp.

INGEBRITSEN, S.E. and SCHOLL, M.A. (1993), *The hydrogeology of Kilauea Volcano,* Geothermics *22,* 255–270.

JACKSON, D.B. and KAUAHIKAUA, J. (1987), *Regional self-potential anomalies at Kilauea volcano.* In *Volcanism in Hawaii* (eds. Decker, R., Wright, T., and Stauffer, P.), U.S. Geological Survey Professional Paper 1350, pp. 947–959.

JAGGAR, T.A., Jr. (1938), *Chain-of-craters crisis,* The Volcano Lett. *459,* 2–4.

KAUAHIKAUA, J. (1993), *Geophysical characteristics of the hydrothermal systems of Kilauea Volcano, Hawaii,* Geothermics *22,* 271–299.

KLEIN, F.W., KOYANAGI, R.Y., NAKATA, J.S., and TANIGAWA, W.R. (1987), *The seismicity of Kilauea's magma system.* In *Volcanism in Hawaii* (eds. Decker, R., Wright, T., and Stauffer, P.), U.S. Geological Survey Professional Paper 1350, pp. 821–825.

MCGEE, K.A., SUTTON, A.J., and SATO, MOTOAKI (1987), *Use of satellite telemetry for monitoring active volcanoes, with a case study of a gas-emission event at Kilauea Volcano, December 1982,* In *Volcanism in Hawaii* (eds. Decker, R., Wright, T., and Stauffer, P.), U.S. Geological Survey Professional Paper 1350, pp. 821–825.

MCGEE, K.A. and GERLACH, T.M. (1998), *Airborne volcanic plume measurements using a FTIR spectrometer, Kilauea volcano, Hawaii,* Geophys. Res. Lett. *25,* 615–618.

MCGEE, K.A. and GERLACH, T.M. (1998), *Annual cycle of magmatic CO_2 in a tree-kill soil at Mammoth Mountain, California: Implications for soil acidification,* Geology *26,* 463–466.

MCPHIE, J., WALKER, G.P.L., and CHRISTIANSEN, R.L. (1990), *Phreatomagmatic and phreatic fall and surge deposits from explosions at Kilauea volcano, Hawaii, 1790 A.D.: Keanakakoi ash member,* Bull. Volcanol. *52,* 334–354.

NORMAN, J.M., GARCIA, R., and VERMA, S.B. (1992), *Soil surface CO_2 fluxes and the carbon budget of grassland,* J. Geophys. Res. *97,* 18,845–18,853.

NORMAN, J.M., KUCHARIK, C.J., GOWER, S.T., BALDOCCHI, D.D., CRILL, P.M., RAYMENT, M., SAVAGE, K., and STRIEGL, R.G. (1997), *A comparison of six methods for measuring soil-interface carbon dioxide fluxes,* J. Geophys. Res. *102,* 28,771–28,777.

ROGIE, J.D., KERRICK, D.M., SOREY, M.L., CHIODINI, G., and GALLOWAY, D.L. (2001), *Dynamics of carbon dioxide emission at Mammoth Mountain, California,* Earth and Planet. Sci. Lett. *188,* 535–541.

SUTTON, A.J., ELIAS, T., GERLACH, T.M., and STOKES, J.B. (2001), *Implications for eruptive processes as indicated by sulfur dioxide emissions from Kīlauea Volcano, Hawaii, 1979—1997,* J. Volcanol. Geotherm. Res. *108,* 283–302.

SYMONDS, R.B., GERLACH, T.M., and REED, M.H. (2001), *Magmatic gas scrubbing: Implications for volcano monitoring,* J. Volcanol. Geotherm. Res. *108,* 303–341.

TILLING, R.I. and JONES, B.F. (1991), *Composition of waters from the research drill hole at summit of Kilauea Volcano and of selected thermal and non-thermal groundwaters, Hawaii,* U. S. Geological Survey Open-file Report 91–133A, 27 pp.

TILLING, R.I., and DVORAK, J.J. (1993), *Anatomy of a basaltic volcano,* Nature *363,* 125–133.

WELLES, J.M., DEMETRIADES-SHAH, T.H., and MCDERMITT, D.K. (2001), *Considerations for measuring ground CO_2 effluxes with chambers,* Chem. Geol. *177,* 3–13.

YANG, X., DAVIS, P.M., DELANEY, P.T., and OKAMURA, A.T. (1992), *Geodetic analysis of dike intrusion and motion of the magma reservoir beneath the summit of Kilauea volcano, Hawaii: 1970–1985,* J. Geophys. Res. *97,* 3305–3324.

ZABLOCKI, C.J. (1978), *Applications of the VLF induction method for studying some volcanic processes of Kilauea volcano, Hawaii,* J. Volcanol. Geotherm. Res. *3,* 155–195.

(Received: January 30, 2003; revised: January 28, 2005; accepted: February 7, 2005)
Published Online First: March 28, 2006

 To access this journal online:
http://www.birkhauser.ch

Pure appl. geophys. 163 (2006) 853–867
0033–4553/06/040853–15
DOI 10.1007/s00024-006-0039-9

© Birkhäuser Verlag, Basel, 2006

❙Pure and Applied Geophysics

Fault-controlled Soil CO_2 Degassing and Shallow Magma Bodies: Summit and Lower East Rift of Kilauea Volcano (Hawaii), 1997

SALVATORE GIAMMANCO,[1] SERGIO GURRIERI,[2] and MARIANO VALENZA[3]

Abstract—Soil CO_2 flux measurements were carried out along traverses across mapped faults and eruptive fissures on the summit and the lower East Rift Zone of Kilauea volcano. Anomalous levels of soil degassing were found for 44 of the tectonic structures and 47 of the eruptive fissures intercepted by the surveyed profiles. This result contrasts with what was recently observed on Mt. Etna, where most of the surveyed faults were associated with anomalous soil degassing. The difference is probably related to the differences in the state of activity at the time when soil gas measurements were made: Kilauea was erupting, whereas Mt. Etna was quiescent although in a pre-eruptive stage. Unlike Mt. Etna, flank degassing on Kilauea is restricted to the tectonic and volcanic structures directly connected to the magma reservoir feeding the ongoing East Rift eruption or in areas of the Lower East Rift where other shallow, likely independent reservoirs are postulated. Anomalous soil degassing was also found in areas without surface evidence of faults, thus suggesting the possibility of previously unknown structures.

Key words: Soil CO_2, Kilauea, volcanic degassing, tectonic structures, geochemical surveying.

1. Introduction

Active tectonic structures in seismogenic and volcanic areas can be pathways for the release of subsurface gases (SUGISAKI *et al.*, 1983; ROSE *et al.*, 1991; KLUSMAN, 1993). On active or quiescent volcanoes, carbon dioxide is the main species in the soil gas released through tectonic structures (BADALAMENTI *et al.*, 1988; PÈREZ *et al.*, 1997; WILLIAMS-JONES *et al.*, 1997), because, after water vapor, CO_2 is the most abundant gas dissolved in magma. Moreover, according to its low solubility in basaltic melts (PAN *et al.*, 1991), CO_2 is one of the first volatile components to be released from magma during its ascent. For these reasons, the output of CO_2 could be a useful indicator of the activity of a volcanic system (GIAMMANCO *et al.*, 1995; DILIBERTO *et al.*, 2002).

Studies recently carried out on Mt. Etna (GIAMMANCO *et al.*, 1998) showed that on such an active composite volcano, old eruptive fissures can be sites of anomalous

[1]INGV, Section of Catania, Piazza Roma 2, I-95123, Italy. E-mail: giammanco@ct.ingv.it
[2]INGV, Section of Palermo, Via La Malfa 153, Palermo, I-90146, Italy
[3]Universitá degli Studi di Palermo, Dipartimento di Chimica e Fisica della Terra ed Applicazioni, via Archirafi 36, Palermo, I-90133, Italy

soil degassing as well as faults. On Mt. Etna, however, this phenomenon was solely observed in about half of the surveyed eruptive fissures, thus suggesting that only those subject to active crustal stress can have soil permeability values high enough to permit deep gas leakage towards the surface (GIAMMANCO *et al.*, 1998, 1999).

On Kilauea (Hawaii), soil CO_2 flux measurements were performed during June 1997 in order to better constrain the relationships between soil degassing and tectonic or volcano-tectonic structures at composite basaltic volcanoes other than Etna. In the case of Kilauea, tectonic structures are those with no evidence of being conduits for magma ascent. On this volcano, carbon dioxide emissions occur mainly through the active summit craters (e.g., GERLACH and GRAEBER, 1985). Few data on other soil gases at Kilauea are available in the current literature (COX, 1983; REIMER, 1987; SIEGEL and SIEGEL, 1987). Our work, therefore, also provided a first relatively large-scale survey of soil CO_2 emissions through the flanks of Kilauea.

The areas investigated on Kilauea were selected on the basis of the density of known eruptive fissures and/or of tectonic structures (Fig. 1). Our investigations were mainly aimed at determining whether a correspondence exists between the location of the soil gas anomalies and the occurrence of tectonic and/or volcano-tectonic structures. We also investigated areas with no field evidence of tectonic structures, but where faults or old eruptive fissures could be postulated based on local structural settings, such as unstable slopes, and the presence of areas covered by recent lavas or

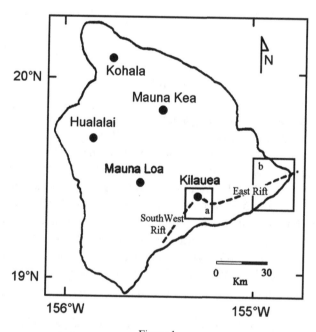

Figure 1
Location of the volcanoes on the island of Hawaii. Inset boxes a and b refer, respectively, to maps of Figures 2 and 3.

tephra. Another important scope of this work was to recognize possible connections between sites of anomalous soil degassing and shallow magmatic intrusions.

2. Geological and Structural Overview

Kilauea volcano is the youngest of five volcanoes that make up the island of Hawaii. It reaches an elevation of about 1230 m a.s.l. and has a subaerial surface area of about 1500 km² (Fig. 1) (HOLCOMB, 1987). Kilauea's eruptive activity is mostly characterized by the effusion of tholeiitic lava flows (WRIGHT and HELZ, 1987). These magmas originate from the mantle at inferred depths of more than 60 km (WRIGHT, 1984; TILLING and DVORAK, 1993). The characteristics of Hawaiian volcanism are typical of an intra-plate hot-spot (HOLCOMB, 1987; WRIGHT and HELZ, 1987). Magma rises from the region of partial melting through almost vertical pathways until it reaches a shallow reservoir beneath Kilauea summit (TILLING and DVORAK, 1993). Recent seismic studies indicate a magma reservoir at a depth of 5–7 km beneath Kilauea (OKUBO et al., 1997) and DAWSON et al. (1999) used seismic data to recognize two shallower magma reservoirs at a depth of 1 to 4 km beneath the southern rim of Kilauea's caldera and the upper East Rift of the volcano. Secondary magma reservoirs may form beneath the rift zones, and are fed by the summit reservoir (TILLING and DVORAK, 1993).

Kilauea's eruptive activity mostly occurs either within its summit caldera or along two rift zones that originate from the summit and extend, respectively, toward the east (Kilauea East Rift Zone, or KERZ) and toward the southwest. Eruptions on the rift zones are fissures eruptions of highly variable duration and intensity. At the time of this writing, an ongoing eruption on the KERZ, that started in 1983, has discharged more than 1×10^9 m³ of this lava (HELIKER et al., 1998).

The whole south flank of Kilauea is subject to deformation induced by pressure of magma that is intruded into shallow reservoirs beneath the rifts (TILLING and DVORAK, 1993). Deformation due to magma intrusion causes faulting in the rocks of this flank of the volcano, with consequent gravitational slumping of whole sections toward the south, where it is not confined by other volcanic edifices (HOLCOMB, 1987; BRYAN and JOHNSON, 1991). For this reason, the southern flank of Kilauea is characterized by a high density of both normal and reverse faults.

The main tectonic structures of Kilauea's south flank are the Koa'e and the Hilina fault systems (HOLCOMB, 1987). The Koa'e system is directed roughly ENE-WSW and links the two rifts zones of the volcano just south of the summit caldera. The faults of the Koa'e system are occasionally sites of eruptions, the last of which occurred in 1973 (TILLING et al., 1987). The Hilina fault system cuts the downhill portion of Kilauea's southern flank. These faults show the largest displacements in the Kilauea area, mostly towards the south. According to stratigraphic studies, the age of these faults was estimated to be at least 23 ka (HOLCOMB, 1987).

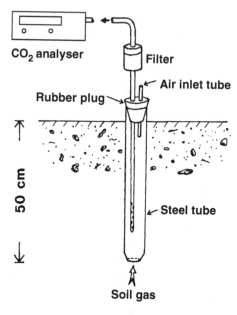

Figure 2
Sketch of the sampling and analysis system used to measure CO_2 fluxes in the soil.

3. Sampling And analytical Methods

Field work on Kilauea in May 1997 consisted of 126 soil measurements of CO_2 flux. Measurements were performed using the method of GURRIERI and VALENZA (1988), also described in GIAMMANCO *et al.* (1995) and DILIBERTO *et al.* (2002). This method (Fig. 2) uses a specially designed probe that is inserted into the soil to a depth of 50 cm. The probe is open at its bottom and with a small tube at the top, which allows air to enter it. By pumping at a constant flow rate, the CO_2 concentration in the mixture of soil gas and air inside the probe reaches a stable value, which depends on the emission rate of CO_2 through the soil. Earlier experiments determined that the values of CO_2 concentration (called *dynamic* because of the way they are determined) are directly proportional to the flux of carbon dioxide through the soil according to the relation $\Phi = k\, C_{dyn}$, where Φ is the flux of soil CO_2 (in g cm^{-2} s^{-1}), C_{dyn} is the *dynamic concentration* of CO_2 (in ppm vol.) and k is an empirical constant (in g ppm^{-1} cm^{-2} s^{-1}). The value of k depends mainly on the geometry of the sampling system and the flow of the pump, which are kept constant, and to a lesser extent on the soil permeability (GURRIERI and VALENZA, 1988; GIAMMANCO *et al.*, 1995). In order to determine the value of k, laboratory tests were carried out (GURRIERI and VALENZA, 1988) where soil degassing was simulated at different known CO_2 fluxes in samples of fine pyroclastic material characterized by permeability ranging between 10 and 60 darcy (about 1 to 6×10^{-11} m^2). It was observed that, within this range of values, soil permeability had a very small impact on the flux measurements (GURRIERI and VALENZA, 1988). Recent developments of this method indicate that the proportionality constant k assumes values that differ appreciably from that given above for soil permeability values lower than 1 darcy. The error induced by the variations of soil permeability of the investigated soils was calculated to be less than 10% (GURRIERI et al., 2000).

Soil gas measurements during the present work were carried out along five sampling profiles (Table 1) in the following areas of Kilauea volcano: i) the summit and upper East Rift of the volcano, including the rim of Halema'uma'u caldera and the upper part of both the Chain of Craters Road and the Hilina Pali Road down to an elevation of about 650 m a.s.l. (Fig. 3); ii) the lower KERZ, roughly bounded by both the villages of Pahoa and Kaimu, and the area of Kapoho (Fig. 4). Measurements were performed under warm and stable weather conditions and no rain fell during the week that preceded our surveys, consequently the effect of meteoric water on soil permeability can be ruled out. This is verified by repeated measurements carried out after one or two days in some randomly selected sites.

Soil CO$_2$ dynamic concentration measurements were performed with a portable fixed-wavelength IR spectrophotometer (Analytical Development Company Limited, U.K.). To obtain the relevant CO$_2$ flux values, a k value of about 7.17×10^{-11} g ppm^{-1} cm^{-2} s^{-1} was used. The instrumental accuracy was within $\pm$ 3%. Such error was obtained from repeated measurements (≥ 3) in each sampling site, and did not affect the results appreciably. In order to eliminate the possible influence of atmospheric pressure changes on soil flux measurements due to elevation effect, the instrument was frequently calibrated during the survey with standard CO$_2$ samples under ambient pressure and temperature conditions. A sampling interval between about 100 and 300 m was chosen as the best compromise between detail of information and number of measurements. The sampling step was deeply affected by recent lava flows, which in some cases impeded our measurements. In general, wider sampling steps increase the chances of missing faults or fissures that might have a CO$_2$ anomaly. However, fault zones and volcanic fissures on Kilauea have generally a width of several tens of meters or more, so the chance that we did not survey even a part of them is fairly low. Also, based on soil CO$_2$ data collected on Mt. Etna, which has similar soil permeability values as well as structural characteristics to Kilauea, the width of soil gas anomalies is generally comparable or larger than that of the tectonic or volcano-tectonic structures (GIAMMANCO et al., 1997, 1998).

The sampling procedure used on Kilauea allowed us to complete each profile in a few hours, during which we assume atmospheric conditions remained constant. This is a necessary requirement for internal consistency of all flux measurements along any given profile. Direct measurements of atmospheric pressure were also frequently carried out to verify the constancy of atmospheric conditions. The observed variations were always less than two millibar, which produce very small effects on soil degassing (DILIBERTO et al., 2002).

Table 1

Soil CO$_2$ flux values measured along the sampling profiles on Kilauea. All flux values are in g cm^{-2} s^{-1}. All distances are in meters from the starting point of each sampling line

Sampling line	Distance	CO$_2$ flux	Sampling line	Distance	CO$_2$ flux	Sampling line	Distance	CO$_2$ flux
A	0	0		2240	6.0×10^{-7}		1840	1.0×10^{-6}
	320	0		2560	1.3×10^{-7}		2400	1.0×10^{-7}
	640	0		2880	1.9×10^{-7}		3520	9.0×10^{-8}
	1120	0		3200	9.0×10^{-8}	D'	4160	7.0×10^{-8}
	1440	0		3520	7.0×10^{-8}	E	0	8.7×10^{-7}
	1920	0		4000	0		160	0
	2240	0		4560	6.0×10^{-8}		320	5.1×10^{-7}
	2560	0		5120	3.0×10^{-8}		480	1.7×10^{-7}
	4080	0		5840	0		640	6.0×10^{-8}
	4480	3.0×10^{-6}	B'	6160	0		800	2.0×10^{-6}
	4800	3.4×10^{-7}	C	0	1.7×10^{-7}		960	3.2×10^{-7}
	5120	4.7×10^{-7}		640	3.9×10^{-7}		1120	8.2×10^{-7}
	5520	3.0×10^{-8}		1120	4.7×10^{-7}		1280	1.4×10^{-7}
	5840	0		1600	2.5×10^{-7}		1440	1.7×10^{-7}
	6400	1.4×10^{-7}		2080	2.9×10^{-7}		1600	2.0×10^{-6}
	6720	0		2640	8.5×10^{-7}		1760	1.0×10^{-8}
	7120	1.0×10^{-8}		2960	0		1920	1.4×10^{-7}
	7600	4.3×10^{-7}		3360	0		2080	0
	7920	9.3×10^{-7}		3680	0		2240	2.2×10^{-7}
	8320	2.4×10^{-7}		4000	9.8×10^{-7}		2400	4.0×10^{-8}
	8720	0		4320	3.4×10^{-7}		2560	0
	9120	0		4560	0		2720	0
	9440	0		4960	1.0×10^{-6}		2880	0
	9760	0		5280	0		3040	0
	9920	1.0×10^{-6}		5560	0		3200	0
	10080	9.0×10^{-8}		6000	0		3360	7.0×10^{-8}
	10320	0		6720	0		3680	0
	10720	0		8320	0		3840	8.5×10^{-7}
	11120	0		5600	1.0×10^{-8}		4000	5.0×10^{-7}
	11440	9.0×10^{-8}		6150	0		4160	2.0×10^{-7}
	11760	0		6550	1.7×10^{-7}		4320	1.1×10^{-7}
	12160	0		6950	0		4480	2.4×10^{-7}
	12560	0		7350	0		4640	1.1×10^{-7}
	13040	0		7550	1.4×10^{-7}		4800	0
A'	13520	0		7850	1.7×10^{-7}		4960	0
B	0	1.4×10^{-7}		8150	5.4×10^{-7}		5120	1.7×10^{-7}
	320	2.0×10^{-7}	C'	8850	9.0×10^{-8}		5280	0
	640	1.0×10^{-8}	D	0	3.0×10^{-8}		5440	3.0×10^{-8}
	960	0		320	0		5600	7.0×10^{-8}
	1280	4.0×10^{-8}		800	0		5760	0
	1680	9.0×10^{-8}		1120	2.9×10^{-7}		6080	0
	1920	0		1440	6.0×10^{-8}	E'	7360	0

Soils at Kilauea are usually thin, but only rarely were we forced to move sampling locations until sufficiently deep soil was found within the above regular sampling interval.

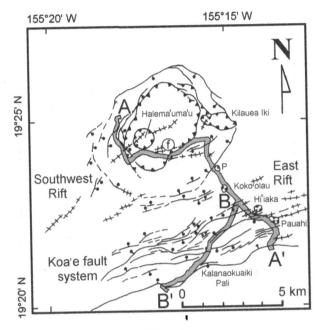

Figure 3

Location of the lines of soil gas measurements on Kilauea's summit area (box a in Fig. 1). A–A' = Crater Rim - Chain of Craters Road; B–B' = Hilina Pali Road. Faults (solid lines with dot on downthrown side; dashed when uncertain) and eruptive fissures (ticked solid lines) are also shown. Structural data are from WOLFE and MORRIS (1996). Letter f indicates the 1974 eruptive fissure. Letter p indicates the Puhimau crater.

4. Results and Discussion of Data

In general, the origin of the CO$_2$ emitted through the soil of active volcanoes, such as Kilauea, can be ascribed to the mixing of two sources: a deep magmatic one and a shallow one linked to organic activity. In contrast with the organic source, the magmatic one is able to sustain high fluxes of gas. KANEMASU et al. (1974) indicate a CO$_2$ flux value of $1.3 \, l \, m^{-2} \, h^{-1}$ (corresponding to $7.36 \times 10^{-8} \, g \, cm^{-2} \, s^{-1}$) as the highest CO$_2$ flux that can be sustained by microbial activity in soil in general. This value will be assumed as a threshold to identify anomalous high CO$_2$ fluxes (presumably caused by CO$_2$ of volcanic origin) from the soils of Kilauea. We believe this value is certainly higher than the soil respiration CO$_2$ flux from the bare volcanic soil of the areas investigated on the Island of Hawaii, but anomalous degassing of magmatic origin through tectonic structures in volcanic environments is usually considerably higher than this threshold value (e.g., GIAMMANCO et al., 1997, 1998; GERLACH et al., 1998; HERNÁNDEZ et al., 1998).

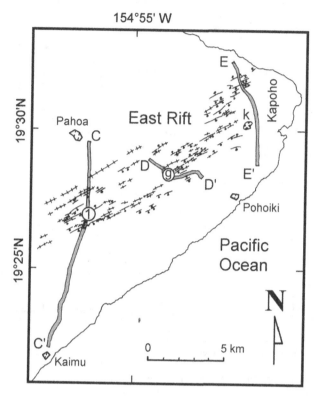

Figure 4

Location of the lines of soil gas measurements on Kilauea's lower East Rift zone (box b in Fig. 1). C–C' = Pahoa–Kaimu Road; D–D' = Pahoa–Pohoiki Road; E–E' = Kapoho profile. Large circle with number inside indicates the site of soil gas sampling for chemical and isotopic analyses; k = Kapoho cone; g = geothermal well. Faults (solid lines with dot on downthrown side; dashed when uncertain) and eruptive fissures (ticked solid lines) are also shown. Structural data are from MOORE and TRUSDELL (1991) and WOLFE and MORRIS (1996).

4.1 Kilauea Summit Area

Measurements in this area (Fig. 3) were carried out along the Crater Rim Road - Chain of Craters Road (line A-A', 35 measurement points) and the Hilina Pali Road (line B-B', 17 measurement points).

Along the Crater Rim Road the highest CO_2 flux value was found close to the western end of the 1974 eruptive fissure (sampling point at 4,480 m in Table. 1 and Fig. 5a). Slightly anomalous soil fluxes were also detected a few hundred meters east of the southeast rim of the caldera, between the eastern edge of the 1974 fissure and two faults related to the caldera-forming collapse. Other major CO_2 anomalies occurred on a fault near the Puhimau crater area (sampling points at 7,920 m in Table 1 and Fig. 5a), which is known for the presence of widespread fumarolic

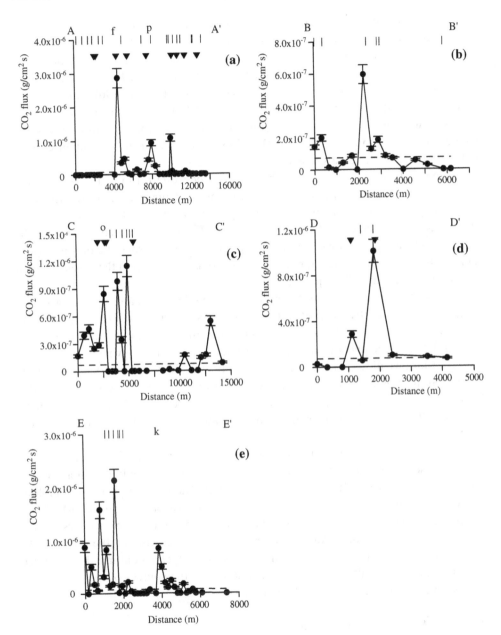

Figure 5

Soil CO$_2$ flux values (g cm^{-2} s^{-1}) measured along the surveyed lines on Kilauea. a) Crater Rim - Chain of Craters Road profile (A–A' in Fig. 2); b) Hilina Pali Road profile (B–B' in Fig. 2); c) Pahoa–Kaimu Road profile (C–C' in Fig. 3); d) Pahoa–Pohoiki Road profile (D–D' in Fig. 3); e) Kapoho profile (E–E' in Fig. 3). Error bars indicate flux values ± 10%. Vertical solid lines indicate intercepted faults; solid triangles indicate intercepted eruptive fissures. The horizontal broken lines indicate the highest value due to organic activity (7.36 × 10^{-8} g cm^{-2} s^{-1}) in the soil (see text for explanation). f = 1974 eruptive fissure; p = Puhimau crater; o = eruptive fissure dated to 400–750 yr B.P. (see text); k = Kapoho Cone.

emissions associated with a weak thermal anomaly in the ground and where previous isotopic data on soil CO_2 ($\delta^{13}C$ values between -4.0 and -1.9 ‰) indicated a clear magmatic origin (FRIEDMAN *et al.*, 1987). The fault intercepted by our profile near Puhimau crater is part of a system that encircles the summit of Kilauea and was the site of eruptive activity as recently as 1974 (WOLFE and MORRIS, 1996). This indicates its connection at that time with the upper KERZ. Points with anomalous CO_2 degassing were also found both on a fault that marks the northernmost limit of the Koa'e fault zone, and very close to the 1973 fissure (northeast of Pauahi crater), although this latter anomaly was barely detectable.

Absence of soil degassing along the faults of the western rim of Kilauea caldera and along some of the upper KERZ may indicate that these tectonic structures are not directly connected to the present pathways used by magma. However, it must be mentioned that in the case of the western rim of Kilauea caldera some parts of the existing faults are buried by layers of hard-packed, altered pyroclastic products (ash, tephra, etc.). The low permeability of these materials may result in the absence of soil degassing in this part of the profile.

An alternative process calls for a generally low level of flank degassing of the volcano during eruptive periods except in the areas affected by faults directly linked to the pathways of magma intrusion beneath the summit caldera and into the KERZ. Such a phenomenon was observed on Mt. Etna, where during large lateral eruptions diffuse degassing is generally very low and restricted to faults connected to the eruptive dike (GIAMMANCO *et al.*, 1995). According to this hypothesis, during non-eruptive periods, including periods of persistent summit activity, diffuse degassing is higher and occurs along most, if not all, of the existing faults.

Along the Hilina Pali road profile, the anomalies of soil CO_2 flux were found on or very close to the northernmost faults belonging to the Koa'e fault system, where they intercept the East Rift zone (Figs. 3 and 5b). It is noteworthy that no anomalous soil degassing was found on the southernmost fault (Kalanaokuaiki Pali) of this system. This behavior was also observed on the same fault where it intersects the Crater Rim – Chain of Craters road profile (Fig. 5a). Also in the case of the Kalanaokuaiki Pali fault, absence of anomalous soil CO_2 emissions may be explained assuming that this fault is not directly connected to the sources of magmatic CO_2 (i.e., volcanic conduits and magma reservoirs).

4.2 Lower Kilauea East Rift Zone

Measurements (Fig. 4) were carried out along the Pahoa-Kaimu road (line C–C', 27 measurement points), along a 4.2 km-long segment on the road connecting Pahoa to Pohoiki, just across the East Rift near the geothermal well (line D-D', 9 measurement points), and along a 7.4 km-long segment on the road crossing the East Rift in the Kapoho area (line E-E', 38 measurement points).

Along the Pahoa–Kaimu profile, the highest CO_2 flux anomalies were found in its northernmost part (Fig. 5c) where most of the rift fractures and faults are intercepted by the profile (Fig. 4). In particular, several contiguous anomalies in soil CO_2 degassing occur just close to Pahoa village and manifest two relative maxima, the highest one near an eruptive fissure dated to 400–750 yr B.P. (WOLFE and MORRIS, 1996). The highest values of CO_2 flux from the soil along this profile were found in two sites located four and five kilometers south of Pahoa, respectively, both at a very close distance from mapped faults. Three other zones of anomalous soil CO_2 fluxes were detected in areas where there is no evidence of faults or fissures; one in the northernmost part of the profile and two in the southernmost part. In these cases, the existence of hidden or buried faults can be postulated to explain such soil gas anomalies. The faults postulated in these zones may be concealed by recent lava flows.

Along the "Geothermal Well" profile, on the Pahoa–Pohoiki road (D–D′ in Fig. 4), CO_2 flux anomalies were found only close to the faults and fissures intercepted by the surveyed line (Fig. 5d).

Along the Kapoho line (E–E′ in Fig. 4) all of the mapped faults are exclusively located in its northern part and were associated with CO_2 flux anomalies (Fig. 5e). Several other anomalous CO_2 flux values were measured just north of the previous anomalies. A further area of anomalous soil CO_2 degassing was found about 2 km south of the southernmost fault, near a large ancient tuff cone (Kapoho Cone, labelled k in Fig. 4). In all of these areas, soil degassing is not associated with visible faults, indicating the possible presence of buried or hidden volcano-tectonic structures.

The $\delta^{13}C(CO_2)$ value from the site with the most intense degassing along this profile (site labelled 1 in Fig. 4) indicates an organic-rich source (−16.3 ‰). Probably, this is due to microbial activity in the uppermost soil horizon (e.g., KANEMASU et al., 1974; HINKLE, 1990). If we assume a two component system with the magmatic end-member of $^{13}C(CO_2)$ at Kilauea having a value of about −3‰ relative to PDB standard (which corresponds to the average value measured at Sulphur Banks, a fumarole field close to the NE edge of the Kilauea caldera; FRIEDMAN et al., 1987; HILTON et al., 1997) and the organic end-member having a value of −28‰ (FAURE, G., 1986), then the magmatic component in the CO_2 emitted from this soil gas site would be about 50%. Such magmatic CO_2 component could originate from degassing magma bodies within the rift. The presence of magma is supported by geophysical and geochemical studies that indicate intrusions occur through the entire length of the rift and such magma bodies may remain molten for relatively long periods of time (probably on the order of several tens of years, e.g., THOMAS, 1987; TILLING and DVORAK, 1993). The magma bodies produce significant thermal anomalies in the ground water (THOMAS, 1987; CONRAD et al., 1997). The heat flow is in places so intense (at least 291 MW over 25×10^6 km^2 of surface on the lower KERZ, according to THOMAS, 1987) as to produce an exploitable geothermal reservoir (tapped by the above-mentioned geothermal well). A direct magmatic source of gas can also be suggested to explain the emissions of CO_2 along the Kapoho line, consistent with the presence of a secondary magma reservoir beneath the area near Kapoho Cone, as postulated by TILLING and DVORAK (1993). However, an apparent discrepancy arises between our data that suggest an active gas-rich magmatic source, and the likely gas-depleted magma present in the secondary reservoir that results from a long residence time. A possible explanation is that envisaged by GERLACH and GRAEBER (1985) and by TILLING and DVORAK (1993): such reservoirs beneath the rift zones are periodically fed by magma that intrudes into the rifts. This mechanism allows replenishment with new magma that carries a higher amount of volatiles than the older magma body. Such volatiles would correspond to those which provide the "type II" gas described by GERLACH and Graeber (1985) as enriched in compounds that have a relatively high solubility in magma (mostly water and halogens), but still have significant amounts of SO_2 and CO_2.

5. Conclusions

Our investigations of diffuse soil degassing, carried out on Kilauea, indicate that degassing of magmatic CO_2 takes place not only through the summit crater of the volcano, but also through faults and old eruptive fissures on its flanks. However, our data do not allow us to quantify the amount of CO_2 released through the soils in the investigated areas. Our findings are consistent with the conclusions of geochemical investigations carried out on Etna (ANZÀ *et al.*, 1993; GIAMMANCO *et al.*, 1997, 1998) as well as other volcanic areas in the world (BADALAMENTI *et al.*, 1988; PÈREZ *et al.*, 1997; WILLIAMS-JONES *et al.*, 1997). It is to be noted, however, that a few years after the end of an eruption, eruptive fissures should not show evidence of degassing. This is due either to obstruction after magma solidification or to sealing from hydrothermal alteration induced by residual magmatic fluids (GIAMMANCO *et al.*, 1999). Therefore, eruptive fissures associated with anomalous soil degassing suggest that tectonic strain is still present. Further, it is reasonable to assume that soil gas anomalies not associated with mapped faults originate from tectonic structures that are either hidden or covered by more recent lava flows. The possibility that some of the hidden structures are eruptive fissures that were covered later by newer volcanic products cannot be ruled out. In any case, these structures should be subject to active tectonic strain that keeps them open.

In contrast with observations on Mt. Etna where all of the surveyed faults and 49% of the surveyed fissures showed anomalous soil CO_2 degassing (GIAMMANCO *et al.*, 1997, 1998), anomalous soil gas emissions on Kilauea were found only over 44% of the tectonic structures and 47% of the volcano-tectonic structures intercepted by our sampling lines. The percentages were obtained by counting the number of anomalous values (i.e., greater than the "organic" threshold of 7.36×10^{-8} g cm^{-2} s^{-1}) on a "true-false" basis (i.e., any value greater than the threshold is considered an anomaly, otherwise it is not). In this computation we did not take into account the intensity of each anomaly. The measured intensity of soil gas anomalies can actually be dependent on the location of sampling points with respect to the maximum gas emission, assuming a log-normal distribution of anomalous values across a degassing fault.

The relatively low number of degassing faults at Kilauea volcano suggests that magmatic gas emissions through Kilauea's flanks occur only along tectonic structures directly connected at depth with the feeding conduits of the volcano or with the magma reservoirs beneath the summit and the rift zones. In addition to this, during periods of rift eruptions such as those when our measurements were carried out, magma is drained towards shallower parts of the volcanic system and is erupted or stored in the rift. In any case, a strong migration of magma occurs and this in turn means a drastic change in the mechanism of magmatic gas release and transport to the surface. Gases exsolved from magma would be carried with it, thus decreasing the gas pressure gradients in the shallow crust towards peripheral areas of the volcano. This causes an increase in the gas pressure gradients towards the eruptive vents and hence along the faults that are in connection with the active magma dike or reservoir. This phenomenon was already observed on Mt. Etna during the voluminous 1991–1993 eruption by GIAMMANCO et al. (1995), who named it "gas-drainage effect".

The "gas–drainage effect" seems to be less marked in the "Geothermal well" and Kapoho lines (respectively, lines D–D$'$ and E–E$'$ in Fig. 5), where 3/4 and 5/6 of tectonic and volcano-tectonic structures, respectively, were associated with anomalous soil degassing. This might support the hypothesis of a local magmatic source of CO$_2$ that is independent of that feeding soil degassing uprift and that might be identified with secondary magma reservoirs (TILLING and DVORAK, 1993).

Acknowledgements

The authors wish to thank the Hawaii Volcanoes National Park for having kindly given the permission to work in protected areas of Kilauea volcano. This work was carried out within the framework of activities coordinated and financially supported by the "Gruppo Nazionale per la Vulcanologia, C.N.R." of Italy.

REFERENCES

ANZÀ, S., BADALAMENTI, B., GIAMMANCO, S., GURRIERI, S., NUCCIO, P. M., and VALENZA, M. (1993), *Preliminary study on emanation of CO$_2$ from soils in some areas of Mount Etna (Sicily)*, Acta Vulcanol. *3*, 189–193.

BADALAMENTI, B., GURRIERI, S., HAUSER, S, PARELLO, F., and VALENZA, M. (1988), *Soil CO$_2$ output in the island of Vulcano during the period 1984–88: Surveillance of gas hazard and volcanic activity*. Rend. Soc. It. Min. Petrog. *43*, 893–899.

BRYAN, C.J. and JOHNSON, C.E. (1991), *Block tectonics of the Island of Hawaii from a focal mechanism analysis of basal slip*, Bull Seismol Soc Am. *81*, 491–507.

CONRAD, M.E., THOMAS, D.M., FLEXSER, S., and VENNEMANN, T.W. (1997) *Fluid flow and water-rock interaction in the East Rift Zone of Kilauea Volcano, Hawaii.* J. Geophys. Res. *102*, 15,021–15,037.

COX, M.E. (1983), *Summit outgassing as indicated by radon, mercury and pH mapping, Kilauea volcano, Hawaii.* J. Volcanol. Geotherm. Res. *16*, 131–151.

DAWSON, P.B., CHOUET, B.A., OKUBO, P.G., VILLASEÑOR, A. and BENZ H.M. (1999), *Three-dimensional velocity structure of the Kilauea caldera, Hawaii,.* Geophys. Res. Lett. *26*, 2805–2808.

FAURE, G. *Principles of Isotope Geology* (New York, John Wiley and Sons 1986).

FRIEDMAN, I. GLEASON, J., and JACKSON, T., *Variation of $\delta^{13}C$ in fumarolic gases from Kilauea volcano.* In, *Volcanism in Hawaii.* (Decker, R.W., Wright, T.L., and Stauffer P.H. eds) Washington, D.C, U.S.G.S. Prof. Paper *1350*, 1987 pp. 805–807.

GERLACH, T.M., and GRAEBER, E.J. (1985), *Volatile budget of Kilauea volcano,* Nature *313*, 273–277.

GERLACH, T.M., DOUKAS, M.P., MC.GEE, K.A., and KESSLER, R. (1998), *Three-year decline of magmatic CO_2 emissions from soils of a Mammoth Mountain tree kill: Horseshoe Lake, CA, 1995-1997.* Geophys. Res. Lett *25*, 1947–1950.

GIAMMANCO, S., GURRIERI, S., and VALENZA, M. (1995), *Soil CO_2 degassing on Mt. Etna (Sicily) during the period 1989-1993: Discrimination between climatic and volcanic influences.* Bull. Volcanol. *57*, 52–60.

GIAMMANCO, S., GURRIERI, S., and VALENZA, M. (1997), *Soil CO_2 degassing along tectonic structures of Mount Etna (Sicily): The Pernicana fault.* Appl. Geochem. *12*, 429–436.

GIAMMANCO, S., GURRIERI, S. and VALENZA, M. (1998), *Anomalous soil CO_2 degassing in relation to faults and eruptive fissures on Mount Etna (Sicily, Italy),* Bull. Volcanol *60*, 252–259.

GIAMMANCO, S., GURRIERI, S., and VALENZA, M. (1999), *Geochemical investigations applied to active fault detection in a volcanic area:The NorthEast Rift on Mt. Etna (Sicily, Italy),* Geophys. Res. Lett. *26*, 2005–2008.

GURRIERI, S., and VALENZA, M. (1988), *Gas transport in natural porous mediums: a method for measuring CO_2 flows from the ground in volcanic and geothermal areas,* Rend. Soc. It. Min. Petrog. *43*, 1151–1158.

GURRIERI, S., CAMARDA, M., RICCOBONO, G. and VALENZA, M. (2000), *Relationships between soil permeability and diffuse degassing in volcanic areas,* Eos, Trans. AGU *81* (48), Fall Meet. Suppl., 2000.

HELIKER, C.C., MANGAN, M.T., MATTOX, T.N., KAUAHIKAUA, J.P., and HELZ, R.T. (1998), *The character of long-term eruptions: inferences from episodes 50-53 of the Pu'u 'O'o-Kupaianaha eruption of Kilauea Volcano,* Bull. Volcanol. *59*, 381–393.

HERNÁNDEZ, P.A., PÉREZ, N.M., SALAZAR, J.M., NAKAI, S., NOTSU, K., and WAKITA, H. (1998), *Diffuse emission of carbon dioxide, methane, and helium-3 from Teide volcano, Tenerife, Canary Islands,* Geophys. Res. Lett. *25*, 3311–3314.

HILTON, D.R., MC.MURTY, G.M., KREULEN, R. (1997), *Evidence for extensive degassing of the Hawaiian mantle plume from helium-carbon relationships at Kilauea volcano.* Geophys. Res. Lett. *24*, 3065–3068.

HINKLE, M.E., *Factors affecting concentrations of helium and carbon dioxide in soil gases.* In, *Geochemistry of Gaseous Elements and Compounds.* (Durrance, E.M., Galimov, E.M., Hinkle, M.E., Reimer, G.M., Sugisaki R., and Augustithis, S.S. eds), (Athens, Theophrastus Publications, 1990) pp. 421–448.

HOLCOMB, R.T., *Eruptive history and long-term behavior of Kilauea volcano.* In *Volcanism in Hawaii* (Decker, R.W., Wright, T.L., Stauffer, P.H. eds), (Washington, DC, U.S.G.S. Prof. Paper 1350 1987) pp.261–350.

KANEMASU, E.T., POWERS, W.L, and SIJ, J.W. (1974), *Field chamber measurements of CO_2 flux from soil surface,* Soil Sci *118*: 233–237.

KLUSMAN, R.W., *Soil Gas and Related Methods for Natural Resource Exploration* (New York, John Wiley and Sons 1993).

MOORE, R.B., and TRUSDELL, F.A. *Geologic Map of the Lower East Rift Zone of Kilauea Volcano, Hawaii: 1: 24,000* (Washington, D.C: U.S. Geological Survey 1991).

OKUBO, P.G., BENZ, H.M., and CHOUET, B.A. (1997), *Imaging the crustal magma source beneath Mauna Loa and Kilauea volcanoes, Hawaii.* Geology *25*, 867–870.

PAN, V, HOLLOWAY, J.R., HERVIG, R.L., (1991), *The pressure and temperature dependence of carbon dioxide solubility in tholeiitic basalt melts.* Geochim. Cosmochim. Acta. *55*, 1587–1595.

PÈREZ, N.M., WAKITA, H, PATIA, H., LOLOK, D., TALAI, B., MC, KEE, C.O. (1997), *Surface geochemical evidence for gas-flow along a seismically active fault zone at Rabaul caldera, Papua New Guinea,* Proc. Gen. Assembly IAVCEI, Puerto Vallarta, Mexico, p. 64.

REIMER, G.M. (1987), *Helium at Kilauea volcano. Part II: Distribution in the summit region.* In *Volcanism in Hawaii.* (Decker, R.W., Wright, T.L., and Stauffer, P.H. eds.) (Washington, D.C, U.S.G.S. Prof. Paper *1350*, 815–819.

ROSE, A.W., HAWKES, H.E., and WEBB, J.S., *Geochemistry in Mineral Exploration.* (London, Academic Press 1991).

SIEGEL, B.Z., and SIEGEL, S.M. (1987), *Hawaiian volcanoes and the biogeology of mercury.* In *Volcanism in Hawaii.* (Decker, R.W., Wright, T.L., and Stauffer, P.H. eds.), (Washington, DC, U.S.G.S. Prof. Paper) *1350*, 827–839.

SUGISAKI, R., IDO, M., TAKEDA, H., ISOBE, Y., HAYASHI, Y., NAKAMURA, N., SATAKE, H., MIZUTANI, Y. (1983), *Origin of hydrogen and carbon dioxide in fault gases and its relation to fault activity,* J. Geology *91,* 3, 239–258.

THOMAS, D.M. (1987) *A geochemical model of the Kilauea east rift zone.* In *Volcanism in Hawaii.* (Decker, R.W., Wright, T.L., and (Stauffer P.H. eds.), (Washington, DC, U.S.G.S. Prof. Paper) *1350*, 1507–1525.

TILLING, R.I., CHRISTIANSEN, R.L., DUFFIELD, W.A, ENDO, E.T., HOLCOMB, R.T., KOYANAGI, R.Y., PETERSON, D.W., and UNGER, J.D. (1987) *The 1972–1974 Mauna Ulu eruption, Kilauea Volcano: an example of quasi-steady-state magma transfer.* In *Volcanism in Hawaii.* (Decker, R.W., Wright, T.L., and Stauffer, P.H. eds.), (Washington, DC, U.S.G.S. Prof. Paper *1350*), 405–469.

TILLING, R.I., and DVORAK, J.J. (1993), *Anatomy of a basaltic volcano.* Nature *363,* 125–133.

WILLIAMS,-JONES, G., HEILIGMANN, M., CHARLAND, A., SHERWOOD, Lollar B., and STIX, J. (1997), *A model of diffuse degassing at three subduction-related volcanoes,* Proc. Gen. Assembly IAVCEI, Puerto Vallarta, Mexico, p. 65.

WOLFE, E.W., and MORRIS, J., *Geologic Map of the Island of Hawaii: 1:100,000* (Washington, D.C, U.S. Geological Survey. 1996).

WRIGHT, T.L. (1984), *Origin of Hawaiian tholeiite: A metasomatic model,* J. Geophys. Res. *89,* 3,233–3,252.

WRIGHT, T.L., and HELZ, R. (1987), *Recent advances in Hawaiian petrology and geochemistry.* In *Volcanism in Hawaii.* (Decker, R.W., Wright, T.L., and Stauffer. P.H. eds.), (Washington, DC, U.S.G.S. Prof. Paper *1350*, 625–640).

(Received: November 2003, revised: January 2005, accepted: January 2005)

To access this journal online:
http://www.birkhauser.ch

Pure appl. geophys. 163 (2006) 869–881
0033–4553/06/040869–13
DOI 10.1007/s00024-006-0038-x

© Birkhäuser Verlag, Basel, 2006

❙Pure and Applied Geophysics

Diffuse Emission of CO_2 from Showa-Shinzan, Hokkaido, Japan: A Sign of Volcanic Dome Degassing

Pedro A. Hernández,[1,2] Kenji Notsu,[2] Hiromu Okada,[3] Toshiya Mori,[2] Masanori Sato,[2] Francisco Barahona,[2,4] and Nemesio M. Pérez[1]

Abstract—Two soil CO_2 efflux surveys were carried out in September 1999 and June 2002 to study the spatial distribution of diffuse CO_2 degassing and estimate the total CO_2 output from Showa-Shinzan volcanic dome, Japan. Seventy-six and 81 measurements of CO_2 efflux were performed in 1999 and 2002, respectively, covering most of Showa-Shinzan volcano. Soil CO2 efflux data showed a wide range of values up to 552 g m^{-2} d^{-1}. Carbon isotope signatures of the soil CO_2 ranged from $-0.9\%_0$ to $-30.9\%_0$, suggesting a mixing between different carbon reservoirs. Most of the study area showed CO_2 efflux background values during the 1999 and 2002 surveys (B = 8.2 and 4.4 g m^{-2} d^{-1}, respectively). The spatial distribution of CO_2 efflux anomalies for both surveys showed a good correlation with the soil temperature, indicating a similar origin for the extensive soil degassing generated by condensation processes and fluids discharged by the fumarolic system of Showa-Shinzan. The total diffuse CO_2 output of Showa-Shinzan was estimated to be about 14.0–15.6 t d^{-1} of CO_2 for an area of 0.53 km^2.

Key words: Showa-Shinzan, volcanic activity, diffuse degassing, carbon dioxide.

Introduction

Carbon dioxide is the major gas species after water vapor in both volcanic hydrothermal fluids and magmas and it is an effective tracer of subsurface magma-degassing due to its low solubility in silicate melts at low to moderate pressure favoring its early exsolution (GERLACH and GRAEBER, 1985; STOLPER and HOLLO-WAY, 1988; PAN et al., 1991). Diffuse degassing studies have shown that volcanoes release high amounts of CO_2 to the atmosphere from both active craters, as plumes

[1]Environmental Research Division, Instituto Tecnológico Y de Energías Renovables (ITER), 38611, Granadilla, S/C de Tenerife, Spain
[2]Laboratory for Earthquake Chemistry, Faculty of Science, The University of Tokyo, Bunkyo-Ku, 113-0033, Tokyo, Japan
[3]Usu Volcano Observatory, Hokkaido University, Sohbetsu-cho, Hokkaido, 052-0103, Hokkaido, Japan
[4]Universidad de El Salvador, San Salvador, El Salvador, Central America

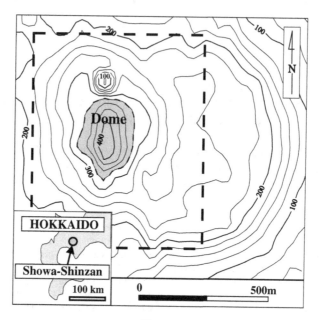

Figure 1
Location and topographic map of Showa-Shinzan volcano. Dashed line indicates the study area.

and fumaroles, and flanks as diffuse soil emanations (ALLARD *et al.*, 1991; BAUBRON *et al.*, 1990; CHIODINI *et al.*, 1996, 1998; SALAZAR *et al.*, 2001; HERNÁNDEZ *et al.*, 1998, 2001a). These emanations occur during both active and repose periods of activity. Therefore, it is difficult to assess the relation between the level of the diffuse discharges and the state of a volcanic system even though several studies have shown a significant correlation of diffuse CO_2 efflux with variations in volcanic activity (GERLACH *et al.*, 1998; GIAMMANCO *et al.*, 1998; HERNÁNDEZ *et al.*, 2001b).

Showa-Shinzan volcanic dome (Fig. 1) was formed during the volcanic eruption of Usu volcano which occurred during 1944 and 1945 as a parasitic cone (dacite lava dome). The dome is located over one of the two parallel zones running NW-SE through the summit of Usu and the northern foot, probably being controlled by the structure of the southern wall of the Toya caldera. It is composed of a circular platform (1,000 m in diameter) and a lava dome (300 m in diameter and 100 m above the platform). The lava domes are hypersthene dacites with about 69% SiO_2 (OBA *et al.*, 1983).

During the 1954–1964 period, Showa-Shinzan has been progressively out-gassed, behaving during the post-1964 period as a meteoric water dominated system (MIZUTANI, 1962) and decreasing in the temperature of the main fumaroles from 1,000 °C to 350 °C. SYMONDS *et al.* (1996) suggested that the close proximity and second degassing of shallow magma due to crystallization causes this prolonged magmatic degassing and that no new magma has been intruded into shallow levels.

The fumaroles also became more air-contaminated (MATSUO et al., 1978) increasing the atmospheric He component with time (NAGAO et al., 1980). NISHIDA and MIYAJIMA (1984) suggested the existence of an intrusive body about 200 m in diameter and several tens of meters below Showa-Shinzan at shallow levels in the base of magnetic measurements. They also reported that some parts of the Showa-Shinzan dome are filled with the lava of high temperature (> 560 °C) although many parts of the dome have cooled down. PÉREZ et al. (2002) reported soil boron and ammonia data from Showa-Shinzan before the 2000 eruption of Usu, finding low boron and relatively high ammonia contents which might indicate the existence of a liquid system vaporized at low temperature and a shallow depth origin.

Here, we present the first detailed study of diffuse CO_2 degassing at Showa-Shinzan volcanic dome with the aim to identify those areas with anomalous levels of diffuse CO_2 degassing and estimate the total output of this gas and its relation with the level of volcanic activity.

Procedures and Methods

Soil CO_2 efflux surveys were carried out in September 1999 and June 2002 under dry and snow-free conditions in order to provide minimal soil gas variations and avoid the influence of soil moisture during the measurements. Soil gas samples were also collected at Showa-Shinzan dome after following carefully considerations of the accessibility and efflux magnitude. For the 1999 and 2002 surveys, 76 and 81 soil CO_2 efflux measurements together with soil gas samples collected at selected locations were carried out in the field, respectively.

Soil CO_2 efflux measurements were performed according to the accumulation chamber method (PARKINSON 1981; CHIODINI et al., 1998). The system consists of a cylindrical chamber opened at the bottom with a fan to improve gas mixing, a NDIR (non-disperse infrared) spectrophotometer (a Dräger Polytron transmitter IR sensor, West Systems) with an accuracy of approximately 5% and a reproducibility of 10% for the range 100–10,000 g m^{-2} d^{-1}. The CO_2 sensor was calibrated in the laboratory using free CO_2 gas (Nitrogen) and a CO_2 standard gas (4% V). We assumed a random error of ± 10% in emission rates, based on the uncertainty from the variability of the measurements carried out in the laboratory. Values of CO_2 efflux (g m^{-2} d^{-1}) were estimated from the rate of concentration increase in the chamber at the observation site, accounting for changes of atmospheric pressure and temperature to convert volumetric concentrations to mass concentrations.

Soil gas samples were collected with a syringe from a probe inserted into the soil at 40–50 cm depth and stored in pre-evacuated vacutainers. The samples were analyzed in the laboratory using a gas chromatograph with a thermal conductivity detector (Ohkura GC 103). Soil CO_2 was analyzed using a Porapaq column with an operating temperature of 30 °C using O_2 as carrier.

Table 1

Summary of descriptive statistics of soil gases and soil temperature at Showa-Shinzan for the 1998 and 2002 surveys

	1999				2002			
	CO_2 efflux (g m^{-2} d^{-1})	CO_2 (%)	δ^{13}C-CO_2 (‰)	Soil Temp (°C)	CO_2 efflux (g m^{-2} d^{-1})	CO_2 (%)	δ^{13}C-CO_2 (‰)	Soil Temp. (°C)
Median	23.2	0.47	−12.67	37.5	9.7	0.19	−11.0	49.0
Minimum	0.14	0.03	−23.5	20.3	0.03	0.07	−30.9	15.2
Maximum	552.3	16.6	−7.9	100.2	288.0	8.69	−0.9	101.3
S.D.	138.3	4.19	5.34	32.6	62.1	1.86	8.20	29.5

Carbon isotope analyses of gases were made with a Finnigan MAT Delta S mass spectrometer. The ^{13}C/^{12}C ratios are reported as δ^{13}C values (±0.1‰) with respect to V-PDB standard.

Results and Discussion

A summary of main descriptive statistics is listed in Table 1. Soil CO_2 efflux values for the 1999 survey ranged from 0.14 up to 552 g m^{-2} d^{-1} with a median of 23.2 g m^{-2} d^{-1}, whereas for the 2002 survey CO_2 efflux values ranged from non-detectable values up to 288 g m^{-2} d^{-1} with a median of 9.7 g m^{-2} d^{-1}. Soil CO_2 concentration of samples for 1999 and 2002 surveys ranged from 0.03 to 16.6% V and 0.07 to 8.7% V, respectively. Isotopic data for carbon (CO_2) ranged from −23.5 to −7.9‰ and −30.9 to −0.9‰, respectively, suggesting different sources for the carbon dioxide in the soil atmosphere. The area covered for both surveys at Showa-Shinzan was approximately 0.5 km^2.

The sampling frequency distribution of the diffuse CO_2 emission rates for both surveys showed a polymodal shape. To check whether the data come from a unimodal or a polymodal distribution, we applied the probability-plot technique (TENNANT and WHITE, 1959; SINCLAIR, 1974) to the entire CO_2 efflux data. At least three distinct modes were found for both 1999 and 2002 data sets: normal I, normal II and the mode comprised between both distributions (Table 2). These three distinct populations are known as background, peak, and intermediate mixed-population, respectively. Results of the 1999 survey showed a background population (normal I) accounting for 44.3% of the total data with a mean of 8.2 g m^{-2} d^{-1} (Figure 2). Peak population (normal II) represented 10.2% of the total data with a mean of 399 g m^{-2} d^{-1}. Regarding to the 2002 survey, background population (normal I) accounted for 66.2% of the total data with a

Table 2

Estimated parameters of partitioned populations in the surveyed area of Showa-Shinzan for 1999 and 2002 surveys

	Population	Mean CO_2 efflux value (g m^{-2} d^{-1})	Proportion (%)	Total diffuse CO_2 output (t d^{-1})
	I	8.2	44.3	2.66
1999	II	67	45.5	12.96
	III	399	10.2	0.06
	I	4.4	66.2	6.39
2002	II	44.2	22.3	6.89
	III	177	11.5	0.77

mean of 4.4 g m^{-2} d^{-1} (Fig. 3). Peak population (normal II) represented 11.5% of the total data with a mean of 177 g m^{-2} d^{-1}.

The existence of three overlapping geochemical populations for soil CO_2 efflux data suggests a close relation of deep perturbations with the volcanic-hydrothermal system on the surface environment. Probability plots show that peak populations for Showa-Shinzan volcanic dome represent a relative low percentage of the total data, having possibly a marked magmatic signature. This observation might indicate that there is an influence of different physical-chemical processes acting at different scales on the diffuse CO_2 degassing at Showa-Shinzan.

CO_2 efflux data from both 1999 and 2002 surveys were used to construct contour maps of Showa-Shinzan using kriging as the interpolation technique which predicts unknown values from the observed data. The variograms for CO_2 efflux (1999 and 2002) and soil temperature (1999), (see Figs. 4, 5 and 6, respectively) were fitted by different functions (linear, exponential, spherical, quadratic, logarithmic and Gaussian) to obtain the best results (JOURNEL, 1988; CRESSIE, 1990; OLIVER and WEBSTER, 1990). Isotropic linear variograms with negligible nugget effect were chosen as the best-fitted to the normal distributed data. The variograms fitted by quadratic, spherical, Gaussian and exponential functions for the 1999 CO_2 efflux data could not be modeled satisfactorily. Figures 4 and 5 show the distribution of soil CO_2 efflux data at Showa-Shinzan volcanic dome for both 1999 and 2002 surveys, respectively. Inspection of the CO_2 efflux distribution map for both surveys shows the following features: (1) Areas of high CO_2 efflux values are characterized by higher temperatures (T > 40°C) whereas those with low CO_2 efflux values correspond to relatively normal soil temperatures (T < 25 °C) and (2) values of CO_2 efflux increase when approaching the fumaroles. Figure 6 shows the spatial distribution of soil temperature for the 1999 survey. The observed positive correlation between high CO_2 efflux and high temperatures in soils (Fig. 7) suggests that part of the energy coming from the remaining heat source beneath Showa-Shinzan volcano is used

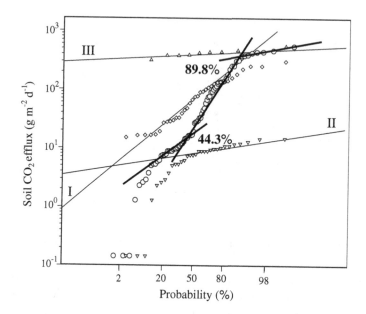

Figure 2
Probability plot of the soil CO_2 efflux values observed during the 1999 survey at Showa-Shinzan volcano.

to heat up the local meteoric water which penetrated through fissures. A good correlation was found between soil CO_2 efflux and soil CO_2 concentration ($r = 0.86$) and between soil temperature and $\delta^{13}C$-CO_2 ($r = 0.87$), whereas a poorer correlation is observed for soil CO_2 concentration and soil temperature ($r = 0.42$).

All the soil anomalies observed in these areas may be related to the degassing of the magma reservoir and upward movement of heated water. This relation also suggests the occurrence of extensive soil degassing generated by condensation processes of fluids originally similar to those discharged by the fumarolic system (CHIODINI *et al.*, 1996). Areas with higher CO_2 concentration levels (> 2 % V) and higher CO_2 efflux values (> 200 g m^{-2} d^{-1}) showed $\delta^{13}C$-CO_2 values between -13.6% and -0.9%, indicating a magmatic contribution for CO_2. Figure 8 shows a biplot of $\delta^{13}C$-CO_2 and soil CO_2 concentration data. Most soil gases from the studied area plot within or near the area defined by a CO_2 concentration from air-like CO_2 (360 ppm V) to 1% V (Box I). This area represents the mixing between the background soil gases which are the result of biogenic and air CO_2 and sites with a contribution of deep-seated CO_2 ($\delta^{13}C$-rich source such as limestone decarbonation or magmatic carbon). Sites of soil CO_2 concentration (> 1% V) show the highest $\delta^{13}C$-CO_2 values ($-6.0\% < \delta^{13}C$-$CO_2 < -0.9\%$), indicating that the efflux of deep-seated CO_2 is restricted to specific areas with higher permeability or associated to

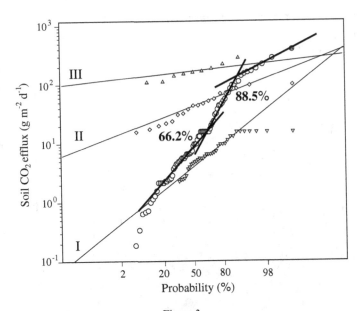

Figure 3
Probability plot of the soil CO₂ efflux values observed during the 2002 survey at Showa-Shinzan volcano.

discrete conduits (Box II). This δ^{13}C-CO₂ range is in good agreement with the δ^{13}C-CO₂ values observed at the fumarolic system, indicating a common origin for the CO₂ in both fumaroles and high CO₂ efflux areas. Probably, fractionation phenomena occurred in the soil due to the fluid condensation contributing to the observed δ^{13}C-CO₂ values.

To estimate the total diffuse CO₂ output released from Showa-Shinzan volcanic edifice, we considered the positive volume enclosed by the soil CO₂ efflux 3-D surface built-up by the kriging in the contouring over the study area, yielding a total of about 15.6 t d^{-1} and 14.0 t d^{-1} for both 1999 and 2002 surveys covering an area of 0.53 km². The estimated output of diffuse CO₂ for Showa-Shinzan is lower than the observed at other active volcanoes such as Etna (55,000 t d^{-1}; ALLARD et al., 1991), Solfatara di Pozzuoli (1500 t d^{-1}, CHIODINI et al., 2001) and Vulcano (700 t d^{-1}, 1.9 km², CHIODINI et al., 1996, 1998) in Italy, Teide (380 t d^{-1}, 0.53 km², HERNÁNDEZ et al., 1998) in Spain, Usu and Miyakejima (39–340 t d^{-1}, 2–2.65 km² and 118 t d^{-1}, 33–41 km², respectively; HERNÁNDEZ et al., 2001a, b) in Japan and Cerro Negro (2,800 t d^{-1}, 0.58 km², SALAZAR et al., 2001) in Nicaragua. The observed differences seem to be related to the dimension of the magmatic source at depth and the level of volcanic activity. To constrain this possibility we have calculated the fraction of deep-seated CO₂ emitted from Showa-Shinzan. Based on the chemical and isotopic analyses of CO₂ we have considered as representative of deep-seated CO₂ efflux the values higher than the

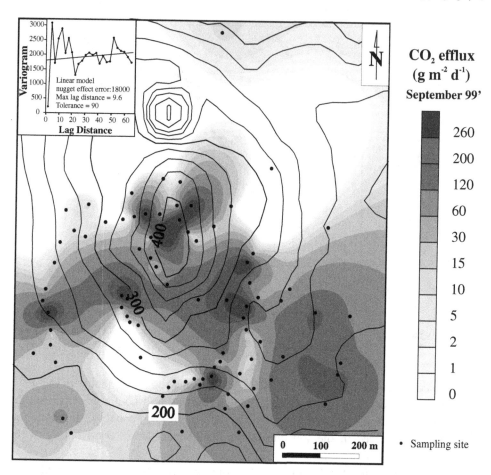

Figure 4
Distribution map of soil CO_2 efflux anomalies (g m^{-2} d^{-1}) and experimental variogram, Showa-Shinzan volcano, September 1999.

background population, which correspond to both the geochemical populations II and III. These values represent the mixing between the biogenic and volcanic CO_2 (magmatic and organic material). Assuming the range of values for populations II and III and using kriging as the interpolation technique, we are able to estimate the fraction of CO_2 considered as volcanic (over the background fraction). Table 2 sets forth the results indicating the CO_2 output for each geochemical population. Under these considerations we have assumed that about 13.0 t d^{-1} and 7.6 t d^{-1} of volcanic CO_2 (83% and 54.5% of the total) have been released as diffuse emanation from Showa-Shinzan during the 1999 and 2002 surveys.

Combining volumes of the extruded portion of the plug dome and the cryptodome it is possible minimally estimate the initial amount of CO_2 in the magma body. Let us

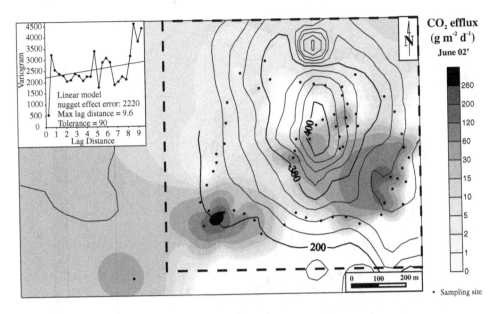

Figure 5

Distribution map of soil CO_2 efflux anomalies (g m^{-2} d^{-1}) and experimental variogram, Showa-Shinzan volcano, June 2002. Dashed line indicates the studied area considered for the estimation of CO_2 output.

assume a simplified cylindrical geometry for both extruded magma and cryptodome, with an average diameter of 300 m and a height of 150 m for the extruded dome and an average diameter of 800–1,200 m and height of 170–200 m for the cryptodome (Fig. 9). The volumes of the extruded dome and cryptodome at Showa-Shinzan are about 0.0106 km^3 and 0.09–0.23 km^3, respectively (SYMONDS *et al.*, 1996). Assuming a dacitic composition with a density between 2.4 and 2.8 g cm^{-3} (MURASE and MCBIRNEY, 1973) and an average CO_2 of 500 ppm (ANDERSON, *et al.*, 1989), the total CO_2 contained initially in this magma body, considering the minimum and maximum calculated volumes and densities, would be approximately 116, 840–357, 291 tons. If we consider the average value of the daily CO_2 output (14.8 t d^{-1}) and as a conceptual model, assume as representative of daily CO_2 output for the period 1944–2002, a total of 313,316 tons are obtained. The similarity in orders of magnitude of both values might be an indicator of the level of degassing of the magmatic source of Showa-Shinzan at depth.

Conclusions

Based on two soil gas surveys, we studied the spatial distribution of diffuse CO_2 emission from the Showa-Shinzan volcano. The chemical and isotopic

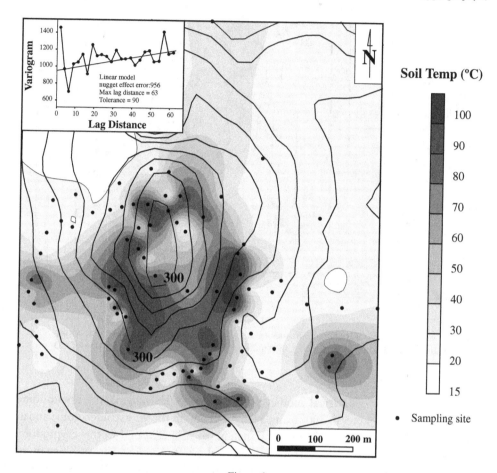

Figure 6
Distribution map of soil temperature anomalies (°C) and experimental variogram, Showa-Shinzan volcano, September 1999.

compositions of the diffuse gas emanations and the CO_2 efflux values indicate that Showa-Shinzan releases relatively low amounts of CO_2 in diffuse form when compared with other volcanic systems. Most of the diffuse CO_2 degassing is concentrated at those areas where anomalous diffuse degassing and soil temperature occur. The good correlations observed between the CO_2 efflux and soil CO_2 concentration and soil temperature indicate an important contribution of advective mechanism controlling the transport of CO_2 to the surface environment. The relatively low CO_2 efflux values together with light $\delta^{13}C(CO_2)$ measured in soil gases, indicate a poor contribution of deep-seated degassing proximal to or incorporated within the volcanic hydrothermal system of Showa-Shinzan. The estimated emission rates of CO_2 for both 1999 and 2002 surveys were 15.6 and 14.0 t d^{-1}, respectively.

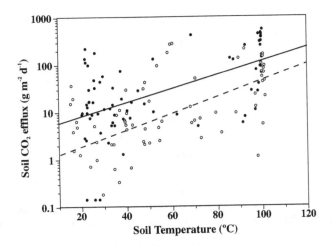

Figure 7

Plot of $\delta^{13}C\text{-}CO_2$ versus soil temperature from the studied area at Showa-Shinzan volcano, September 1999. Black circles and open circles represent 1999 and 2002 surveys, respectively. Solid and dashed lines represent the correlation lines between $\delta^{13}C\text{-}CO_2$ and soil temperature for both 1999 and 2002 surveys.

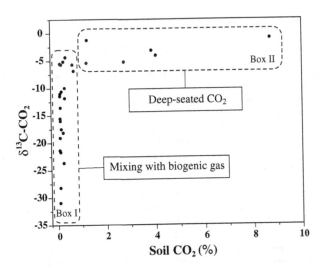

Figure 8

Plot of $\delta^{13}C\text{-}CO_2$ versus soil CO_2 concentration from the studied area at Showa-Shinzan volcano. Boxes I and II represent the mixing between the background soil gases and the sites with anomalous CO_2 efflux values (volcanic gas), respectively.

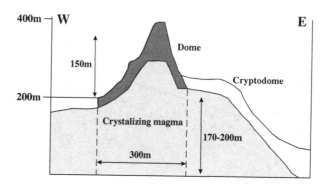

Figure 9

Schematic W-E cross section through Showa-Shinzan dome, showing the extrusive dome and the accompanying lava, the cryptodome and speculative subsurface magma (from SYMONDS *et al.*, 1996).

Acknowledgements

We thank Ana Martinez Soriano for her assistance in the field work. We are indebted to the staff of Usu Volcano Observatory and Yoichi Shimoike for their help and assistance. This research was carried out with grant from the JSPS Program in Japan (P.H.P.).

REFERENCES

ALLARD, P., CARBONELLE, J., DAJLEVIC, D., LE BRONCE, J., MOREL, P., ROBE, M.C., MAURENADS, J.M., FAIVRE-PIERRET, R., MARTIN, D., SABROUX, J.C., and ZETTWOOG, P. (1991), *Eruptive and diffuse emissions of CO_2 from Mount Etna,* Nature *351,* 387–391.

ANDERSON, A.T., NEWMAN, S., WILLIAMS, S.N., DRUITT, T.H., SKIRIUS, C., and STOLPER, E. (1989), *H_2O, CO_2, Cl and gas in Plinian and ash-flow Bishop rhyolite,* Geology *17,* 221–225.

BAUBRON, J.C., ALLARD, P., and TOUTAIN, J.P. (1990), *Diffuse volcanic emissions of carbon dioxide from Vulcano Island, Italy,* Nature *344,* 51–53.

CHIODINI, G., FRONDINI, F., and RACO, B. (1996), *Diffuse emission of CO_2 from the Fossa crater, Vulcano Island (Italy),* Bull. Volcanol. *58,* 41–50.

CHIODINI, G., CIONI, R., GUIDI, M., RACO, B., and MARINI, L. (1998), *Soil CO_2 efflux measurements in volcanic and geothermal areas,* Appl. Geochem. *13,* 543–552.

CHIODINI, G., FRONDINI, F., CARDELLINI, C., GRANIERI, D., MARINI, L., and VENTURA, G. (2001), *CO_2 degassing and energy release at Solfatara Volcano, Campi Flegrei, Italy,* J. Geophys. Res. *106* (B8), 16213–16221.

CRESSIE, N.A.C., (1990), *The origins of kriging,* Mathemat. Geology. *22,* 239.

GERLACH, T., DOUKAS, M., MCGEE, K., and KESSLER, R. (1998), *Three-year decline of magmatic CO_2 emission from soils of a Mammoth Mountain tree kill: Horseshoe Lake, CA, 1995–1997,* Geophys. Res. Lett. *25,* 1947–1950.

GERLACH, T.M.J. and GRAEBER, E.J. (1985), *Volatile budget of Kilauea volcano,* Nature *313,* 273–277.

GIAMMANCO, S., INGUAGGIATO, S., and VALENZA, M. (1998), *Soil and fumarole gases of Mount Etna: Geochemistry and relations with volcanic activity,* J. Volcanol. Geotherm. Res. *81,* 297–310.

HERNÁNDEZ, P.A., PÉREZ, N.M., SALAZAR, J.M., NOTSU, K., and WAKITA, H. (1998), *Diffuse emission of carbon dioxide, methane, and helium-3 from Teide volcano, Tenerife, Canary Islands,* Geophys. Res. Lett. *25,* 3,311–3,314.

HERNÁNDEZ, P.A., NOTSU, K., SALAZAR, J.M., MORI, T., NATALE, G., OKADA, H., VIRGILI, G., SHIMOIKE, Y., SATO, M., and PÉREZ, N.M. (2001a), *Carbon dioxide degassing by advective flow from Usu volcano, Japan,* Science *292,* 83–86.

HERNÁNDEZ, P.A., SALAZAR, J.M., SHIMOIKE, Y., MORI, T., NOTSU, K., and PÉREZ, N.M. (2001b), *Diffuse emission of CO$_2$ from Miyakejima volcano, Japan,* Chem. Geol. *177,* 175–185.

JOURNEL, A., *Principles of Environment Sampling* (American Chemical Society), (Washington, DC 1988), pp. 45–72.

MATSUO, S., SUZIKI, M., and MIZUTANI, Y. (1978), *Nitrogen to argon ratio in volcanic gases,* In: *Terrestrial Rare Gases.* (E.C. Alexander, Jr. and M. Ozima, eds.), (Cent Acad Pub Japan, Tokyo, 1978), pp.17–25.

MIZUTANI, Y. (1962), *Origin of lower temperature fumarolic gases at Show Shinzan,* J. Earth Sci. Nagoya Univ. *10,* 135–148.

MURASE, T. and McBIRNEY, A.R. (1973), *Properties of some common igneous rocks and their melts at high temperatures,* Geol. Soc. Am. Bull. *84,* 3,563–3,592.

NAGAO, K., TAKAOKA, N., MATSUO, S., MOZITANI, Y., and MATSUBAYASHI, Y. (1980), *Change in rare gas composition of the fumarolic gases from the Show Shinzan volcano,* Geochem. J. *14,* 139–143.

NISHIDA, Y. and MIYAJIMA, E. (1984), *Subsurface structure of Usu volcano, Japan as revealed by detailed magnetic survey,* J. Volcanol. Geotherm. Res. *22,* 271–285.

OBA, T., KATSUI, Y., KURASAWA, H., IKEDA, Y., and UDA, T. (1983), *Petrology of historic rhyolite and dacite from Usu volcano, North Japan,* J. Fac. Sci, Hokaido Univ., Ser. IV, *20,* 274–290.

OLIVER, M.A. and WEBSTER, R. (1990), *Kriging: A method of interpolation for geographical information system,* Int. J. Geograph. Info. Systems *4* (3), 313–332.

PAN, V., HOLLOWAY, J.R., and HERVIG, R.L. (1991), *The pressure and temperature dependence of carbon dioxide solubility in tholeiitic basalts melts,* Geochim. Cosmochim. Acta *55,* 1587–1595.

PARKINSON, K.J. (1981), *An improved method for measuring soil respiration in the field,* J. Appl. Ecol. *18,* 221–228.

PÉREZ, N.M., HERNÁNDEZ, P.A., CASTRO, L., SALAZAR, J.M., NOTSU, K., MORI, T., and OKADA, H. (2002), *Distribution of Boron and Ammonia in soil around Usu volcano, Japan, Prior to the 2000 eruption,* Bull. Volcanol. Soc. Japan *47,* 347–351.

SALAZAR, J.M., HERNÁNDEZ, P.A., PÉREZ, N.M., MELIAN, G., ALVAREZ, J., and NOTSU, K. (2001), *Diffuse volcanic emissions of carbon dioxide from Cerro Negro volcano, Nicaragua,* Geophys. Res. Lett. *28,* 4275–4278.

SINCLAIR, A.J. (1974), *Selection of thresholds in geochemical data using probability graphs,* J. Geochem. Explor. *3,* 129–149.

STOLPER, E. and HOLLOWAY, J.R. (1988), *Experimental determination of the solubility of carbon dioxide in molten basalt at low pressure,* Earth. Plan. Science Lett. *87,* 397–408.

SYMONDS, R.R., MIZUTANI, Y., and BRIGGS, P.H. (1998), *Long-term geochemical surveillance of fumaroles at Showa-Shinzan dome, Usu volcano, Japan,* J. Volcanol. Geotherm. Res. *73,* 177–211.

TENNANT, C.B. and WHITE, M.L. (1959), *Study of the distribution of some geochemical data,* Econ. Geol. *54,* 1281–1290.

(Received: May 23, 2003; revised: May 15, 2005; accepted: May 30, 2005)

To access this journal online:
http://www.birkhauser.ch

Pure appl. geophys. 163 (2006) 883–896
0033–4553/06/040883–14
DOI 10.1007/s00024-006-0050-1

© Birkhäuser Verlag, Basel, 2006

❚ Pure and Applied Geophysics

Anomalous Diffuse CO_2 Emission prior to the January 2002 Short-term Unrest at San Miguel Volcano, El Salvador, Central America

NEMESIO M. PÉREZ,[1] PEDRO A. HERNÁNDEZ,[1] ELEAZAR PADRÓN,[1]
RAFAEL CARTAGENA,[2] RODOLFO OLMOS,[2] FRANCISCO BARAHONA,[2] GLADYS MELIÁN,[1]
PEDRO SALAZAR,[1] and DINA L. LÓPEZ[3]

Abstract—On January 16, 2002, short-term unrest occurred at San Miguel volcano. A gas-and-steam-ash plume rose a few hundred meters above the summit crater. An anomalous microseismicity pattern, about 75 events between 7:30 and 10:30 hours, was also observed. Continuous monitoring of CO_2 efflux on the volcano started on November 24, 2001, in the attempt to provide a multidisciplinary approach for its volcanic surveillance. The background mean of the diffuse CO_2 emission is about 16 g m^{-2} d^{-1}, but a 17-fold increase, up to 270 g m^{-2} d^{-1}, was detected on January 7, nine days before the January 2002 short-term unrest at San Miguel volcano. These observed anomalous changes on diffuse CO_2 degassing could be related to either a sharp increase of CO_2 pressure within the volcanic-hydrothermal system or degassing from an uprising fresh gas-rich magma within the shallow plumbing system of the volcano since meteorological fluctuations cannot explain this observed increase of diffuse CO_2 emission.

Key words: San Miguel, volcanic activity, diffuse degassing, carbon dioxide.

Introduction

The symmetrical cone of San Miguel, one of the most active volcanoes in El Salvador (Central America) rises from near sea level to 2,132 meters of elevation. This basaltic volcano is located in the eastern part of the country and lies on the southern fault of the Central American graben at the intersection with NW-SE trending faults (Fig. 1). San Miguel volcano has erupted at least 30 times since 1699, all events classified with a VEI of 1 or 2. Many of the earlier eruptions occurred at flank vents, but since 1867 all have taken place at the summit (MEYER-ABICH, 1956; SIMKIN and SIEBERT, 1994). San Miguel volcano has had active fumaroles in the summit region at

[1]Environmental Research Division, Instituto Tecnológico y de Energías Renovables (ITER), 38611, Granadilla, S/C de Tenerife, Spain
[2]Instituto de Ciencias de la Tierra, Universidad de El Salvador, El Salvador, Central America
[3]Department of Geological Sciences, 316 Clippinger Laboratories Ohio University, Athens, OH, 45701, USA

least since 1964, and their variable intensity sustains usually a faint plume emitted from its summit crater. Since 1970, weak explosions took place four times, the last in January 2002 (GVN BULLETIN, 2002).

The January 2002 short-term volcanic unrest at San Miguel was characterized by the emission of a gas-and-steam plume of 100 metric tons/day of sulfur dioxide containing a slight amount of ash rising with a mushroom-like profile a few hundred meters above the summit crater of San Miguel (GVN BULLETIN, 2002). In addition, the event was accompanied by anomalous microseismicity with about 75 events between 7:30 and 10:30 hours, on January 16, 2002. This seismic swarm was followed by progressively decreasing seismic activity during the next days. This style of increased seismicity and gas emission is within the range of normal activity at San Miguel according to SNET (Servicio Nacional de Estudios Territoriales; the National organization in-charge of the volcano monitoring at El Salvador) since intermittent periods of vigorous steam-and-gas emission from San Miguel have been commonly reported in recent years (GVN BULLETIN, 2002). The population at risk from a San Miguel eruption with significant ashfall is a mix of urban and rural residents. The city of San Miguel, at the foot of the NE flank of the volcano, has a population of ~150,000, and the rural zone that would likely be affected has a population of ~100,000 (GVN BULLETIN, 2002).

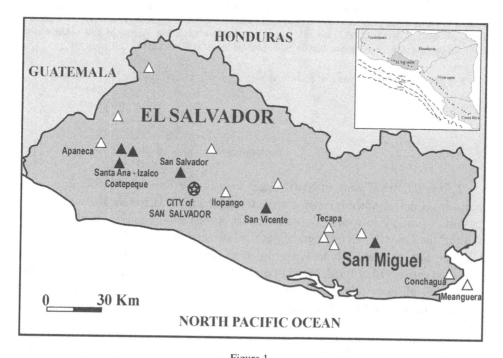

Figure 1
San Miguel volcano located in the eastern part of El Salvador, Central America. Closed triangles represent those active volcanic systems with a continuous geochemical monitoring station of CO_2 efflux.

Scientists have long recognized that gases dissolved in magma provide the driving force of volcanic eruptions, but only recently geochemical techniques permitted continuous measurement of different types and manifestations of volcanic gases released into the atmosphere. Carbon dioxide is the major gas species after water vapor in both volcanic fluids and magmas and it is an effective tracer of subsurface magma-degassing, due to its low solubility in silicate melts (GERLACH and GRAEBER, 1985). Since this degassed CO_2 travels upward by advective-diffusive transport mechanisms and manifests itself at the ground surface, soil CO_2 efflux pattern changes over time provide information about subsurface magma movement (HERNÁNDEZ et al., 2001a; CARAPEZZA et al., 2004).

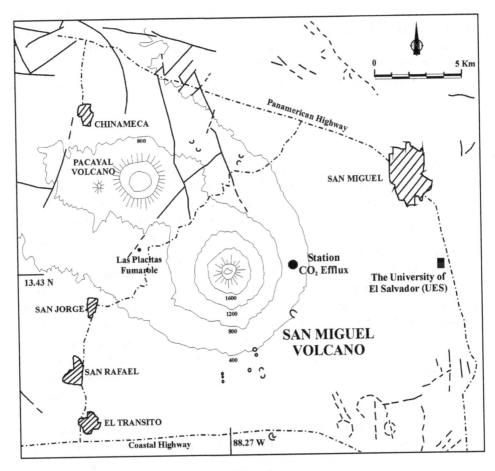

Figure 2

Simple morphological and structural map of San Miguel volcano and its surroundings. This map shows major inhabited areas in the well as the location of the Las Placitas' fumarole and the geochemical station (closed circle). Solid lines represent faults and dashed lines represent inferred fault systems. Dotted lines represent major highways.

Extensive work on diffuse CO_2 degassing has been performed at volcanic and geothermal areas over the last 13 years, suggesting that even during periods of quiescence, volcanoes release large amounts of carbon dioxide in a diffuse form (BAUBRON *et al.*, 1990; ALLARD *et al.*, 1991; FARRAR *et al.*, 1995; CARTAGENA *et al.*, 2004; CHIODINI *et al.*, 1996, 1998; HERNÁNDEZ *et al.*, 1998, 2001b; GERLACH *et al.*, 1998; SOREY *et al.*, 1998; GIAMMANCO *et al.*, 1998; PÉREZ *et al.*, 1996, 2004; ROGIE *et al.*, 2001; SALAZAR *et al.*, 2001). However, few works related to continuous monitoring of the diffuse CO_2 degassing have been published (ROGIE *et al.*, 2001; SALAZAR *et al.*, 2002, 2004; MORI *et al.*, 2002; GRANIERI *et al.*, 2003; CARAPEZZA *et al.*, 2004; BRUSCA *et al.*, 2004).

Conventional geophysical methods are operating for the volcano monitoring program of San Miguel (e.g., volcano seismicity). In order to provide a multidisciplinary approach to volcanic surveillance of San Miguel, the Spanish Aid-Agency (AECI) provided the financial-aid to set up a geochemical station for the continuous monitoring of diffuse CO_2 degassing. Similar monitoring stations were installed at Santa Ana-Izalco-Coatepeque, San Salvador and San Vicente volcanic systems in 2001.

Procedures and Methods

With the aim of detecting anomalous temporal variations in diffuse CO_2 emission rates related to changes of volcanic activity at San Miguel, an automatic geochemical station (WEST Systems, Italy) was installed on November 24, 2001. This station continuously monitors diffuse CO_2 degassing on the eastern flank of San Miguel volcano (594 m a.s.l., Fig. 2). The station is equipped with an on-board microcomputer as well as sensors to measure the air CO_2 gas concentration, wind speed and direction, air temperature and relative humidity (1 m above the ground). The accumulation chamber is lowered on to the ground for several minutes every hour, and during this period the gas is continuously extracted from the chamber, sent to the IR spectrophotometer, and then injected again into the chamber. The latter is equipped with a mixing device in order to improve gas mixing. Soil CO_2 efflux is then measured according to the accumulation chamber method by means of a NDIR (non-dispersive infrared) spectrophotometer (Dräger Polytron IR transmitter) with a double-beam IR detector with solid state sensor compensated in temperature. Accuracy of 3% is acquired for a reading at 350 ppm. The automatic geochemical station is powered with a solar cell panel and a backup battery. Values of CO_2 efflux ($g\ m^{-2}\ d^{-1}$) are estimated from the rate of concentration increase in the chamber at the observation site, accounting for changes of atmospheric pressure and temperature to convert volumetric concentrations to mass concentrations. The chamber rim is designed to be set properly on the ground in order to eliminate the input of atmospheric air, which could cause significant errors, especially on windy days. The reproducibility for the range 10–20,000 $g\ m^{-2}\ d^{-1}$ is ± 10%. We assumed a random

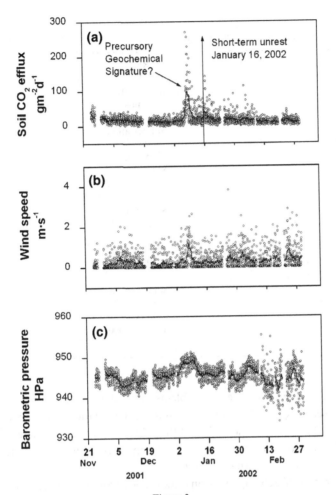

Figure 3
Time series of (a) diffuse CO_2 emission rate, (b) wind speed and (c) barometric pressure on the eastern flank of San Miguel Volcano from November, 2001 to March, 2002.

error of $\pm 10\%$ in the emission rates of CO_2, based on variability of the replicate measurements carried out on known CO_2 efflux rates in the laboratory. All measurements were performed during the dry season in El Salvador.

Results and Discussion

A time series of 2,326 hourly observations of diffuse CO_2 emission rate, wind speed and barometric pressure from 17:00 hours of November 24, 2001, to 14:00 hours of March 1, 2002, as well as their moving averages of 36 hours are shown in Figure 3. A total of 12.4% of missing data throughout the observation

Table 1

Summary of results from the geochemical station at San Miguel Volcano

CO_2 efflux (g m^{-2} d^{-1})	Range	< 0.25–270
	Mean	29.4
	Std deviation	33.2
Air relative humidity (%)	Range	12.5–97.3
	Mean	54.2
	Std deviation	19.3
Air temperature (°C)	Range	15.9–34.5
	Mean	23.9
	Std deviation	4.2
Bar pressure (hPa)	Range	932.1–959.0
	Mean	945.3
	Std deviation	2.7
Wind speed (m s^{-1})	Range	0.0–13.7
	Mean	1.1
	Std deviation	1.6
Wind direction (°)	Range	0.8–341.7
	Mean	202.1
	Std deviation	90.5
Tidal ( )	Range (x1000)	−999.3–2241.7
	Mean (x 1000)	189.7
	Std dev (x1000)	754.2

period occurred due to telemetry problems. The effect of the missing data on the overall data set is not likely to produce spurious correlation results because most of the missing data were recorded when diffuse CO_2 degassing on the eastern flank of San Miguel volcano was stationary. Table 1 summarizes the result of the CO_2 efflux data and the observed external variables.

The observed diffuse CO_2 emission data ranged from negligible values up to 270 g m^{-2} d^{-1}. These time series may be separated into four distinct set intervals: (i) from the beginning of the observation until January 6, 2002, characterized by a low variance of the CO_2 efflux and relatively low diffuse CO_2 emission rates of about 16 g m^{-2}, d^{-1}; (ii) from January 6 to 16, 2002, characterized by a short-term transient in the soil CO_2 degassing emission $\sim$ 9 days before the short-term unrest with CO_2 efflux values reaching up to 270 g m^{-2} d^{-1}; (iii) from January 16 to February 9, 2002, characterized by wider scatter of efflux data and a relatively high CO_2 efflux value (143 g m^{-2} d^{-1}) occurring at the onset of an anomalous seismic pattern at San Miguel volcano in January 16, 2002, and (iv) a period from February 9 to March 1, 2002, characterized by CO_2 efflux values similar to those observed within period (i).

Wind speed values ranged from 0 to 3.8 m/s. Average wind speed was 0.18 m/s. However, wind speed time series showed a simultaneous increase with the transient of soil CO_2 efflux in period (ii). Barometric pressure ranged from 938 to 949 HPa within periods (i) to (iii). Barometric pressure time series reflected the passing of low

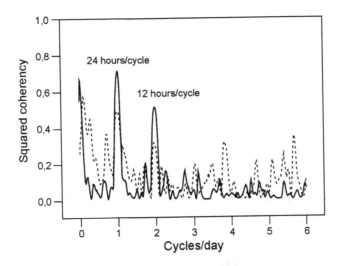

Figure 4

Cross-spectral analysis of the soil CO$_2$ efflux and wind speed (solid line) and soil CO$_2$ efflux and barometric pressure (dashed line) time series.

and high pressure fronts over the observation site. A high pressure atmospheric regime started on January 2 and ended January 12, 2002. Both wind speed and barometric pressure displayed high values during period (ii), where a short-term transient in the soil CO$_2$ degassing was observed. These observations drive the question whether wind speed and/or barometric pressure may be responsible for the observed changes in the soil CO$_2$ degassing at San Miguel volcano. At this aim, time domain classical techniques were applied to get the autocorrelation functions of the three original time series, showing the presence of cycles of the same length.

Because various types of serial dependence (trends, cycles, and autoregressive serial dependence) within time series can give rise to spurious or misleading correlations between time series (WARNER, 1998), soil CO$_2$ efflux, wind speed and barometric pressure time series were standardized and prewhitened before spectral analysis was performed. Soil CO$_2$ efflux time series was log-transformed and differenced one time to obtain a much more stationary time series. Wind speed and barometric pressure time series only required one order differencing before getting stationarity. Then spectral coherences between the prewhitened soil CO$_2$ efflux and wind speed, as well as soil CO$_2$ efflux and barometric pressure time series were computed (Fig. 4). Cross-spectral analysis of the prewhitened time series shows that a high percentage of shared variance between both sets of time series occur at one and two cycles per day, which correspond's to periods of 24 and 12 hours, respectively, suggesting a causal influence or coupling of wind speed and barometric pressure on the observed soil CO$_2$ efflux values at diurnal and semi-diurnal frequencies. These results suggest a close relationship between the temporal variations of the soil CO$_2$ efflux and those of meteorological variables, showing

24- and 12-hour periodicities (i.e., wind speed, barometric pressure, solar radiation, air temperature, etc.).

Once the main cyclic components of the prewhitened soil CO_2 efflux time series have been identified, it was possible to model the original soil CO_2 efflux signal using the harmonic analysis (Fig. 5a), based on the following equation

$$\Phi(t) = \mu + \sum_{i=1}^{2} \{A_i \cos(\varpi_i t) + B_i \sin(\varpi_i t)\} + \varepsilon,$$

where $\Phi(t)$ and μ are the observed soil CO_2 efflux time series and its mean, respectively, ω_i (i = 1,2) are the main identified frequencies corresponding to the main periodic components of the soil CO_2 efflux signal, A_i and B_i are model

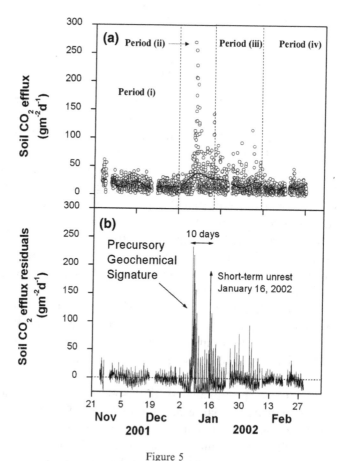

Figure 5
(a) Original soil CO_2 efflux signal (closed circles) and fitted harmonic model (solid line) with two major periodic components; (b) residual of harmonic analysis showing the persistence of a large amplitude transient in the soil CO_2 time series.

parameters to adjust during the fitting process, t is the observation time and, finally, ϵ are the residuals uncorrelated with the cos and sin terms. Using this simple deterministic model, the autoregressive behavior of the soil CO_2 efflux time series, which is linked to the periodic behavior of the meteorological variables (wind speed and barometric pressure), can be removed from the original soil CO_2 efflux signal. As a result of this filtering procedure, a large amplitude transient in the soil CO_2 degassing still remains within the residuals (Fig. 5b) temporally related with the onset of the anomalous seismic pattern at San Miguel volcano.

Low amplitude and almost universal short-term CO_2 efflux changes driven by meteorological fluctuations have been observed in others volcanic systems (SALAZAR et al., 2000, 2002; PADRÓN et al., 2001; ROGIE et al., 2001; GRANIERI et al., 2003). This significant transient increase of CO_2 efflux rate observed on January 6 and 7, 2002, cannot be explained in terms of meteorological fluctuations such as barometric pressure and wind speed temporal variations recorded at the station site.

Since meteorological fluctuations cannot explain the observed relative increase of the CO_2 efflux rate on January 6 and 7, 2002 (Fig. 3), it is evident that the observed changes of diffuse CO_2 emission rate on the eastern flank of San Miguel volcano should be related to either a sharp increase of CO_2 pressure within the volcanic-hydrothermal system or degassing from an uprising fresh gas-rich magma within the shallow plumbing system of the volcano as it was explained for the observed changes on CO efflux prior to the Stromboli 2002–2003 eruptive events (CARAPEZZA et al., 2004; BRUSCA et al., 2004). As an alternative procedure to isolate the joint impulse-response function of the CO_2 efflux and determine the effects of variations in the hydrothermal system on diffuse CO_2 efflux, we used multivariate regression analysis (MRA) to delineate the relations between CO_2 efflux and external factors and then use these relations to filter out the effects of these factors on the measured efflux history. In this case, we considered in the analysis the most relevant parameters to explain the variability of the CO_2 efflux, barometric pressure, wind speed, tidal, air temperature, air relative humidity, wind direction and power supply. MRA is commonly used to predict a dependent variable, y, as a function of relevant

Table 2

Stepwise forward results From the MRA analysis

	Multiple R	F to ENTER	p-level
Air temperature	0.132472	36.331	0.0000
TIDAL	0.178912	30.371	0.0000
Barometric pressure	0.191681	9.981	0.0016
Wind Speed	0.199183	6.201	0.0128
Power supply	0.200368	1.001	0.3170
Wind direction	0.200887	0.440	0.5071
Relative humidity	0.201147	0.221	0.6381

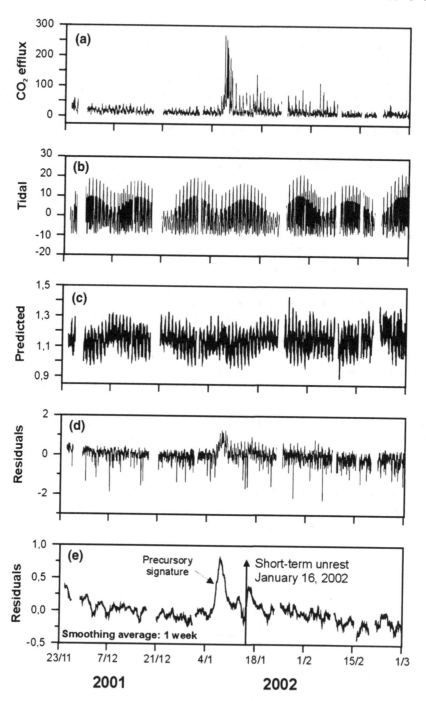

Figure 6

Observed (a) and predicted (c) CO_2 efflux time series of San Miguel Volcano. Figures 6d and 6e show the total residuals and smoothed residual of the MRA for the same data series.

explanatory variables, x1, x2,..., xn. Results of MRA provide an understanding of the percentage of the variability in y that is explained by the selected set of x variables. A forward stepwise regression analysis (Table 2) was performed to identify the most efficient set of x variables that would explain most of the variability in y and also to assess whether the relationship between y and the selected set of x variables is statistically significant, with the retained explanatory variables. A F to enter criterion of 1.0 and F to remove 0.0 for entering or removing an explanatory variable were used. The correlation coefficient (r) between the CO_2 efflux and each explanatory variable was calculated. The highest simple correlation was between the CO_2 efflux and air temperature (r = 0.13). The statistical significance level of the regression model (p value) was less than 0.0001 with air temperature and TIDAL as independent variables, indicating a highly significant model. The square of the multiple regression coefficients (r^2) for the regression equation was 0.20, which means that about 20% of the variability in the estimated CO_2 efflux is explained by the regression model. The time evolution of the residuals of MRA filtering is shown in Figure 6. Again, the filtered trend and residual components showed the anomalous increase on the diffuse CO_2 degassing 10 days before the onset of the short-term unrest at San Miguel volcano.

Fluid pressure fluctuations in volcanic systems might be also related to stress/strain changes in the subsurface due to either magma rising beneath the volcano or the occurrence of a relatively high magnitude earthquake in the vicinity of the volcano. The last rationale was recently described by SALAZAR et al. (2002) at San Vicente volcano (El Salvador, Central America) where significant CO_2 efflux rate fluctuations were measured relative to a 5.1 magnitude earthquake, which occurred to 25 km away from the observation site on May 8, 2001. A significant increase in CO_2 efflux rate was mainly driven by strain changes prior to this earthquake.

Since the seismic activity recorded in the vicinity of San Miguel volcano at the end of 2001 and early 2002 do not reflect the occurrence of a relatively high magnitude earthquake, magma movement at depth seems the most reasonable scenario for the observed increase on the diffuse CO_2 emission rate at San Miguel volcano on January 6 and 7, 2002. This geological phenomena should be responsible for a relative increment of the internal pressure of magmatic volatiles due to magma rising within the volcanic-hydrothermal system; therefore, driving an increase of diffuse CO_2 emission in the surface environment of San Miguel volcano prior to the observed vigorous steam-and-gas emission through the plume and microseismicity changes on January 16, 2002.

After the observed precursory geochemical signature of diffuse CO_2 emission rate at San Miguel volcano, period (iii), CO_2 efflux values showed a higher dispersion than those observed before January 6, 2002. This scattered behavior on the CO_2 efflux rate could mainly be due to long-period earthquakes, volcanic tremor, and explosion events which were recorded at San Miguel in late January and February (GVN BULLETIN, 2002).

Conclusions

An increase up to 17-fold the background mean value of diffuse CO_2 emission rate was observed before short-term unrest at San Miguel volcano on January 16, 2002. These CO_2 efflux changes were not driven by fluctuations of meteorological variables such as wind speed and barometric pressure and seem clearly associated to a relative increment of CO_2 pressure within the volcanic-hydrothermal system due to magma movement at depth beneath San Miguel volcano. These results showed the potential of applying continuous monitoring of soil CO_2 efflux to improve and optimize the detection of early warning signals of future volcanic crisis at San Miguel as well as in other active volcanic systems. Further observations, after fixing the radiotelemetry system are needed to verify the existence of a close relationship between diffuse CO_2 emission rate and volcanic activity.

Acknowledgements

The authors are grateful to the Office of the Spanish Agency for International Cooperation (AECI) in El Salvador, the Spanish Embassy in El Salvador, The Ministry of Environment and Natural Resources of the Government of El Salvador, and Geotérmica Salvadoreña (GESAL) for their assistance during the fieldwork. We thank José M. Salazar for constructive suggestions to improve the paper and Tomás Soriano for help in the field work. We are indebted to the GRP of El Salvador's Civil National Police for providing security during our stay in El Salvador. This research was mainly supported by the Spanish Aid Agency (Agencia Española de Cooperación Internacional – AECI), and additional financial-aid was provided by the Cabildo Insular de Tenerife, CajaCanarias (Canary Islands, Spain), the European Union, and The University of El Salvador.

REFERENCES

ALLARD, P., CARBONELLE, J., DAJLEVIC, D., LE BRONCE, J., MOREL, P., ROBE, M.C., MAURENADS, J.M., FAIVRE-PIERRET, R., MARTIN, D., SABROUX, J.C., and ZETTWOOG, P. (1991), *Eruptive and diffuse emissions of CO_2 from Mount Etna*, Nature *351*, 387–391.
BAUBRON, J.C., ALLARD, P., and TOUTAIN, J.P. (1990), *diffuse volcanic emissions of carbon dioxide from Vulcano Island, Italy*, Nature *344*, 51–53.
BRUSCA, L., INGUAGGIATO, S., LONGO, M., MADONIA, P., and MAUGERI R. (2004), *The 2002–2003 eruption of Stromboly (Italy): Evaluation of the volcanic activity by means of continous monitoring of soil temperature, CO_2 flux, and meteorological parameters*, Geochem., Geophys. Geosyst. *5*, Q12001, doi:10.1029/2004GC000732.

CARAPEZZA, M. L., INGUAGGIATO, S., BRUSCA, L., and LONGO, M. (2004), *Geochemical precursors of the activity of an open-conduit volcano: The Stromboli 2002–2003 eruptive events*, Geophys. Res. Lett. *31*, L07620, doi:10.1029/2004GL019614.

CARTAGENA, R., OLMOS, R., LÓPEZ, D., BARAHONA, F., SORIANO, T., HERNÁNDEZ, P.A., and PÉREZ, N.M. (2004), *Diffuse degassing of carbon dioxide, radon and mercury at San Miguel Volcano, El Salvador, Central America*, Bull. Geol. Soc. America *375*, 203–212.

CHIODINI, G., FRONDINI, F., and RACO, B. (1996), *Diffuse emission of CO$_2$ from the Fossa Crater, Vulcano Island (Italy)*, Bull. Volcanol. *58*, 41–50.

CHIODINI, G., CIONI, R., GUIDI, M., RACO, B., and MARINI, L. (1998), *Soil CO$_2$ flux measurements in volcanic and geothermal areas*, Appl. Geochem. *13*, 543–552.

FARRAR, C.D., SOREY, M.L., EVANS, W.C., HOWLE, J.F., KERR, B.D., KENNEDY, B.M., KING, Y., and SOUTHON, J.R. (1995), *Forest-killing diffuse CO$_2$ emission at Mmammoth Mountain as a sing of magmatic unrest*, Nature *376*, 675–678.

GERLACH, T.M.J. and GRAEBER, E.J. (1985), *Volatile budget of Kilauea Volcano*, Nature *313*, 273–277.

GERLACH, T.M.J., DOUKAS, M.K., McGEE and KESSLER, R. (1998), *Three-year decline of magmatic CO$_2$ emission from soils of a Mammoth Mountain tree kill: Horseshoe Lake, CA, 1995–1997*, Geophys. Res. Lett. *25*, 1947–1950.

GIAMMANCO, S., GURRIERI, S., and VALENZA, M. (1998), *Anomalous soil CO$_2$ degassing in relation to faults and eruptive fissures on Mount Etna (Sicily)*, Bull. Volcanol. *69*, 252–259.

GRANIERI, D., CHIODINI, G., MARZOCCHI, W., and AVINO, R. (2003). *Continuous monitoring of CO$_2$ soil diffuse degassing at Phlegraean Fields (Italy): Influence of environmental and volcanic parameters*, Earth Planet. Sci. Lett. *212*, 167–179.

GVN BULLETIN (2002), *Minor gas-and-ash emission in January 2002; Summary of earlier activity*, Bull. Global Volcan. Network *27*, 02.

HERNÁNDEZ, P.A., PÉREZ, N.M., SALAZAR, J.M., NOTSU, K., and WAKITA, H. (1998), *Diffuse emission of carbon dioxide, methane, and helium-3 from Teide Volcano, Tenerife, Canary Islands*, Geophys. Res. Lett. *25*, 3,311–3,314.

HERNÁNDEZ, P.A., SALAZAR, J.M., SHIMOIKE, Y., MORI, T., NOTSU, K., PÉREZ, N.M. (2001b), *Diffuse emission of CO$_2$ from Miyakejima Volcano, Japan*, Chem. Geology *177*, 175–185.

HERNÁNDEZ, P.A., NOTSU, K., SALAZAR, J.M., MORI, T., NATALE, G., OKADA, H., VIRGILI, G., SHIMOIKE, Y., SATO, M., and PÉREZ, N.M. (2001a), *Carbon dioxide degassing by advective flow from Usu Volcano, Japan*, Science *292*, 83–86.

ISHIGURO, M., SATO, T., TAMURA, Y., and OOE, M. (1984), *Tidal data analysis: An introduction to Bbaytap*, Proc. Institute of Statist. Math. *32*, 71–85.

MEYER-ABICH, H. (1956), *Los Volcanes Activos De Guatemala Y El Salvador, (América Central)*, Anales Del Servicio Geológico Nacional De El Salvador, pp. 3–129, Ministerio De Obras Públicas. República De El Salvador.

MORI, T., NOTSU, K., HERNÁNDEZ, P.A., SALAZAR J, M.L., PÉREZ, N.M., VIRGILI, G., SHIMOIKE, Y., and OKADA, H. (2002), *Continuos monitoring of soil CO^2 efflux from the summit region of Usu Volcano, Japan*, Bull. Volcanol. Soc. Japan *47*, 339–345.

PADRÓN, E., SALAZAR, J.M.L., HERNÁNDEZ, P.A., and PÉREZ, N.M. (2001), *Continuous monitoring of diffuse CO$_2$ degassing from Cumbre Vieja Volcano, La Palma, Canary Islands*, EOS, Trans. Am. Geophys. Union *82*, F1132.

PÉREZ, N.M., WAKITA, H., LOLOK, D., PATIA, H., TALAI, B., and MCKEE, C.O. (1996), *Anomalous soil gas CO$_2$ concentrations and relation to seismic activity at Rabaul Caldera, Papua New Guinea*, Geogaceta *20*, 1000–1003.

PÉREZ, N.M., SALAZAR, J.M.L., HERNÁNDEZ, P.A., SORIANO T., LOPEZ, K., and NOTSU, K. (2004), *Diffuse CO$_2$ and ^{222}rn dcegassing from San Salvador Volcano, El Salvador, Central America*, Bull. Geol. Soc. Am. *375*, 227–236.

ROGIE, J.D., KERRICK, D.M., SOREY, M.L., CHIODINI, and GALLOWAY, D.L. (2001), *Dynamics of carbon dioxide emission at Mammoth Mountain, California*, Earth Planet. Sci. Lett. *188*, 535–541.

SALAZAR, J.M.L., PÉREZ, N.M., and HERNÁNDEZ, P.A. (2000), *Secular variations of soil CO$_2$ flux levels at the summit cone of Teide Volcano, Tenerife, Canary Islands*, EOS, Trans. Am. Geophys. Union *81*, F1317.

SALAZAR, J.M.L., HERNÁNDEZ, P.A., PÉREZ, N.M., MELIÁN, G., ÁLVAREZ, J., SEGURA, F., and NOTSU, K. (2001), *Diffuse emission of carbon dioxide from Cerro Negro Volcano, Nicaragua*, Geophys. Res. Lett. *28*, 4275–4278.

SALAZAR, J.M.L., PÉREZ, N.M., HERNÁNDEZ, P.A., SORIANO, T., BARAHONA, F., OLMOS, R., CARTAGENA, R., LÓPEZ, D., LIMA, N., MELIÁN, G., GALINDO, I., PADRÓN, E., SUMINO, H. and NOTSU, K., (2002), *Precursory diffuse carbon dioxide degassing signature related A 5.1 magnitude earthquake in El Salvador, Central America*, Earth Planet. Sci. Lett. *205*, 81–89.

SALAZAR, J.M.L., HERNÁNDEZ, P.A., PÉREZ, N.M., OLMOS, R., BARAHONA, F., CARTAGENA, R., SORIANO, T., LOPEZ, K., and NOTSU, K. (2004), *Spatial and temporal variations of diffuse CO_2 degassing at Santa Ana-Izalco-Coatepeque Volcanic Complex, El Salvador, Central America*, Bull. Geolog. Soc. Am. *375*, 135–146.

SIMKIN, T., and SIEBERT, L., *Volcanoes of The World, 2nd Ed.* (Geosci. Press. Tuscon, Arizona 1994), 349 pp.

SOREY, M.L., EVANS, W.C., KENNEDY, B.M., FARRAR, C.D., HAINSWORTH, L.J., and HAUSBACK, B. (1998), *Carbon dioxide and helium emissions from a reservoir of magmatic gas beneath Mammoth Mountain, California*, J. Geophys. Res. *103*(B7), 15,303–15,323.

WARNER, REBECCA, M. *Spectral analysis of time-series data* (Guilford Press 1998), 225 pp.

(Received: December 1, 2004; revised: November 18, 2005; accepted: November 28, 2005)

To access this journal online:
http://www.birkhauser.ch

Pure appl. geophys. 163 (2006) 897–914
0033–4553/06/040897–18
DOI 10.1007/s00024-006-0045-y

© Birkhäuser Verlag, Basel, 2006

❚Pure and Applied Geophysics

In situ Permeability Measurements Based on a Radial Gas Advection Model: Relationships Between Soil Permeability and Diffuse CO_2 Degassing in Volcanic Areas

MARCO CAMARDA,[1] SERGIO GURRIERI,[1] and MARIANO VALENZA[2]

Abstract—In this paper we have developed a new method for measuring *in situ* soil permeability, which is based on the theory of radial gas advection through an isotropic porous medium. The method was tested in the laboratory and at several locations on the island of Vulcano (Aeolian Islands, Italy). It consists of a special device which generates a gas source at a depth of 50 cm and it permits measurement of the relative induced pressure in nearby soil at different depths. The characteristic error of the method was less than 10%.

Furthermore, soil permeability measurements were carried out in the island of Vulcano during different periods of the year (between May 2000 and June 2001). A strong decrease in permeability in the upper layers of the soil during and after rainfall was noted, with very poor correlations between the spatial distributions of soil CO_2 flux and shallow soil permeability.

Key words: Gas soil permeability, volcanic areas, radial gas advection, soil degassing.

1. Introduction

Gas transport in porous media is due either to diffusion and/or advection processes (ETIOPE and MARTINELLI, 2002). The former process is induced by a concentration gradient while the latter is driven by a pressure gradient. In both cases, the physical and chemical properties of the medium and of the fluid strongly influence the transport process. If these properties are known, quantitative models for gas transport through porous media can be developed. One important parameter which must be known in these models is soil permeability (L^2). This parameter depends on the properties and the specific conditions of the soil such as porosity, structure, tortuosity, specific surface, air saturation etc. (MOLDRUP et al., 1998). Soil permeability can be defined as the ease with which a fluid can pass through the soil. Laboratory methods, employed in measuring soil gas permeability, consist of special filtration devices in which the soil samples are traversed in

[1] Istituto Nazionale Geofisica e Vulcanologia, Sezione di Palermo, V. Ugo La Malfa 153, 90146 Palermo, Italy. E-mail: s.gurrieri@pa.ingv.it
[2] Dipartimento Chimica e Fisica della Terra ed Applicazioni, V. Archirafi 36, 90100 Palermo, Italy

one direction by a gas flux. Generally, treatment of soil samples with these methods (sample collection, transport and insertion inside the measuring device) profoundly modify all soil properties and the resulting permeability values could be affected by serious errors (EVANS and KIRKHAM, 1949). In order to solve this problem, various empirical methods were developed to measure soil permeability directly in the field (EVANS and KIRKHAM, 1949; GROVER, 1955; FISH and KOPPI, 1994). The devices employed consist of an inverted chamber placed on the soil and connected to a gas tank. The gas pressure reached in the chamber is proportional to gas flux and soil permeability, according to an empirical model of gas advection through a homogeneous porous medium.

A new method for measuring shallow soil permeability *in situ* and the effects of this parameter on soil degassing in a volcanic areas (Vulcano, Italy) will be discussed in this paper. This method is based on the theory of radial gas advection through an isotropic porous medium. The model shows the relationship between the permeability of the medium and the pressure gradients induced by a radial and continuous gas source. The method was tested in the laboratory and successively used to measure soil permeability in various areas on Vulcano, a small volcanic island in the Aeolian archipelago which is located in the southern Tyrrhenian Sea, close to the north Sicilian coast. The island is an active volcano, which last erupted in the period 1888–90, and, at present, is characterized by solphataric activity. Soil permeability surveys were performed both in the summer and winter periods and the results obtained were compared with soil gas fluxes, which were measured at the same time and in the same locations.

2. Generalities on Fluid Advection

Advective fluid transport in a natural porous medium is governed by the Darcy's law. This highlights the relationship between velocity (v) of the fluid and the pressure gradient (∇P):

$$v = -\frac{k}{\mu}\nabla(P + \rho g z),$$

where g is the acceleration gravity constant, ρ and μ are the density and the viscosity of the fluid respectively, and k is the *intrinsic permeability* of the soil. Regarding these fluids, the term $\rho g z$ in the following calculations will be ignored because it is three orders of magnitude less than the P term. Theoretically, *intrinsic permeability* only depends on the properties of the porous medium and not on the permeating fluid (CARMAN, 1956). This assumption is generally true when fluids do not interact with soil. In other cases (especially water and other liquids), the fluid can interact with soil, thereby changing its properties. *Intrinsic permeability* could, thereby, strongly depend on the characteristics of the fluid (FISH and KOPPI, 1994; MICHAELS and LINN, 1954).

In this work we shall consider the advection of air and CO_2 through a porous medium. k is the *intrinsic permeability* of the medium for these gases which we call *gas permeability*, as suggested by several authors (MOLDRUP *et al.*, 1998; MCCARTHY and BROWN, 1992).

Combining Darcy's law with the continuity equation [$\text{div}(\rho v) + n\partial\rho/\partial t = 0$], in which n is the porosity of the porous medium, the fundamental equation for gas transfer through porous media is obtained (SCHEIDEGGER, 1974):

$$\text{div}\left(\rho\frac{k}{\mu}\nabla P\right) = n\frac{\partial\rho}{\partial t}$$

or

$$\nabla^2 \rho P = 0 \tag{1}$$

if the system is under steady state conditions, at constant temperature (μ = constant) and the porous medium is homogeneous (k = constant). Moreover, assuming ideal behavior regarding the gas involved in the transfer process and constant temperature:

$$\rho = cP$$

where $c = M/RT$, M is the molecular weight of a generic gas, R is the universal gas constant and T is the absolute temperature. Substituting this expression in Equation (1), Laplace's equation is obtained:

$$\nabla^2\chi = 0 \tag{2}$$

where $\chi = cP^2$ Equation (2) can be used to determine the gas pressure spatial variation for any gas advection model.

Modeling a radial gas pressure distribution through a porous medium (Fig. 2), it was useful to express Laplace's equation using the polar spherical coordinates r, θ and γ:

$$\frac{\partial}{\partial r}\left(r^2\frac{\partial\chi}{\partial r}\right) = 0$$

in which we do not consider the terms θ and γ, in light of a hypothesis of a homogeneous porous medium. The solution of this differential equation is:

$$\chi = a + b/r \tag{3}$$

where a and b depend on boundary conditions. Considering two spherical shells with radii of R_1 and R_2, which are concentric with the gas source and assuming that the pressure of these two shells is equal to P_1 and P_2 respectively, the expression of the constant a and b can be easily found:

$$a = cP_1^2 - \frac{b}{R_1} \quad \text{and} \quad b = \frac{c(P_2^2 - P_1^2)R_1R_2}{R_1 - R_2}.$$

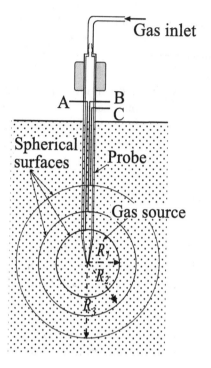

Figure 1
Schematic representation of the probe used in making soil permeability measurements and graphical representation of the isobar surfaces produced from the soil; A, B, C are tubes which measure the pressure and R_1, R_2, R_3 are the radii of the spherical shells. The tubes are connected to external pressure sensors (see Fig. 2).

Substituting a and b terms into equation (3) and solving for the pressure, we obtain the following equation:

$$P = \sqrt{P_1^2 - \frac{(P_2^2 - P_1^2)R_2}{R_1 - R_2}\left(\frac{R_1}{r} - 1\right)}, \tag{4}$$

which shows the variation of the gas pressure (P) as a function of the radius (r), generated by a radial gas source inside a homogeneous porous medium. Substituting the first derivative of the equation (4) at $r = R_1$ in Darcy's law, the equation for the volumetric gas flux across a spherical shell of radius R_1 is obtained:

$$\varphi_{r=R_1} = \frac{2k\pi}{\mu}\frac{R_1 R_2}{R_1 - R_2}\frac{(P_2^2 - P_1^2)}{P_1}. \tag{5}$$

Equation (5) predicts that the gas flux crossing a spherical shell of radius R_1, when a radial gas source is generated in a homogeneous porous medium of permeability k and the gas pressure at R_1 and R_2 shells is equal to P_1 and P_2 respectively.

3. Laboratory and Field Tests

The permeability values discussed in this paper were calculated according to equation (5) and the measurements of gas pressure for different concentric spherical shells. The radial gas source was generated by using the probe shown in Figure 2 (diameter = 2 cm, length = 75 cm), which was inserted in the soil at a depth of 50 cm. The gas pressure gradients are measured by three thin tubes, externally connected to water manometers and located inside the probe; the tubes were opened at depths of 35, 40 and 45 cm (A, B and C in the Figure 1). Figure 2 shows the equipment used for measuring soil permeability. The gas flow was generated by a membrane pump connected with a flowmeter, and both the flux and pressure measurements were carried out at steady state (reached in a few minutes).

The method was tested in the field (the island of Vulcano) and in the laboratory. The permeability values measured in the field, together with relative confidence intervals, have been reproduced in Table 1. Three permeability values, $k_{1,2}$, $k_{2,3}$ and

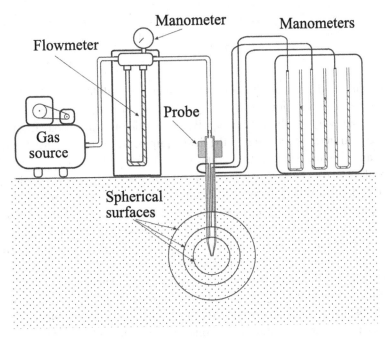

Figure 2

Permeability equipment used for field measurements, as discussed in this paper. The apparatus consists of: an external pump, a flux meter, a probe with which to supply gas to the soil, three water manometers for reading the soil gas pressure at different depths.

Table 1

Soil permeability measurements and the relative confidence interval relative to different soil portions and different air fluxes. The sites with the same number and different subscripts refer to permeability measurements carried out a few meters from one another. Discrepancy values $\delta(k_{1,2} - k_{1,3})$ expressed in % between the $k_{1,2}$ and $k_{1,3}$ permeability values are also shown. The permeability $k_{1,2}$ and $k_{1,3}$ are very similar and confirms the assumption that the porous medium can be considered quite homogeneous. The air flux values reported here were corrected by the output air pressure values

Site	Flux ($cm^3 s^{-1}$)	$k_{1,2}$ (μm^2)	$k_{2,3}$ (μm^2)	$k_{1,3}$ (μm^2)	$\delta(k_{1,2}\text{-}k_{1,3})$
1_a	260	14.6 ± 0.8	13 ± 2	13.9 ± 0.6	7
1_b	230	12.3 ± 0.6	16 ± 3	12.6 ± 0.5	1
101	230	18 ± 1	19 ± 2	18 ± 1	0
4_a	280	50 ± 5	47 ± 12	48 ± 3	4
4_b	178	48 ± 6	45 ± 15	46 ± 3	4
4_c	230	43 ± 5	27 ± 10	37 ± 3	15
30_a	60	10 ± 2	10 ± 5	10 ± 2	0
30_b	185	7.5 ± 0.3	12 ± 2	7.9 ± 0.3	5
30_c	255	6.4 ± 0.2	5.5 ± 0.3	5.9 ± 0.2	8
6	275	63 ± 6	70 ± 23	62 ± 4	2
43_a	260	6.6 ± 0.2	8.6 ± 0.8	6.5 ± 0.2	2
43_b	265	7.1 ± 0.2	7.0 ± 0.8	6.9 ± 0.2	3
41	275	34 ± 3	28 ± 7	31 ± 2	9
8_a	305	35 ± 3	31 ± 8	33 ± 3	6
8_b	310	33 ± 2	31 ± 6	32 ± 1	3
12_a	275	9.7 ± 0.3	17 ± 4	10.5 ± 0.3	8
12_b	295	6.8 ± 0.2	16 ± 3	7.0 ± 0.2	3
40_a	290	9.5 ± 0.3	11 ± 1	9.5 ± 0.3	0
40_b	265	10.9 ± 0.4	13 ± 2	10.9 ± 0.4	0

$k_{1,3}$ were measured for each site, each value referring to different spherical soil shells with inner and outer radii of $R_1 - R_2$, $R_2 - R_3$ and $R_1 - R_3$ respectively. The confidence interval was obtained by applying the rule of error propagation (TAYLOR, 2000) to equation (5). According to the field equipment, the uncertainty regarding the pressure and flux measurements is equal to ± 10 Pa and ± 4 cm^3 s^{-1} respectively. The absolute error in the permeability measurements depends on the pressure difference relative to the spherical shells under consideration. According to equation (5), soil gas pressure rapidly decreases with increasing distance from the gas source (Figure 3). Therefore, the most distant shells were characterized as possessing the lowest difference in pressure. Considering the uncertainty regarding the water manometer, any error in the pressure difference measured between the R_2 and R_3 shells is significantly higher than with the other shells (Table 1).

Gas flux values and soil permeability also influence the error made in taking soil permeability measurements. Low flux and high soil permeability yield small differences in gas pressure and, as a consequence, permeability measurements will be affected by larger errors.

Table 1 also reports discrepancies $\delta(k_{1,2} - k_{1,3})$, expressed as percentages between the permeability values measured, considering the spherical shells of soil with inner

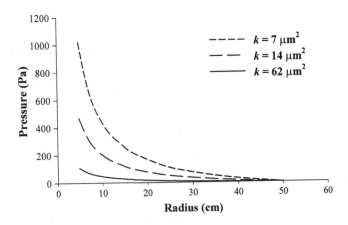

Figure 3

Theoretical variation in soil gas pressure as a function of the radius for three different permeability values (7, 14 and 62 μm^2 respectively). The pressure produced in the soil significantly decreases with increasing distance from the gas source (0 cm).

and outer radii of $R_1 - R_2$ and $R_1 - R_3$. The calculated permeability produced very similar results with a discrepancy of less than 10 %, except in one case (15 % for the site 4). The agreement between the permeability values $k_{1,2}$ and $k_{1,3}$ calculated for the same sites confirms our assumption that the porous medium can be considered relatively homogeneous.

Finally, the method was also tested comparing the *in situ* measurements with the values determined by using standard laboratory procedures. Several soil samples were collected in sites where soil permeability had been previously measured with the method described in this paper. The four selected sites (1, 4, 6 and 8) were characterized as having no cohesive soils with different permeability values (15–90 μm^2). The soil samples were collected with a shovel from a hole approximately 50 cm deep; in many cases the extracted soils were disturbed. The permeability of the soil samples was measured in the laboratory by a gas permeameter (LOOSVELDT *et al.*, 2002) and the values compared with the *in situ* measurements (Table 2). The laboratory results were

Table 2

Comparison between laboratory and in situ soil permeability values. The discrepancies between these values are also shown. The in situ data are the mean values of the permeability measurements reported for each site in Table 1

Sample	k (μm^2) (laboratory)	k (μm^2) (*in situ*)	$\delta(\%)$
4	48.5	45.3	7
6	85.5	62.5	31
1	16.9	13.5	22
8	41.2	33.3	21

always higher than the *in situ* measurements with a discrepancy ranging from 7 to 31 % (Table 2). These results can be explained by considering that the sample procedure and laboratory treatments destroyed the original soil structure and caused an increase in porosity. Moreover, during the waiting period prior to taking the measurements, samples were subjected to a continuous decrease in soil moisture content which resulted in an increase in gas permeability.

4. Field data and discussion

4.1 Field data collection

Five surveys of soil permeability measurements on Vulcano Island (Aeolian Archipelago, Italy) were conducted during the period May 2000–June 2001. The

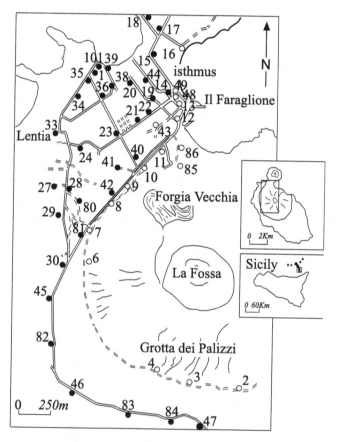

Figure 4
Location of the 48 permeability and flux measurement sites. k_m is the mean value of the soil permeability. The empty circles indicate the 17 sites used to investigate the effect of rainfall on soil permeability.

Table 3

Permeability values calculated in different periods of the year in the Island of Vulcano

Site	May-26-2000 k (μm^2)	January-18-2001 k (μm^2)	March-28-2001 k (μm^2)	May-05-2001 k (μm^2)	June-16-2001 k (μm^2)
2	64.7	n.d	70.1	49.4	49.4
3	56.1	n.d	44.0	33.2	28.4
4	60.1	27.4	60.1	44.0	52.5
6	60.1	20.8	28.4	24.8	24.1
7	41.8	22.0	34.6	25.7	29.5
8	30.6	23.3	64.7	39.7	31.9
9	60.1	39.7	64.7	70.1	60.1
10	30.6	23.3	34.6	24.1	26.6
11	18.4	22.0	28.4	31.9	41.8
12	41.8	18.8	44.0	22.0	22.0
13	46.5	28.4	49.4	44.0	49.4
14	46.5	19.7	27.4	4.5	30.6
15	28.4	33.2	44.0	44.0	20.8
16	70.1	34.6	64.7	52.5	60.1
17	26.6	17.1	28.4	18.8	28.4
18	21.4	39.7	23.3	20.8	8.3
19	24.1	44.0	60.1	52.5	24.1
20	18.4	13.4	34.6	49.4	46.5
21	76.4	44.0	70.1	41.8	76.4
22	44.0	28.4	25.7	19.7	22.0
23	60.1	26.6	56.1	44.0	46.5
24	30.6	17.1	33.2	25.7	30.6
26	24.8	26.6	24.8	16.7	15.4
27	15.7	11.4	20.8	17.1	21.4
28	24.8	24.1	24.8	20.8	22.0
29	44.0	24.3	36.1	24.8	28.4
30	14.2	10.6	14.8	12.5	12.3
33	49.4	n.d	76.4	28.4	26.6
34	33.2	18.4	18.8	17.5	16.7
35	60.1	28.4	60.1	39.7	34.6
36	60.1	29.5	44.0	24.1	24.8
38	15.1	7.3	18.4	8.2	20.8
39	64.7	49.4	70.1	60.1	64.7
41	34.6	49.4	46.5	60.1	76.4
42	31.9	24.8	30.6	25.7	28.4
43	20.8	24.8	22.0	14.2	15.1
45	41.8	10.5	39.7	60.1	17.9
46	24.1	20.8	39.7	20.8	34.6
47	41.8	24.1	36.1	31.9	28.4
48	37.9	44.0	64.7	60.1	76.4
49	11.3	18.4	20.8	18.4	18.8
80	28.4	19.2	18.8	16.4	15.4
81	23.3	17.9	30.6	17.1	17.1
82	60.1	34.6	70.1	44.0	10.6
83	33.2	18.4	30.6	26.6	33.2
84	60.1	20.8	44.0	19.2	28.4
85	44.0	39.7	56.1	20.8	24.1
86	27.4	12.3	22.0	17.9	26.6

explored area (Figure 4) covers about 2.2 square kilometers and is located on the western flank of La Fossa cone, between the Vulcanello isthmus (on the north) and Grotta dei Palizzi (on the south). The soil permeability measurements were performed in the same 48 fixed sampling sites in which soil CO_2 flux measurements are periodically carried out for volcanic surveillance purposes (DILIBERTO *et al.*, 2002). The measured permeability values ranged from between 5 and 80 μm^2 and the obtained results are reported in Table 3.

4.2 Temporal Variation in Soil Permeability

Figure 5 presents the permeability maps of the investigated area relating to five different periods in a one-year period; in general, the most permeable areas are located in the southernmost part of the area (Figure 5, Grotta dei Palizzi) under investigation. Other deposits characterized by high permeability ($> 60 \ \mu m^2$) are located either on the isthmus and the Faraglione. In general, these soils are essentially composed of pyroclastic debris devoid of any vegetation. The lowest permeability values (4–30 μm^2) were found south of the Lentia area where the soils are characterized by a higher degree of cohesion often covered with vegetation.

Soil permeability measurements taken in the surveyed area have also been discussed in a recent work by CARAPEZZA and GRANIERI (2004). Regarding the measurements taken by the authors of this paper, CARAPEZZA and GRANIERI obtained a similar permeability range (2—60 μm^2) however the frequency distributions were very different. For example, the mean value obtained by CARAPEZZA and GRANIERI (2004) is significantly lower than those found with our investigations (see Figure 5). The reason for such discrepancies is mainly due to the different methodologies used to measure soil permeability. The values obtained by Carapezza and Granieri were obtained in the laboratory by a variable load water permeameter and they indicate the ease with which water can pass through the soil. They cannot, therfore, be used in studies regarding soil gas transport, such as discussed in this paper. As documented by MICHAELS and LINN (1954) and FISH and KOPPI (1994), the permeability value of a soil sample depends on the nature of the fluids used to measure permeability. Very high differences have been found between water permeability and gas permeability and, in particular, the former was found to be generally lower than the latter (LOOSVELDT *et al.*, 2002). According to these arguments, the use of water permeability does not seem to be a suitable choice in soil gas transport studies.

As highlighted by the maps shown in Figure 5, soil permeability is characterized by time variations while the spatial position of relative maximum and minimum values is relatively constant. The maximum permeability variations were recorded in January 2001 when the values were approximately 30% lower than the mean value in the other surveys. This significant decrease is certainly due to the abundant rainfall which fell prior to the January survey (3 days of rainfall with more than 20 mm of

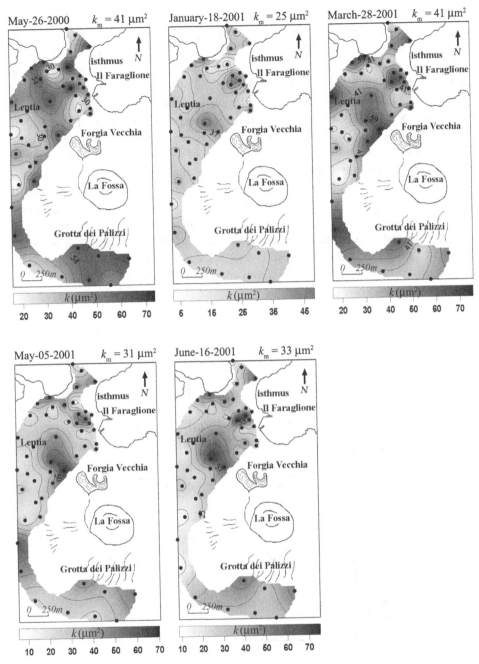

Figure 5

Soil permeability maps of the investigated area. Location of sites from where the measurements were taken are indicated by black dots.

H_2O per day). In order to evaluate the effect of rain on soil permeability, 5 surveys in a 17-site sampling grid (the empty circles in Figure 4) were carried out before and after a period of heavy rainfall. The first survey took place before a period of rainfall (May–05-2001, 11:30 am), after a long dry period, and the other surveys were undertaken 1^h, 18^h, 41^h and 1 month after the end of the period of rainfall. Plots a, b, c and d in Figure 6 show the permeability values obtained after the rainfall event versus the permeability values measured before the period of rainfall. In general, the best fitting slopes have a value less than one because the soil permeability values measured up to 18 and 41 hours after the rainfall were lower than those measured before. On the contrary, the permeability values measured 1 month later are very close to the values measured before the rainfall event and, obviously, the relative slope $\cong 1$ (Figure 6). Temporal variation in soil permeability, therefore, strictly depends on soil moisture content. Once the rainfall ceases, evaporation phenomena, drainage of water from the upper soil down to the water table and lack of rainfall result in a progressive decrease in soil water content and in a corresponding increase in gas permeability up to the pre-rainfall conditions (Figure 7).

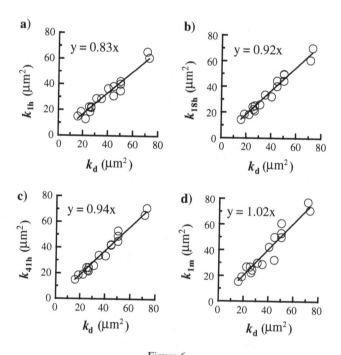

Figure 6
Soil permeability values carried out before the period of rainfall (k_d) versus those taken 1^h (k_{1h}), 18^h (k_{18h}), 41^h (k_{41h}) and 1 month (k_{1m}) after the rainfall has ceased.

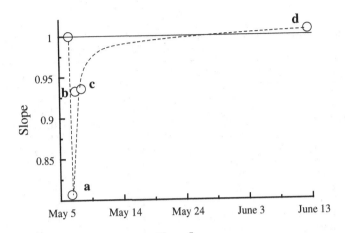

Figure 7

Slopes of the lines of Figures 6a, 6b, 6c and 6d versus time. This diagram shows the significant decrease in soil permeability immediately after the period of rainfall.

4.3 Relationship between Soil Permeability and Diffuse CO_2 Degassing

In order to evaluate the relationship between soil permeability and CO_2 soil degassing, five surveys regarding both parameters were performed and the values obtained were compared. According to a method proposed by GURRIERI and VALENZA (1988), the CO_2 flux measurements were taken from exactly the same sites as the permeability measurements. The flux values obtained (Table 4), ranging between 1.0×10^{-6} and 5.4×10^{-3} cm s^{-1}.

The flux maps (Figure 8) show the spatial distribution of the CO_2 emissions from the soil. For all field measurements, the highest CO_2 fluxes were located in the areas of Grotta dei Palizzi and Faraglione, while lower fluxes were recorded in the areas to the north of the Grotta dei Palizzi and the area adjacent to the Lentia Complex. This distribution seems to reflect regional tectonic structures in the area. The highest exhaling areas are aligned along a N-S direction, together with the main fumarolic areas located at La Fossa crater, the Forgia Vecchia craters on the eastern beach of the isthmus and around Il Faraglione (VENTURA et al., 1999).

A very low numerical correlation was found between the permeability values measured in each period and the corresponding values of soil CO_2 flux (Table 6). According to these data, low soil permeability seems to have a secondary influence on the spatial distribution of diffuse soil gas emissions. High soil CO_2 fluxes were found in areas with low as well as high permeability values (Figs. 5 and 8). Several studies on soil degassing carried out in volcanic geothermal areas (BAUBRON, 1996; GIAMMANCO et al., 1998; DILIBERTO et al., 2002) have shown that the presence of a high degassing area is strictly related to the occurrence of active tectonic structures, which are high permeable channels capable of driving deep gases toward the surface.

Table 4

Soil CO$_2$ fluxes measured in different periods of the year in the Island of Vulcano. The fluxes values shown in this table are expressed in cm s^{-1}

Site	May-26-2000	January-18-2001	March-28-2001	May-05-2001	June-16-2001
2	3.4E-04	n.d.	1.5E-04	4.2E-04	1.3E-04
3	2.7E-04	n.d.	6.1E-05	2.7E-04	2.0E-04
4	3.1E-03	3.1E-03	1.1E-03	5.4E-03	1.1E-03
6	1.0E-06	1.0E-06	5.7E-06	3.8E-06	3.8E-06
7	7.6E-05	7.6E-05	3.0E-05	2.3E-05	5.1E-05
8	1.0E-05	1.0E-05	7.6E-06	1.3E-05	5.7E-06
9	7.2E-05	7.2E-05	3.0E-05	5.7E-05	3.0E-05
10	1.1E-05	1.1E-05	2.9E-05	3.0E-05	3.0E-05
11	1.7E-05	1.7E-05	2.7E-05	3.8E-05	3.0E-05
12	4.6E-04	4.6E-04	4.2E-04	4.6E-04	2.4E-04
13	1.5E-03	1.5E-03	1.3E-03	3.0E-03	1.1E-03
14	8.7E-04	8.7E-04	3.0E-04	1.1E-05	8.4E-05
15	1.3E-04	1.3E-04	1.6E-04	1.4E-03	3.4E-05
16	6.1E-05	6.1E-05	2.7E-05	3.8E-05	3.4E-05
17	2.0E-03	2.0E-03	1.5E-05	2.3E-05	2.7E-05
18	2.8E-05	2.8E-05	1.7E-05	2.3E-05	1.1E-05
19	6.1E-05	6.1E-05	2.7E-04	5.7E-05	3.2E-04
20	5.3E-05	5.3E-05	4.6E-05	4.6E-05	1.4E-04
21	1.3E-04	1.3E-04	7.2E-05	5.7E-05	9.1E-05
22	1.5E-05	1.5E-05	7.6E-06	2.7E-05	7.6E-06
23	5.3E-05	5.3E-05	4.6E-05	5.7E-05	4.2E-05
24	8.0E-06	8.0E-06	1.5E-05	2.3E-05	2.3E-05
26	2.7E-05	2.7E-05	9.5E-06	1.9E-05	1.9E-05
27	6.0E-06	6.0E-06	1.1E-05	1.9E-05	1.1E-05
28	2.2E-05	2.2E-05	1.5E-05	2.3E-05	1.5E-05
29	3.8E-05	3.8E-05	3.4E-05	4.2E-05	1.1E-04
30	3.0E-05	3.0E-05	1.7E-05	2.3E-05	3.4E-05
33	2.6E-05	n.d.	2.3E-05	1.9E-05	1.9E-05
34	1.5E-05	1.5E-05	3.8E-06	5.7E-06	1.1E-05
35	1.0E-04	1.0E-04	1.2E-04	1.2E-04	8.0E-05
36	4.1E-05	4.1E-05	4.2E-05	3.8E-05	4.6E-05
38	1.5E-05	1.5E-05	1.1E-05	1.5E-05	1.5E-05
39	9.1E-05	9.1E-05	5.7E-05	5.3E-05	5.3E-05
41	5.7E-05	5.7E-05	8.6E-05	8.0E-05	5.7E-05
42	4.6E-05	4.6E-05	3.2E-05	1.5E-05	3.4E-05
43	6.0E-06	6.0E-06	7.6E-06	2.3E-05	1.1E-05
45	4.6E-05	4.6E-05	3.0E-05	5.3E-05	1.7E-05
46	6.4E-05	6.4E-05	5.1E-05	4.9E-05	8.0E-05
47	5.7E-05	5.7E-05	2.7E-05	1.9E-05	3.4E-05
48	1.5E-03	1.5E-03	2.6E-03	9.0E-04	2.2E-03
49	2.0E-04	2.0E-04	3.2E-04	6.1E-05	3.6E-04
80	1.5E-05	1.5E-05	7.6E-06	3.8E-06	3.4E-05
81	1.1E-05	1.1E-05	1.9E-05	1.9E-05	1.9E-05
82	9.1E-05	9.1E-05	5.1E-05	4.9E-05	2.1E-04
83	2.6E-05	2.6E-05	1.9E-05	2.5E-05	3.8E-05
84	8.7E-05	8.7E-05	1.3E-05	3.0E-05	3.4E-05
85	1.0E-05	1.0E-05	1.5E-05	1.1E-05	1.9E-05
86	6.8E-05	6.8E-05	6.5E-05	3.8E-05	6.1E-05

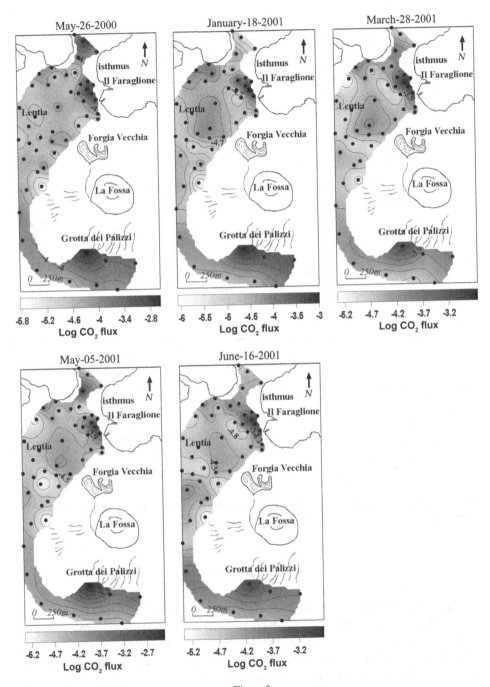

Figure 8

CO_2 flux maps of the investigated area. Location of sites from where the measurements were taken are indicated by black dots.

Table 5

Permeability values measured at the island of Vulcano in a fixed sampling grid of 17 points before (May 5, 11:30 am) and at different times after rainfall

Site	k_d (before rain) (μm^2)	k_{1h} (after 1^h) (μm^2)	k_{18h} (after 18^h) (μm^2)	k_{41h} (after 41^h) (μm^2)	k_{1m} (after 1^m) (μm^2)
2	49.3	n.d	49.4	48.4	49.4
3	34.6	n.d	33.2	33.6	28.4
4	49.4	41.8	44.0	44.4	52.5
6	26.6	22.0	24.8	24.1	24.1
7	30.6	28.4	25.7	25.7	29.5
8	44.0	30.6	39.7	41.5	31.9
9	40.5	60.1	70.1	70.1	60.1
10	24.8	22.0	24.1	24.1	26.5
11	39.7	36.1	31.9	33.2	41.8
12	25.7	18.4	22.0	22.0	22.0
13	44.0	37.9	44.0	41.8	49.4
16	49.4	39.7	52.5	52.5	60.1
43	15.7	14.8	14.2	14.8	15.1
48	70.1	64.7	60.1	64.7	76.4
49	18.4	18.4	18.4	18.4	18.8
85	26.6	21.4	20.8	21.4	24.1
86	22.0	12.7	17.9	18.8	26.5

Table 6

Values of correlation coefficient relative to the relation between permeability and CO_2 flux data obtained for each survey

Permeability and CO_2 flux survey	R
May-05-200	0.14
January-18-2001	0.10

Shallowest cover could only slightly influence the shape and size of the soil degassing anomalies.

5. Conclusion

Applying a physical model to gas radial advection through porous media, a new method has been developed with which to determine *in situ* soil permeability by taking simple measurements of gas pressure at different depths. To perform this method, a special probe has been developed and tested, both in the field and the laboratory. A comparison between permeability values measured in the same soils by an *in situ* radial gas advection method and other classical laboratory procedures both indicate that the laboratory values are higher than those obtained *in situ* (up to 30 % higher in our tests).

Permeability measurements were performed with this new method over a large sector on the island of Vulcano (Aeolian Islands, Italy) and they identified a range of values of between 4–80 μm^2. Permeability values were compared with soil CO_2 fluxes, which had been measured at the same site and in the same time, and a very low correlation between these two parameters was found. This result suggests that the permeability in the upper layers of the soil is not the main factor in determining the spatial distribution of soil gas emissions.

Finally, several permeability surveys were undertaken at different times on a fixed sampling grid before and after a heavy period of rainfall. The values obtained have identified a strong decrease in soil permeability, of up to 30 %, after an intense period of rainfall.

Acknowledgements

This study was supported entirely by the G.N.V. - I.N.G.V. within the ambit of the following research project: *Multidisciplinary Investigations into the Mass and Energy Budgets concerning active Italian volcanoes* – Task: *Studies on Diffuse Soil Degassing and Tectonics*. We would also like to thank Dr. R. Favara, Dr. S. I. Diliberto and S. Francofonte for their collaboration in the field.

REFERENCES

BAUBRON, J. C., *Prospection, caractérisation et variabilité temporelle d'émanations gazeuses diffuse a l'Etna (Sicile, Italie)*, (Années 1993 et 1994. Contract EV5V–CT92–0177, Rapport BRGM R 38820 dr/hgt 1996) pp. 1–74.

CARMAN, P.C., *Flow of Gases through Porous Media* (Academic Press, New York, 1956) pp. 182.

DILIBERTO, I.S., GURRIERI, S., and VALENZA, M. (2002), *Relationships between diffuse CO_2 emissions and volcanic activity on the island of Vulcano (Aeolian Islands, Italy) during the period 1984–1994*, Bull. Volcanol. *64*, 219–228.

ETIOPE, G. and MARTINELLI, G. (2002), *Migration of carrier and trace gases in the geosphere: an overview*, Phys.Earth Planet. Inter. *129*, 185–204.

EVANS, D.D. and KIRKHAM, D. (1949), *Measurement of the air permeability of soil in situ*, Soil. Sci. Soc. Am. Proc. *14*, 65–73.

FISH, A.N. and KOPPI, A.J. (1994), *The use of a simple air permeameter as a rapid indicator of functional soil pore space*, Geoderma *63*, 255–264.

GIAMMANCO, S., GURRIERI, S., and VALENZA, M. (1998), *Anomalous CO_2 degassing in relation to faults and eruptive fissures on Mount Etna (Sicily, Italy)*, Bull. Volcanol. *60*, 252–259.

GROVER, B.L. (1955), *Simplified air permeameters for soil in place*, Soil Sci. Soc. Am. Proc. *19*, 414–418.

GURRIERI, S. and VALENZA, M. (1988), *Gas transport in natural porous mediums: A method for measuring CO_2 flows from the ground in volcanic and geothermal areas*, Rend. Soc. Ital. Min. Petrog. *43*, 1151–1158.

LOOSVELDT, H., LAFHAJ, Z., and SKOCZYLAS, F. (2002), *Experimental study of gas and liquid permeability of a mortar*, Cement and Concrete Res. *32*, 1357–1363.

McCARTHY, K. P.and BROWN, K.W. (1992), *Soil gas permeability as influenced by soil gas-filled porosity*, Soil. Sci. Soc. Am. J. *56*, 997–1003.

MICHAELS, A. S. and LINN, C. S. (1954), *The permeability of Kaolinite*, Ind. Eng. Chem. *45*, 139–1246.

MOLDRUP, P., POULSEN, T. G., SCHJONNING, P., OLESEN, T., and YAMAGUCHI, T. (1998), *Gas permeability in undisturbed soils: Measurement and predictive models,* Soil Science *163,* 180–189.

PINAULT, J. L. and BAUBRON, J. C. (1996), *Siganal processing of soil gas radon, atmospheric pressure, moisture, and soil temperature data: A new approach for radon concentaration modeling,* J. Geophy. Res. *101,* 3157–3171

SCHEIDEGGER, A. E., *The Physics of Flow through Porous Media,* (University of Toronto Press, 1974).

TAYLOR, J., *Introduzione all'analisi degli errori: lo studio delle incertezze nelle misure fisiche,* (Zanichelli, 2000).

VENTURA, G., VILARDO, G., MOLANO, G., and PINO, N. A. (1999), *Relationships among crustal structure, vulcanism and strike-slip tectonics in the Lipari-Vulcano Volcanic Complex (Aeolian Islands, Southern Tyrrhenian Sea, Italy),* Phys. Earth and Planet. Inter. *116,* 31–52.

(Received: June 19, 2003; revised: July 6, 2004; accepted: July 23, 2003)

To access this journal online:
http://www.birkhauser.ch